Mechanical Science and Engineering

Mechanical Science and Engineering

Editor: Roman Fritz

NY RESEARCH PRESS
New York

Published by NY Research Press
118-35 Queens Blvd., Suite 400,
Forest Hills, NY 11375, USA
www.nyresearchpress.com

Mechanical Science and Engineering
Edited by Roman Fritz

International Standard Book Number: 978-1-63238-533-8 (Hardback)

Cataloging-in-publication Data

Mechanical science and engineering / edited by Roman Fritz.
p. cm.
Includes bibliographical references and index.
ISBN 978-1-63238-533-8
1. Mechanical engineering. 2. Mechanical movements. 3. Machinery. I. Fritz, Roman.
TJ145 .M43 2017
621--dc23

Printed in the United States of America.

Contents

Preface

Mechanical Engineering is an interdisciplinary field of study which combines techniques and principles from a range of disciplines including physics, materials science, etc. It includes the designing, maintenance and analysis of mechanical systems. This book provides comprehensive insights into the field of mechanical engineering. It elucidates new techniques and their applications in a multidisciplinary approach. It also sheds light on the concepts and innovative models around prospective developments with respect to mechanical engineering. Students, researchers, experts and all associated with this field will benefit alike from this book. It will help the readers in keeping pace with the rapid changes in this field.

This book has been an outcome of determined endeavour from a group of educationists in the field. The primary objective was to involve a broad spectrum of professionals from diverse cultural background involved in the field for developing new researches. The book not only targets students but also scholars pursuing higher research for further enhancement of the theoretical and practical applications of the subject.

It was an honour to edit such a profound book and also a challenging task to compile and examine all the relevant data for accuracy and originality. I wish to acknowledge the efforts of the contributors for submitting such brilliant and diverse chapters in the field and for endlessly working for the completion of the book. Last, but not the least; I thank my family for being a constant source of support in all my research endeavours.

Editor

1

On the positioning error of a 2-DOF spherical parallel wrist with flexible links and joints – an FEM approach

G. Palmieri

Università degli Studi eCampus, Via Isimbardi 10, 22060 Novedrate (CO), Italy

Correspondence to: G. Palmieri (giacomo.palmieri@uniecampus.it)

Abstract. This paper deals with an elasto-static analysis of a 2-DOF (degrees of freedom) spherical parallel wrist where the links and the joints are considered flexible. Theoretically, the mobile platform of the wrist rotates around a fixed point which is the intersection of all the joint axes. However, if the flexibility of the limbs is taken into account, while the base platform (BP) and the mobile platform (MP) are assumed rigid, two centers can be identified: one for the BP and the other for the MP. In general such points are not coincident; as a result, the positioning of the MP is affected by inaccuracies, which can be evaluated in terms of displacement and orientation errors. A finite-element method (FEM) approach is used to analyze the problem in a series of configurations of the wrist; the results are elaborated in order to obtain continuous maps of the errors over the workspace of the machine.

1 Introduction

A spherical parallel wrist is a manipulator formed by at least two kinematic chains (limbs) connecting a base platform (BP) to a mobile platform (MP) which is characterized by a spherical motion with 2 or 3 degrees of freedom (DOFs). Furthermore, all the links of the kinematic chains can have only a spherical motion; thus, in spherical parallel manipulators with revolute joints, each limb is composed of links connected by revolute joints whose axes intersect at a common point which represents the center of the spherical motion. Several examples of 3-DOF spherical parallel wrists used to orient objects, machine-tool beds or cameras can be found in the literature (Asada and Granito, 1985; Gosselin and Angeles, 1989; Gosselin and Hamel, 1994; Di Gregorio, 2004); furthermore, 2-DOF wrists used as pointing devices have been proposed in Gosselin and Caron (1999) and Kong (2011). As described in Wu et al. (2014), two characteristic points can be identified in a spherical parallel wrist: the point at the intersection of the axes of the actuated joints at the base is the BP center, while the point at the intersection of the axes of the joints on the mobile platform is the MP center. Theoretically, the BP center and the MP center overlap with the spherical center. In reality, due to manufacturing errors and tolerances and to the deformation of the bodies under external loads, the joint axes do not intersect at a common point. If it is assumed that the base and the mobile platforms are perfectly rigid and manufactured with high-precision tools, we can still identify the BP and MP centers; however, they are shifted and the motion of the MP cannot be spherical.

The evaluation of the positioning error of a manipulator is a fundamental step in the design process if accuracy and stiffness are the guidelines for the mechanical project, as is the case for machining or assembly manipulators. Stiffness analysis of manipulators has been widely studied in recent years, mainly by means of analytic methods; examples include the studies of Majou et al. (2007), Pashkevich et al. (2009), Cammarata et al. (2012), Dong et al. (2014) and Wu et al. (2014). In most of these works of these works, the analytic model is exploited to evaluate the stiffness of the manipulator in a series of configurations inside the workspace. As the computational time of such models is quite low, a large number of configurations can be analyzed, obtaining, as a result, maps of the stiffness indices over the workspace of the machine. On the other hand, some simplifying assumptions are necessary to formulate the problem, and results may be affected by inaccuracies. The implementation of a finite-element model is usually done in order to validate the analytic model comparing the results on a single or few configurations of the manipulator.

This paper investigates how the elastic deformation of the links and joints of a spherical parallel wrist generates an error in the positioning of the MP of the machine. Deformations are generated by an external load applied to the MP which simulates the interaction force exerted by an external manipulator during a cooperative assembly task. Recent advances in the field of flexible-multibody simulation tools today allow for a new approach to the stiffness analysis of manipulators to be pursued (Wu et al., 2014; Palmieri et al., 2014b): a configuration-dependent finite-element model, built as a result of searching for an optimum compromise between accuracy and computational effort, is directly used to perform the elasto-static analysis of the wrist in a series of configurations inside the workspace.

The remainder of the paper is organized as follows. Section 2 describes the kinematics and the mechanics of the parallel spherical wrist studied in this work; specifications and requirements of the machine are summarized and reference systems are fixed. After that, in Sect. 3, implementation of the finite-element model is discussed. Data obtained from the FEM analysis are then processed by means of a procedure, described in Sect. 4, which returns the positioning error of the MP in terms of displacement and orientation errors. Finally, results are discussed in Sect. 6.

2 Kinematics and mechanical design of the spherical wrist

The wrist studied in this paper is a non-overconstrained mini pointing device characterized by a parallel kinematics, which provides it with 2 DOFs of rotation. As described in Palmieri et al. (2014a), the mini wrist is conceived to be integrated in a mini assembly cell together with a second manipulator responsible for the Cartesian positioning; the final goal is to realize a fully automated mini cell for "general purpose" assembly tasks with accuracy and repeatability of a few microns and overall dimensions of a few centimeters. In this context, the emphasis given to the stiffness analysis and to the accuracy evaluation of the manipulators is fully motivated.

The functional and mechanical design of the device are described in Palmieri et al. (2015). The main characteristics of the wrist are a workspace of $\pm 54.7^\circ$ about every axis lying on the horizontal plane, a payload of 100 g, an overall size contained within a cube with a side of 150 mm, a circular MP with a 50 mm diameter, and a resolution on the order of $10^{-2\circ}$. The kinematic structure of the wrist is based on the very common scheme of the parallel five-bar linkage (Gosselin and Caron, 1999; Kong, 2011). This well-known mechanism allows the platform to rotate with 2 DOFs around a point $\mathbf{O}$ located at the intersection of all the axes of the revolute joints. Starting from the basic scheme of the spherical five-bar linkage, in which all the joint axis are orthogonal,

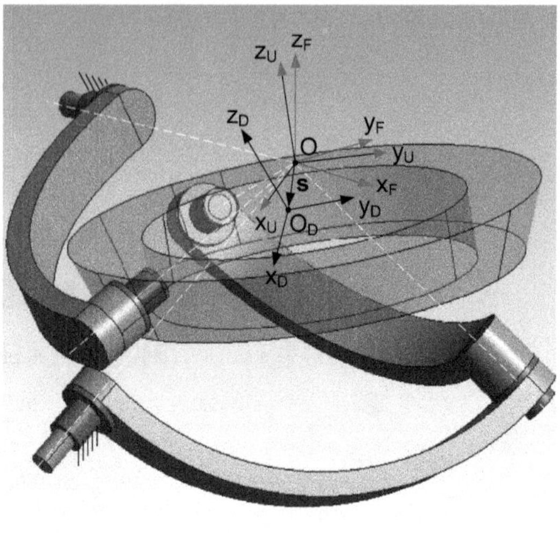

Figure 1. Frames attached to the BP $\{\mathcal{F}\}$, to the MP in the undeformed configuration $\{\mathcal{U}\}$ and to the MP in the deformed configuration $\{\mathcal{D}\}$.

a series of modifications have been introduced in order to guarantee the required workspace and to move the spherical center $\mathbf{O}$ of the wrist up from the supporting surface of the MP (Palmieri et al., 2015). The resulting kinematic scheme is shown in Fig. 1: in the hypothesis of rigid limbs, $\mathbf{O}$ is the intersection of all the joint axes and coincides with the BP center and with the MP center; the bushings marked with a hatched area are fixed to the frame and the rotations of the supported shafts are the motor angles. In the same figure, three different frames are introduced. The first frame, $\{\mathcal{F}\}$, is the frame of the BP with origin at the intersection of the axes of the two actuated joints. The frame associated with the MP in the undeformed configuration is $\{\mathcal{U}\}$, with the origin in common with $\{\mathcal{F}\}$ in $\mathbf{O}$; the relative rotation between $\{\mathcal{U}\}$ and $\{\mathcal{F}\}$ is due to the kinematics of the machine. When an external load is applied, the limbs are subjected to elastic deformations. As a result, the MP moves from its theoretical position to a new position identified by the frame $\{\mathcal{D}\}$. In this case, the center of the MP is shifted from $\mathbf{O}$ to $\mathbf{O}_D$ by a vector called $\boldsymbol{s}$, while the relative orientation between $\{\mathcal{U}\}$ and $\{\mathcal{D}\}$ is described by a rotation matrix called ${}^{D}_{U}\mathbf{R}$.

The mechanical design of the wrist has been carried out until the realization of the prototype shown in Fig. 2. All of the links composing the limbs are made of steel, such as the frame, while the MP is made of aluminum alloy. An auxiliary plate and a pyramidal frame are shown in the figure, but they are not considered part of the wrist in the remainder of this study; in fact, such components constitute a particular design conceived for a demo task presented in Palmieri et al. (2015). According to Al-Widyan et al. (2011), all passive joints of the kinematic chain, which are ideally revolute joints, have been realized with bushings which also allow for small axial displacements; however in practice, they realize

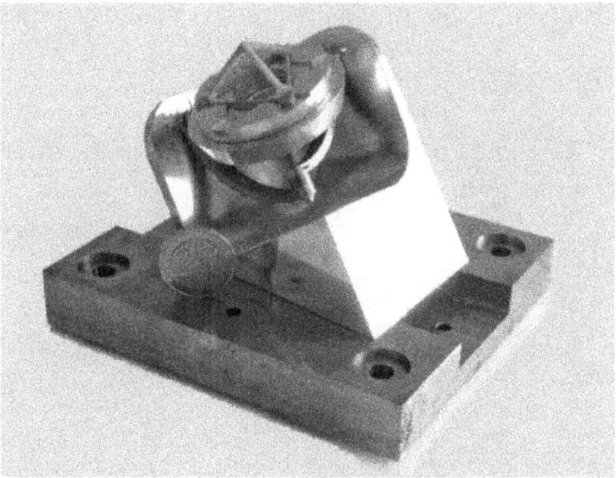

Figure 2. Prototype of the parallel spherical wrist.

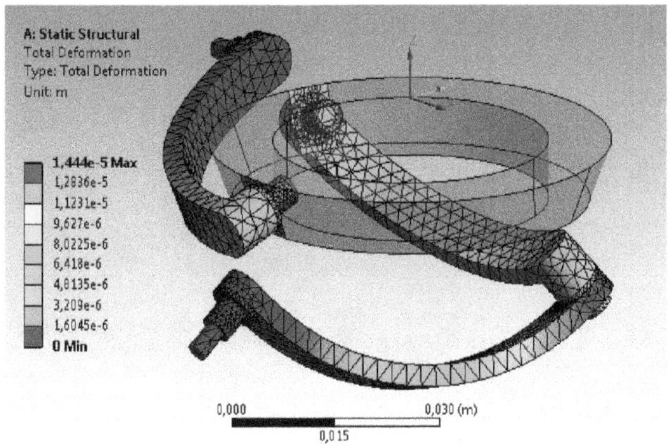

Figure 3. Mesh of the flexible components of the wrist and map of the total displacement for a load equal to $F = -1$ N along the z axes applied to the center of the MP.

cylindrical joints instead of revolute joints. With this expedient, the wrist behaves like a non-overconstrained mechanism that is able to cope with the manufacturing errors: machining tolerances lead to joint axes that inevitably do not intersect at a common point; thus the use of revolute joints would give a hyperstatic structure subjected to deformation when the links are assembled. On the other hand, the addition of extra DOFs in the passive joints introduces an undesired mobility which affects the accuracy and the stiffness of the machine. Due to the small dimensions of the wrist components and to the necessity of reduced friction, miniaturized polymeric bushings have been used; the low Young's modulus of such components introduces deformations concentrated at joints, which have been taken into account in the subsequent elasto-static analysis.

3 Finite-element model

The finite-element model was created in Ansys Workbench using the static structural module. A simplified CAD model of the wrist was imported, eliminating small features like chamfers and fillets and small components like screws and fasteners. The frame of the machine and the MP are considered perfectly rigid, while links and bushings at the joints are flexible. The material assigned to the links is steel (Young's modulus $E = 200$ GPa), while bushings are modeled in polymeric material with a Young's modulus of 2.4 GPa. The mesh is characterized by tetrahedral elements with refinements at the contact surfaces for a total of about 16 000 elements and 28 000 nodes (Fig. 3). The size of the elements was set by a sensitivity analysis: the detail level of the mesh was refined until results stabilized. Kinematic pairs have been introduced in order to replicate the isostatic kinematic structure of the wrist: all the bushings of the passive joints are fixed with the inner pin, while they are connected to the outer hub by a cylindrical pair; furthermore, frictionless surface contacts have been introduced between the flanges of the bushings and the corresponding surfaces of the connected bodies. Figure 4 shows the behavior of one of the modeled joints, where the translation of the cylindrical joint, in addition the rotation, is evident. The two bushings which support the pins connected to the motors are fixed to ground; each one of these bushings is connected to the corresponding pin by a revolute joint, and the relative corresponds to the motor rotation. By varying the motor angles, by means of the joint configuration tool, it is possible to configure the wrist in different positions in which the elasto-static analysis is then performed. A unitary vertical downward force is applied at the center of the MP. Such force simulates the interaction with an external manipulator during an assembly operation: the wrist orients an object that is aligning one of its surfaces with the horizontal plane, and the manipulator pushes against it in the normal direction. The magnitude of the force, set at 1 N, corresponds to the payload of the wrist, equal to 100 g. The deformation of the wrist is evaluated for each configuration by reading the displacement of seven points fixed to the MP; such data will be used in a subsequent procedure for the estimation of MP positioning error.

4 Positioning error estimation

Given a series of n points fixed to the mobile platform, the position of the ith point in the undeformed configuration with respect to the global fixed frame is called $\mathbf{P}_{i,U} = \left[x_{i,U}\ y_{i,U}\ z_{i,U}\ 1\right]^T$; the homogeneous representation is used. After the load is applied, each one of those points moves according to the rigid body motion of the MP due to the elastic deformation of the limbs; its position in the deformed configuration is $\mathbf{P}_{i,D} = \left[x_{i,D}\ y_{i,D}\ z_{i,D}\ 1\right]^T$, again with respect to the global fixed frame. The following set of equations must be satisfied:

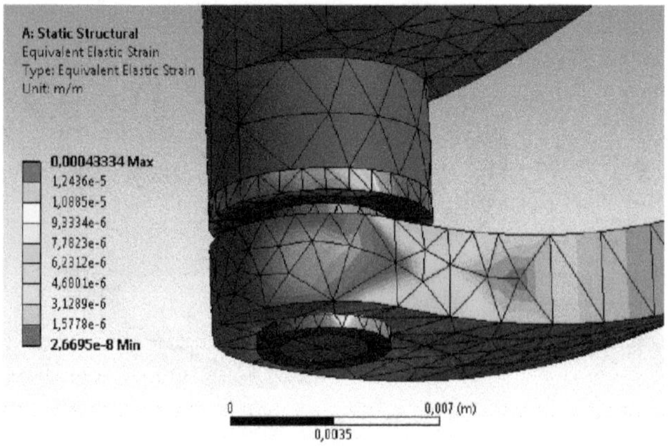

Figure 4. Modeling of the bushings: plot of the equivalent elastic strain on the deformed configuration (scale 100 ×).

$$\mathbf{P}_{i,D} = {}^{D}_{U}\mathbf{T}\mathbf{P}_{i,U} \quad i = 1, 2, \ldots, n. \tag{1}$$

${}^{D}_{U}\mathbf{T}$ in Eq. (1) is the 4 × 4 homogeneous matrix representing the MP rigid motion of roto-translation due to the deformation of the limbs. As known, ${}^{D}_{U}\mathbf{T}$ is composed of a rotation matrix ${}^{D}_{U}\mathbf{R}$ and a vector $s = (\mathbf{O}_D - \mathbf{O})$, representing the relative orientation and the displacement between the origins of the undeformed $\{\mathcal{U}\}$ and the deformed $\{\mathcal{D}\}$ frames, respectively:

$$ {}^{D}_{U}\mathbf{T} = \left[\begin{array}{c|c} {}^{D}_{U}\mathbf{R} & s \\ \hline \mathbf{0}^T & 1 \end{array}\right]. \tag{2}$$

The coordinates of a series of n points in the undeformed and deformed configuration are known from the results of the FEM analysis. They can be collected in the following matrices of dimension $4 \times n$:

$$\mathbf{A}_U = \left[\mathbf{P}_{1,U} \cdots \mathbf{P}_{n,U}\right] \quad \mathbf{A}_D = \left[\mathbf{P}_{1,D} \cdots \mathbf{P}_{n,D}\right]. \tag{3}$$

Thus, on the basis of Eq. (1), a first estimation of the roto-translation matrix ${}^{D}_{U}\mathbf{T}$ can be obtained as the result of an unconstrained least-squares problem:

$$ {}^{D}_{U}\widetilde{\mathbf{T}} = \mathbf{A}_D\mathbf{A}_U^{\dagger}, \tag{4}$$

where † stands for the pseudo-inverse operator[1] and $\widetilde{\ }$ is referred to first-estimate entities. Due to the numerical approximations of the FEM analysis, ${}^{D}_{U}\widetilde{\mathbf{T}}$ does not match the constraints of a rigid body motion: its 3 × 3 submatrix ${}^{D}_{U}\widetilde{\mathbf{R}}$, in particular, does not satisfy the orthogonality condition. A proper rotation matrix ${}^{D}_{U}\mathbf{R}$ can be found as the orthogonal factor of the polar decomposition[2] (Baron and Angeles, 2000; Higham, 1986) of the first unconstrained estimate ${}^{D}_{U}\widetilde{\mathbf{R}}$:

$$ {}^{D}_{U}\widetilde{\mathbf{R}} = \mathbf{Q}\mathbf{W}{}^{D}_{U}\mathbf{R} = \mathbf{Q}. \tag{5}$$

Once ${}^{D}_{U}\mathbf{R}$ is known, it is possible to rewrite the system of Eq. (1) in the form

$$\mathbf{A}_D = \left[\begin{array}{c|c} {}^{D}_{U}\mathbf{R} & \mathbf{0} \\ \hline \mathbf{0}^T & 1 \end{array}\right]\mathbf{A}_U + \mathbf{S}, \tag{6}$$

where

$$\mathbf{S} = \begin{bmatrix} s_1 & \cdots & s_n \\ 1 & & 1 \end{bmatrix}. \tag{7}$$

Finally,

$$\mathbf{S} = \mathbf{A}_D - \left[\begin{array}{c|c} {}^{D}_{U}\mathbf{R} & \mathbf{0} \\ \hline \mathbf{0}^T & 1 \end{array}\right]\mathbf{A}_U. \tag{8}$$

In the theoretical assumption of a rigid body motion, all vectors s_i obtained from Eq. (8) are equal. In the actual case, they differ because of numerical approximations; thus, a least-squares problem has to be solved in order to find the translation vector s that minimizes the error function e:

$$e = \frac{1}{n}\sum_{i=1}^{n} \| s - s_i \|^2 \underset{s}{\rightarrow} \min. \tag{9}$$

For data with a Gaussian noise distribution, it can be demonstrated (Lenarcic and Parenti-Castelli, 2001) that the result of the least-squares minimization is coincident with the mean value of the scattered samples. Thus, in our case, it is reasonable to assume that

$$s = \frac{1}{n}\sum_{i=1}^{n} s_i. \tag{10}$$

The norm of s represents the magnitude of the displacement of the MP center as a consequence of the elastic deformation. In order to also describe the orientation error by means of a scalar quantity, an axis-angle representation of the rotation matrix ${}^{D}_{U}\mathbf{R}$ is used; the angle ϕ obtained from such a representation represents the magnitude of the orientation error of the MP.

The previously outlined procedure has been carried out for 25 different configurations of the wrist, which form an equally spaced grid over the joint space. Figures 5 and 6 show the plot of the displacement and orientation error, respectively, in the workspace of the wrist. Maps are obtained as interpolation of the data grid using biharmonic spline functions. A black solid line borders a subspace, called operative workspace, which represents the required range of motion for the particular application the wrist is designed for (Palmieri et al., 2015). The maximum shift s of the MP center can reach up to 70×10^{-3} mm at the borders of the workspace, while the maximum value of ϕ is of about $5.5 \times 10^{-2\circ}$. Considering the data confined inside the operative workspace,

[1] The Moore–Penrose pseudo-inverse of a $m \times n$ rectangular matrix $\mathbf{A}$ with $m < n$ can be directly calculated as $\mathbf{A}^{\dagger} = \mathbf{A}^T\left(\mathbf{A}\mathbf{A}^T\right)^{-1}$.

[2] The polar decomposition of a square matrix $\mathbf{A}$ is a matrix decomposition of the form $\mathbf{A} = \mathbf{Q}\mathbf{W}$, where $\mathbf{Q}$ is an orthogonal matrix and $\mathbf{W}$ is a positive semi-definite Hermitian matrix.

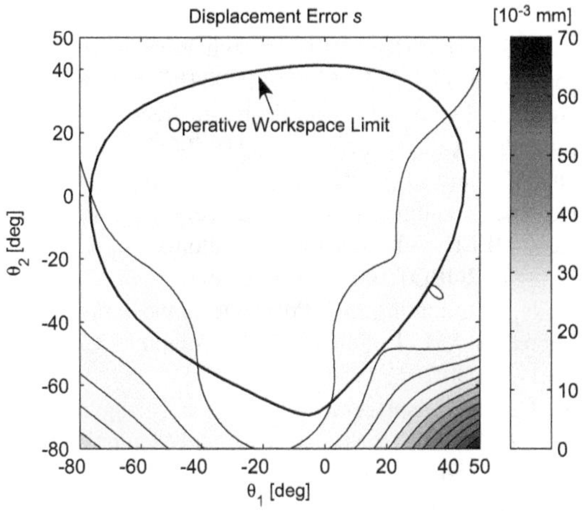

Figure 5. Displacement error s of the MP mapped in the joint space.

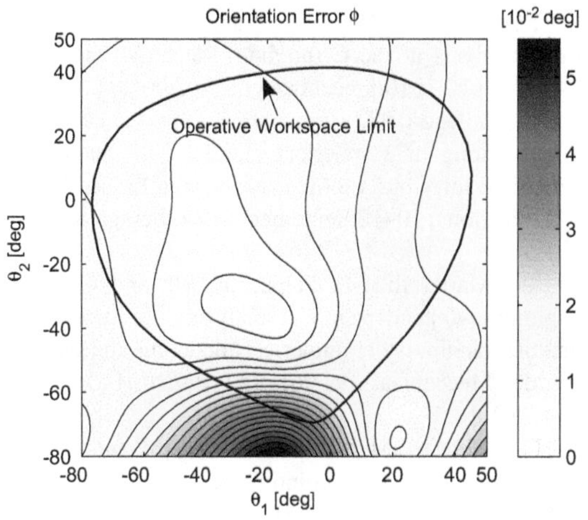

Figure 6. Orientation error ϕ of the MP mapped in the joint space.

the positioning errors are much lower: the mean, minimum and maximum values of s and ϕ are shown in Table 1; the mean value of the orientation error ($1.29 \times 10^{-2\circ}$), in particular, is comparable with the angular resolution of the wrist ($1.5 \times 10^{-2\circ}$) related to the resolution of the encoders. The positions assumed by the center of the MP in the configurations analyzed inside the operative workspace generate a cloud of points whose convex hull is represented in Fig. 7. The location and the dimension of the hull can be evaluated on the basis of the statistical values of the components of the vector $\boldsymbol{s}$ with respect to the frame $\{\mathcal{F}\}$, which are shown in Table 2.

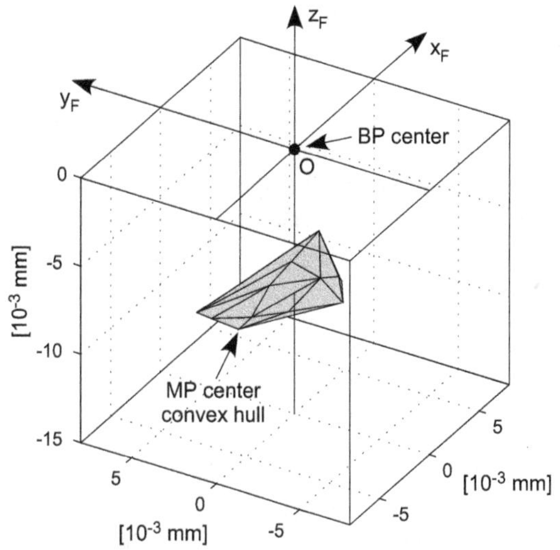

Figure 7. Convex hull of the positions of the MP center for a set of configurations of the wrist inside the operative workspace.

Table 1. Statistical values of the displacement error s and orienting error ϕ inside the operative workspace.

		Mean	Min	Max
s	$[10^{-3}\,\text{mm}]$	8.73	5.50	14.78
ϕ	$[10^{-2\circ}]$	1.29	0.64	3.61

Table 2. Statistical values of the components of the displacement error vector $\boldsymbol{s}$ inside the operative workspace.

		Mean	Min	Max
s_x	$[10^{-3}\,\text{mm}]$	1.17	−2.06	5.12
s_y	$[10^{-3}\,\text{mm}]$	0.53	−2.12	5.63
s_z	$[10^{-3}\,\text{mm}]$	−8.39	−13.51	−5.11

5 Conclusions

The elasto-static analysis of a 2-DOF spherical parallel wrist where the links and the joints are considered flexible is described in this paper. An FEM approach has been proposed: starting from a simplified CAD of the manipulator, a reconfigurable finite-element model of the wrist has been implemented in Ansys Workbench. An external load is applied to the MP, simulating a cooperative assembly operation, and the consequent elastic deformation is evaluated. The displacements of a set of points fixed to the MP are then processed in a least-squares minimization procedure aimed to find the roto-translation that best fits the motion of the MP. Once the displacement and orientation errors are found for a series of configurations of the wrist, continuous maps over the workspace of the machine are obtained by interpolation. From examination of the operative workspace, it can be seen that the average orientation error is on the order of $10^{-2\circ}$, which is compliant with the requirements of the wrist; moreover, the wrist exhibits a displacement error with an average value on the order of 10^{-2} mm.

The main contribution of the work is the development of a novel methodology that allows for configuration-dependent stiffness analyses of manipulators with flexible limbs to be performed. The methodology is based on the interaction between multibody and FEM simulation tools and represents an alternative, or better an integration, to simplified analytic methods. Besides the evaluation of the positioning error, the proposed method can also be applied to perform an elastodynamic analysis of the wrist once a specific end-effector design is defined and mass properties of all components are known.

Acknowledgements. The present research was developed in the framework of the project PRIN2009 MM & A, funded by the Italian Ministry for Education, Universities and Research (MiUR) and jointly carried out by the Polytechnic University of Marche, the University of Brescia, University of Bergamo and the Institute of Industrial Technologies and Automation of the Italian National Research Council.

References

Al-Widyan, K., Ma, X. Q., and Angeles, J.: The robust design of parallel spherical robots, Mech. Mach. Theory, 46, 335–343, doi:10.1016/j.mechmachtheory.2010.11.002, 2011.

Asada, H. and Granito, J. A. C.: Kinematic and static characterization of wrist joints and their optimal design, in: vol. 2, IEEE T. Robot. Automat., 2, 244–250, 1985.

Baron, L. and Angeles, J.: The direct kinematics of parallel manipulators under joint-sensor redundancy, IEEE T. Robot. Automat., 16, 12–19, doi:10.1109/70.833183, 2000.

Cammarata, A., Condorelli, D., and Sinatra, R.: An Algorithm to Study the Elastodynamics of Parallel Kinematic Machines With Lower Kinematic Pairs, J. Mech. Robot., 5, 011004, doi:10.1115/1.4007705, 2012.

Di Gregorio, R.: The 3-RRS Wrist: A New, Simple and Non-Overconstrained Spherical Parallel Manipulator, J. Mech. Design, 126, 850–855, doi:10.1115/1.1767819, 2004.

Dong, G., Song, Y., Sun, T. S., and Lian, B.: Elasto-dynamic Analysis of a Novel 2-DoF Rotational Parallel Mechanism, in: Proceedings of 2014 Workshop on Fundamental Issues and Future Research Directions for Parallel Mechanisms and Manipulators 7–8 July 2014, Tianjin, China, 2014.

Gosselin, C. and Angeles, J.: The Optimum Kinematic Design of a Spherical Three-Degree-of-Freedom Parallel Manipulator, J. Mech. Design, 111, 202–207, doi:10.1115/1.3258984, 1989.

Gosselin, C. and Caron, F.: Two degree-of-freedom spherical orienting device, United States patent US 5966991, 1999.

Gosselin, C. and Hamel, J.-F.: The agile eye: a high-performance three-degree-of-freedom camera-orienting device, IEEE T. Robot. Automat., 1, 781–786, doi:10.1109/ROBOT.1994.351393, 1994.

Higham, N.: Computing the Polar Decomposition - with Applications, SIAM J. Scient. Stat. Comput., 7, 1160–1174, doi:10.1137/0907079, 1986.

Kong, X.: Forward Displacement Analysis and Singularity Analysis of a Special 2-DOF 5R Spherical Parallel Manipulator, J. Mech. Robot., 3, 024501, doi:10.1115/1.4003445, 2011.

Lenarcic, J. and Parenti-Castelli, V.: A method for determining movements of a deformable body from spatial coordinates of markers, J. Robot. Syst., 18, 731–736, doi:10.1002/rob.8111, 2001.

Majou, F., Gosselin, C., Wenger, P., and Chablat, D.: Parametric stiffness analysis of the Orthoglide, Mech. Mach. Theory, 42, 296–311, doi:10.1016/j.mechmachtheory.2006.03.018, 2007.

Palmieri, G., Callegari, M., Carbonari, L., and Palpacelli, M.: Design and testing of a spherical parallel mini manipulator, in: IEEE/ASME Int. Conf. on Mechatronic and Embedded Systems and Applications, 10–12 September 2014, Senigallia, AN, Italy, 2014a.

Palmieri, G., Martarelli, M., Palpacelli, M., and Carbonari, L.: Configuration-dependent modal analysis of a Cartesian parallel kinematics manipulator: numerical modeling and experimental validation, Meccanica, 49, 961–972, doi:10.1007/s11012-013-9842-4, 2014b.

Palmieri, G., Callegari, M., Carbonari, L., and Palpacelli, M.: Mechanical design of a mini pointing device for a robotic assembly cell, Meccanica, doi:10.1007/s11012-015-0132-1, in press, 2015.

Pashkevich, A., Chablat, D., and Wenger, P.: Stiffness analysis of overconstrained parallel manipulators, Mech. Mach. Theory, 44, 966– 982, doi:10.1016/j.mechmachtheory.2008.05.017, 2009.

Wu, G., Bai, S., and Kepler, J.: Mobile platform center shift in spherical parallel manipulators with flexible limbs, Mech. Mach. Theory, 75, 12–26, doi:10.1016/j.mechmachtheory.2014.01.001, 2014.

2

Flexibility oriented design of a horizontal wrapping machine

H. Giberti[1] and A. Pagani[2]

[1]Politecnico Di Milano, Dipartimento di Meccanica, Campus Bovisa Sud, via La Masa 1, 20156, Milano, Italy
[2]Fpz S.p.a., Via Fratelli Cervi, 18, 20049 Concorezzo (MB), Italy

Correspondence to: H. Giberti (hermes.giberti@polimi.it)

Abstract. Flexibility and high production volumes are very important requirements in modern production lines. In most industrial processes, in order to reach high production volumes, the items are rarely stopped into a production line and all the machining processes are executed by synchronising the tools to the moving material web. "Flying saw" and "cross cutter" are techniques widely used in these contexts to increase productivity but usually they are studied from a control point of view.

This work highlights the kinematic and dynamic synthesis of the general framework of a flying machining device with the emphasis on the driving system chosen and the design parameter definition, in order to guarantee the required performance in terms of flexibility and high production volumes. The paper develops and applies a flexibility oriented design to an horizontal wrapping machine.

1 Introduction

In modern production systems it is increasingly important to increase productivity and at the same time ensure high flexibility levels with respect to the change of product or the size thereof. These requirements are by definition antithetical (Sethi and Sethi, 1990; Shewchuk and Moodie, 1998; Matthews et al., 2006). It is difficult for high production machines to elaborate a range of highly diversified goods. On the other hand it is difficult for flexible machines to reach high production levels.

In most industrial processes, in order to reach high production volumes, the items pass through the production line in a continuous way. Thus the items or process are rarely stopped and all the machining processes are executed with the items in movement. Therefore the tools have to be synchronised to the moving material web and after the machining process, those tools have to be positioned at the starting point for the next cycle.

Processes such as welding, embossing, printing, cutting, sealing, gripping, etc., normally found in a production line, are by their very nature not continuous. In these cases the manufacturing processes have to be executed when the item is stopped. Thus the production line works in an intermittently way. To eliminate the wasting of time in stopping and restarting the line it is necessary that the tools follow the items.

Regardless of the industrial field, when the tool moves along a rectilinear trajectory, the application is generally called "flying saws" (Diekmann and Luchtefeld, 2008) and the tool is mounted on a slide that moves together with the piece to be worked. After the machining process has been completed, the tool returns to its original position ready for the next work cycle. Alternatively if the tool moves along a closed trajectory, usually a circular one, the flying tool is referred to as a cross cutter (Diekmann and Luchtefeld, 2008). These kind of manufacturing processes are generally referred to as "flying machining" and several devices have been developed to perform these in various industrial fields.

Regardless of the industrial sector and the flying machining solution chosen, the design set of problems and the methods of controlling the system are the same. As shown in Strada et al. (2012) in fact flying saw and cross cutter systems could been parametrized and studied in an analogous way. Obviously technical solutions developed to move the tools are different but the methodology to synthesise the system could be considered similar.

It is possible to find several papers that have been published regarding flying machining but none of these deals with the problem in general: each one regards specific cases. Most of them are about the control problem. In Varvatsoulakis (2009), for example, a new digital control system has been designed and implemented in order to replace the existing obsolete one in a cutting system into a production line of STAHL-37 steel tubes. In this case the existing hardware of the cutting system (motor, drive, mechanical equipment) has been maintained. A similar approach is presented in Bebic et al. (2012), but in this case the authors suggest substituting the drive and control systems in order to improve the performance of the cross cuter in the paper or board production line. For these purposes a close examination of the characteristics and requirements of basic subsystems of the paper-board cross cutter from the control system perspective is done.

The control system is studied in depth in Wu et al. (2014). In that work the authors propose a control architecture based on ARM and FPGA to reach high-speed, high-precision, high dynamic, high rigidity performance in a flyng shear cutting system. In view of the increase in demand in the face of the increasing of wrapper machine request for wrapping machines, particularly in the Chinese market (Wu, 2010), the authors of the paper Shao et al. (2012) show a synchronizing servo motion and an iterative learning control useful for horizontal flow wrapper. Also in this case the focus of the work is on the control system and on the architecture whereby one can obtain good cutting accuracy and eliminate the repeatable position error. The control problems have been widely studied since the second half of the last century Shepherd, 1964. With the spread of new electronic devices the control approach changed shifting from analog solutions to digital ones Visvambharan, 1988 up to the more modern approaches mentioned above.

These studies address the control system in reaching the required performance and no analysis is addressed on the layout of the cutting tool. In Peric and Petrovic (1990) an optimal control system is considered in order to minimize the driving torque. In this case kinematic and dynamic are taken into account but without a detailed study on the effects that the design parameters have in terms of attainable productivity. A proposal for the revision of the cross cutter system layout is presented in Hansen et al. (2003). The authors suggest operating the cutter by separately controlled servo drives but, also in this case the focus remains on how to control the cutter position.

This work highlights the kinematic and dynamic synthesis of a general flying machining device. Particular attention is paid to the choose of the driving system and the design parameters, so as to guarantee the required performance in terms of flexibility and high production volumes. By virtue of the generalisation set up in Strada et al. (2012) the design method is refers to the cross cutter solution which is widespread in food packaging systems.

The focus of this study is on a flexibility oriented design procedure which takes into account the input parameters necessary to avoid limitations and constraints to the potential of the machine. A general framework is provided, allowing the designer to assess different possible motor-reducer solutions and design parameter combinations, taking into account the various advantages or limitations in term of flexibility. This new approach satisfies two requirements. The first one can verify, theoretically the cutting flexibility in an existing cutting machine. The second can design a new cutting machine capable of reaching a much higher production flexibility level.

This work is organized as follows. In Sect. 2, the flying cutting machine is described. In Sect. 3 the motion laws adopted to perform the cutting operation are set out and analysed while their effects on the dynamic loads are set out in Sect. 4. In Sect. 5 case study simulations and results are presented. In conclusion in Sect. 6 the final considerations are summarized.

2 Flow-pack systems: horizontal wrapping machine

A particularly lively industry in which the flying machining is used is the packaging field. A packaging machine is a system used to cover wholly or partially single items or collected group of them with a flexible material. Wrapping machine is a kind of packaging machine that is used to wrap small items with paper or plastic film. The first noted wrapping machine was developed by William and Henry Rose, in England at the end of the nineteenth century (Hooper, 1999).

The typical layout of a flow-pack machine is depicted in Fig. 1. A specific wrapping machine is taken as an example in order to support the theoretical background with a numeric example. It is worth noting that the following considerations are general and not related to a specific flying cutting technology application. The purpose of this kind of machine is to weld and cut the double plastic film that will form the package, while the product is already between them. The plastic film is unreel by the film feed roller and passes through the forming box that folds it in the final configuration. It is important to note that the product arrives on a conveyor-belt and the plastic film is bent around it. The product moves forward to the unit that package it. A couple of rotating heads are used to execute these operations.

Usually, they are synchronous, having the same motor and control unit, even if some attempts to adopt an asynchronous control strategy have been made (Hansen et al., 2003). On their external circumferences, n tools are mounted with the double purpose to weld and cut the packages. In fact, each tool is constituted by a central saw profile to cut each package, whose ends are simultaneously welded by heat-seals units fitted on the side of the saw profile.

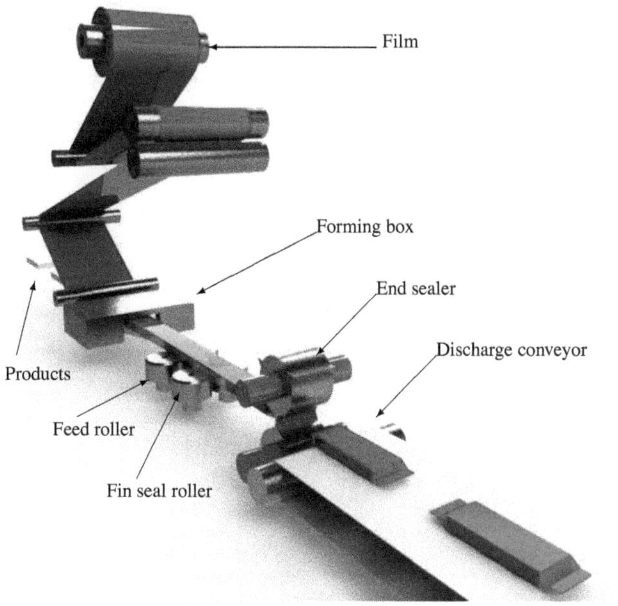

Figure 1. Wrapping machine.

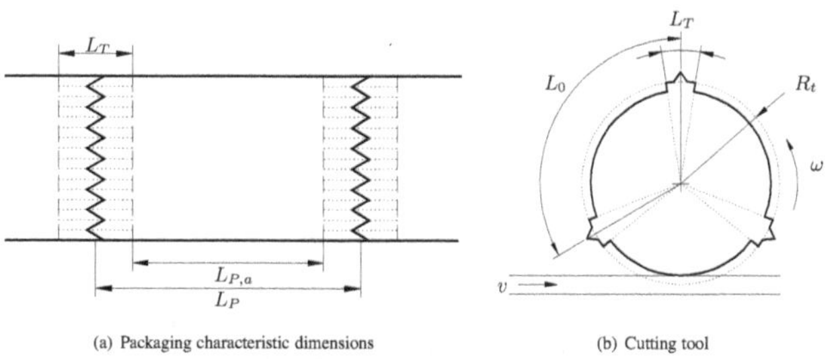

Figure 2. Characteristic dimensions.

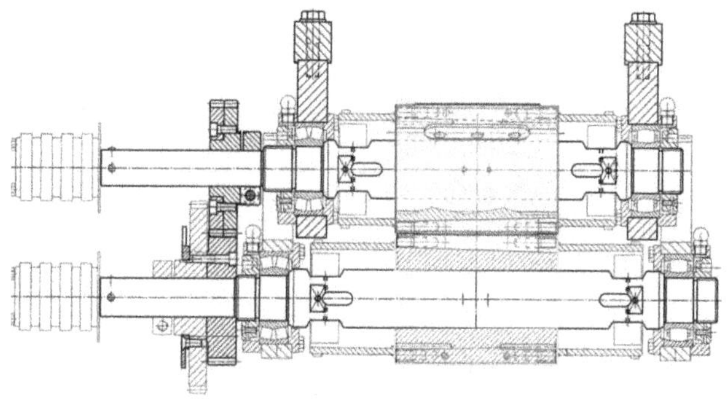

Figure 3. Rotating heads.

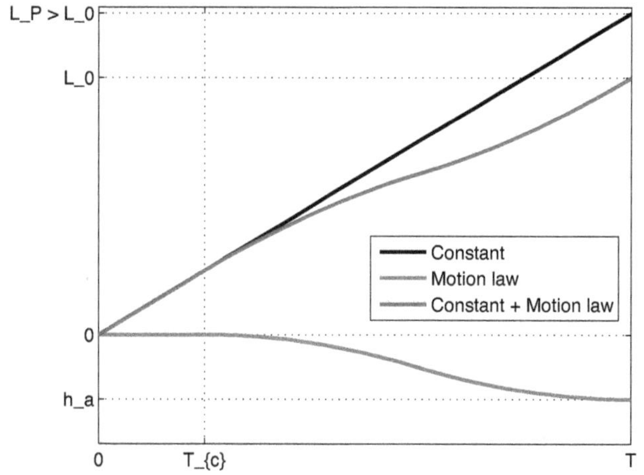

Figure 4. Motion law superposition.

2.1 Design framework

Each package is composed by three parts as sketched in Fig. 2a: two welded terminals ($L_T/2 + L_T/2$) and the central part where the object to package lies ($L_{P,a}$). It is important to highlight that the expression "product length" L_P used in this work refers to the total length of the packaged unit and not only to the length of the object to be packaged (L_a).

Thus, the dimensional parameters L_T and L_P are the starting dimensions to design the machine.

The more suitable working condition is to have a constant angular speed in order to have negligible dynamic loads and thus this is the nominal working condition. It correspond to an established product length defined as "base" or "design" length L_0. In every other cases, if the product length is different from the design one, acceleration or deceleration are required in order to account for the imposed target product length. Typically, the base length L_0 is provided by the costumer because it represents the most common length and thus the target of the designer is to set up a machine which shows the best performance in this configuration.

Thus, the radius R_t of the rotating head (Fig. 2b) is defined in order to obtain a circumference which length is proportional to the design length itself:

$$2\pi R_t = N L_0$$

The integer ratio $N = 2\pi R_t / L_0$ corresponds to the number of cutting tools to be installed onto the rotating head. The dimension of the rotating head R_t is usually bounded by the layout configuration of the machine (Fig. 3).

3 Laws of motion

One of the main characteristics of an automated machine is its productivity: it represents the starting point to define the kinematic link between each part of the whole mechanism. To satisfy the assigned productivity P of a product with a length L_P, the conveyor-belt has to maintain a constant velocity v equal to:

$$v = \frac{L_p \cdot P}{60}$$

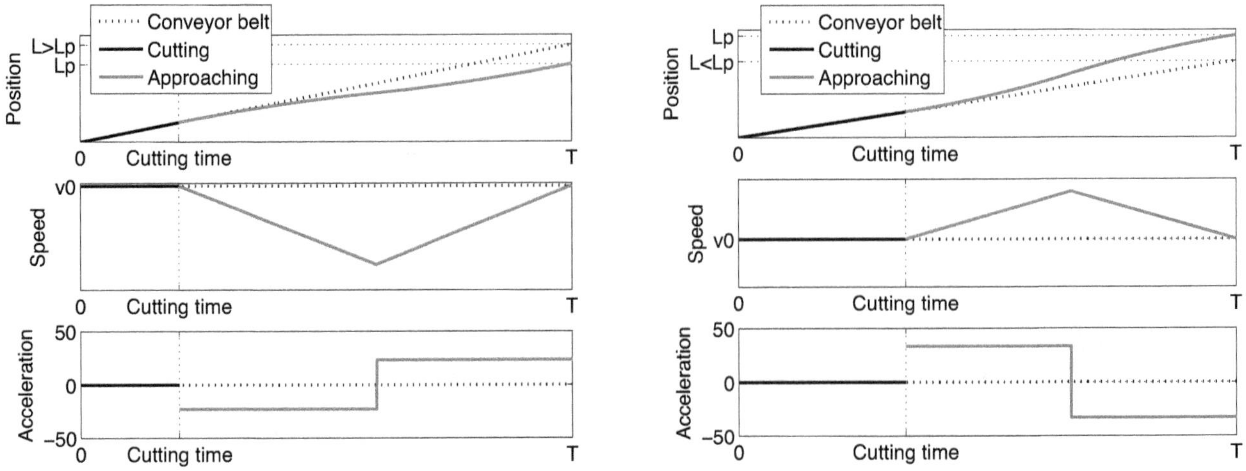

(a) (b)

Figure 5. Motion sequence of the sealing process (L_P L_0).

P is usually expressed in pieces $\min^{-1}$, the cycle time is equal to $T = 60/P$. The total time T is defined as the sum of the duration of two phases:

$$T = T_t + T_a$$

- the cutting phase T_t. It is the part of the cycle dedicated to weld and cut the packaging of the product,
- the approaching phase between two cuts. It corresponds to the time T_a from the finish of a cut and the beginning of the following one.

3.1 Cutting phase

During the cutting phase the angular velocity of the rotating head is kept constant to cut and weld the packaging properly. The tangential velocity of the rotating head has to be equal to the one of the conveyor-belt v, resulting in a null relative velocity between them. Thus, the conveyor-belt velocity v can be also defined as:

$$v = \frac{L_t}{T_t}$$

because during the cutting phase T_t the conveyor-belt shift of a distance equal to L_t. This condition allows to define the angular velocity ω_t of the rotating head during the cutting phase:

$$\omega_t = \frac{v}{R_t} = \frac{L_t}{R_t T_t}$$

Usually the length of the welded part of the packaging is defined a-priori and does not depend to the product length L_P. Thus, the length L_t is not a design parameter for the law of motion because it is imposed by the dimension of the cutting tools and it is typically defined by the costumer.

3.2 Approaching phase

As mentioned above, if the product length is equal to the design one ($L_P = L_0$) the rotating head maintains during the approaching phase a constant angular velocity ω_a equal to the one of the cutting phase (ω_t). If the product length is greater or smaller than the design one, the angular velocity of the rotating head during the approaching phase must decrease or increase to properly repositioning the tool for the next cutting phase. A smart approach to generalize the problem is to describe the angular velocity ω as the sum of two contributes:

- ω_t: the constant angular velocity that allows to cut the L_0 length,
- $\omega_a = \omega_t + \Delta\omega$ where $\Delta\omega$ the variation of angular velocity needed to get the tools in the correct position to execute the next cut.

The variation of the angular speed $\Delta\omega$ depends on the product length and the design length. It is null only if the product length is equal to the design one. In the other cases to define its value it can be convenient to consider the equivalent linear path of the tool as a function of time. Using this different point of view it is possible to define the law of motion of the tool as the superimposition of the path corresponding to the constant angular speed and of the "Δ" path needed to reach at the correct position and time the package to process

(Fig. 4), considering that its duration is equal to the one of the approaching phase one and it correspond to a linear distance equal to $h_a = L_O - L_t$.

Two conditions can be reached:

- $L_P < L_O$. The approaching length h_a is greater than the required one: $h = L_P - L_t$. The rotating heads must accelerate to recover this additional length.
- $L_P > L_O$. The approaching length h_a is smaller than the required one: the rotating heads must decelerate. In extreme cases, it must rest or reverse the rotation direction.

A motion law with a total lift equal to $h = L_P - L_t$ and a duration time equal to $T_a = T - T_t$ is adopted to perform the modulation of the rotating heads velocity. In Fig. 5 both the cases above described are shown. The dotted line represents the feed of the conveyor-belt. Being its speed constant, as a function of time, it has a linear trend, starting from zero and ending at the processed length L_P. During the cutting phase, the feed of the rotating heads is the same of the one of the conveyor-belt, being null the relative velocity between them. If the product length is longer than the design one, the rotating head have to slow down (Fig. 5a). If it is smaller than the design one, the rotating head must increase its angular velocity in order to recover the length deficit as reported in Fig. 5b.

3.3 Dimension-less design of motion laws

Named $y(t)$ the path of the rotating head during the approaching phase, it is important to note that its "shape" is not defined a priori. In fact, some different laws of motion, even if they result in very similar behavior in the positioning, differ in relevant ways if the corresponding accelerations are analyzed as shown for three different motion laws in Fig. 6.

Each law of motion can be expressed using a dimension-less space and time parameters:

$$\zeta = \frac{y(t)}{h} \xi = \frac{t}{T_a}$$

The results is that the law of motion is totally describable using the corresponding $\zeta = \zeta(\xi)$ function, with $0 \leq \zeta \leq 1$ and $0 \leq \xi \leq 1$. The velocity and the acceleration are obtainable using the following differential relations:

$$\dot{y} = \frac{\mathrm{d}y}{\mathrm{d}t} = \frac{\mathrm{d}(h\zeta)}{\mathrm{d}(T_a\xi)} = \zeta' \frac{h}{T_a}$$

$$\ddot{y} = \frac{\mathrm{d}\dot{y}}{\mathrm{d}t} = \frac{h}{T_a}\frac{\mathrm{d}\zeta'}{\mathrm{d}t} = \frac{h}{T_a}\frac{\partial \zeta'}{\partial \xi}\frac{\mathrm{d}\xi}{\mathrm{d}t} = \zeta'' \frac{h}{T_a^2}$$

being ζ' and ζ'' the dimensionless expressions of velocity and acceleration, respectively.

Every law of motion must satisfy null speed both at the starting and at the ending time instants ($\zeta'(\xi = 0) = \zeta'(\xi = 1) = 0$), while it must provide the correct lift starting from $\zeta(\xi = 0) = 0$ reaching $\zeta(\xi = 1) = 1$ at the end. As a consequence, it can be demonstrated that the only constrains on the dimensionless acceleration ζ'' are:

$$\int_0^1 \zeta''(\xi)\mathrm{d}\xi = 0 \qquad \int_0^1 \zeta''(\xi)\zeta\mathrm{d}\xi = -1$$

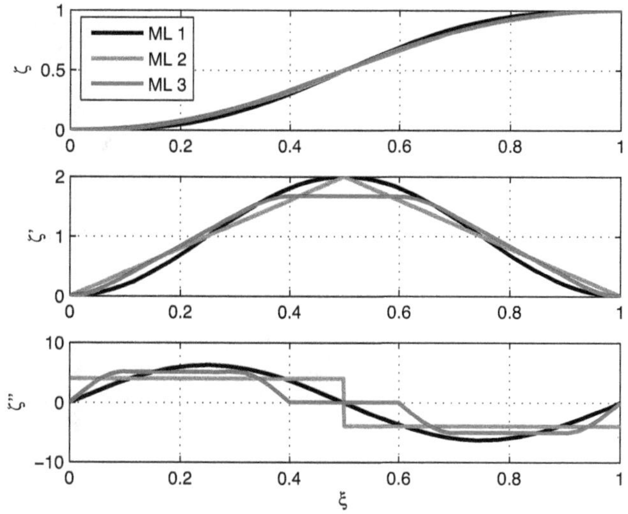

Figure 6. Comparison between different motion laws.

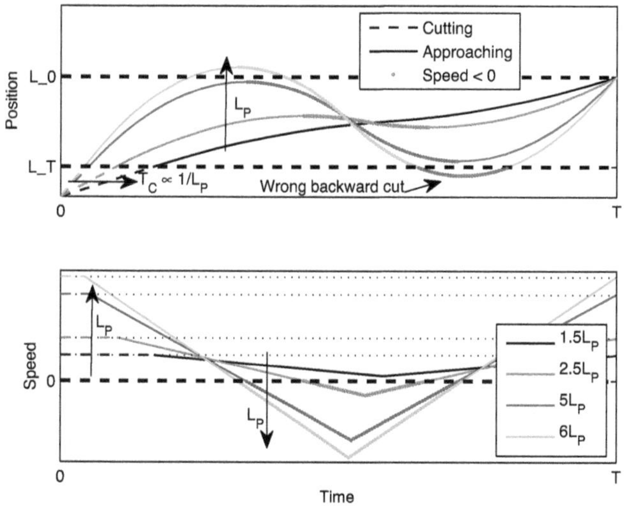

Figure 7. Backward cut increasing the product length.

Using the dimensionless form to describe the laws of motion, some coefficients can be defined to capture several of their notable properties. Thus, it is possible to define the dimension-less speed coefficients C_v, that is useful to take into account the peak value of the speed, defined as:

$$C_v = \dot{y}_{max} \frac{h}{t_a}$$

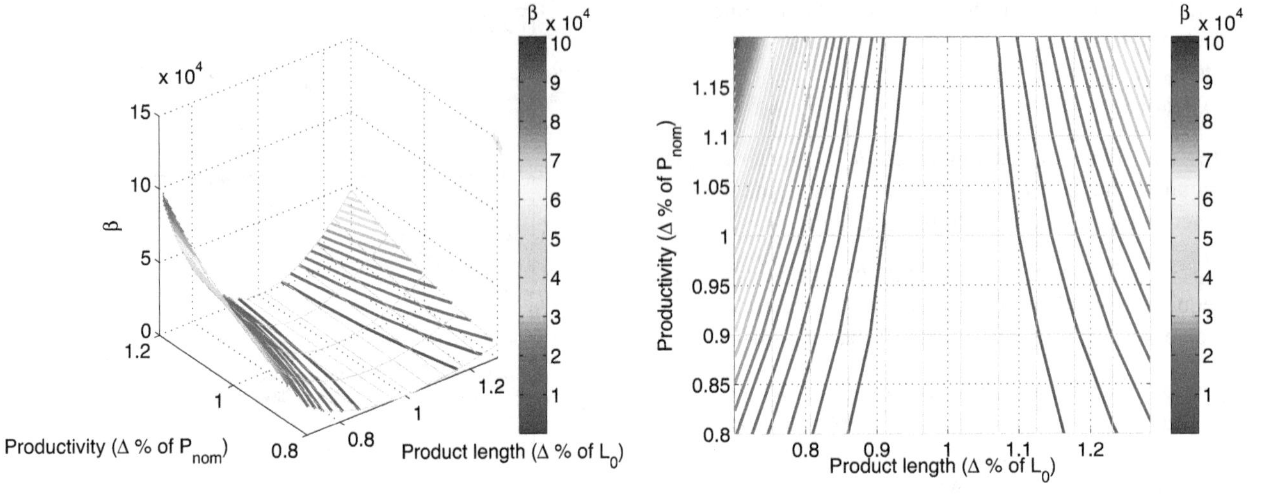

(a) (b)

Figure 8. β as a function of product length and productivity.

Table 1. Dimension-less r.m.s. acceleration and speed coefficients.

Motion law	C_a	$C_{a,rms}$	C_v
Acc const symm	4	4	2
1/3-1/3-1/3	4.5	3.67	1.5
Cubic	6	3.46	1.5
Cycloidal	2π	4.44	2

Furthermore, dealing with acceleration, it is possible to define the dimension-less acceleration coefficient C_a and the dimension-less root mean square (r.m.s.) acceleration coefficient $C_{a,rms}$ defined respectively as:

$$C_a = \ddot{y}_{max}\frac{h}{t_a^2} C_{a,rms} = \ddot{y}_{rms}\frac{h}{t_a^2}$$

A comparative collection of C_a, $C_{a,rms}$ and C_v is provided (Table 1) in order to highlight their effectiveness in describing and comparing the properties of different laws of motion.

The advantage of using the dimension-less form to deal with the different laws of motion is that they are quickly comparable referring to the coefficients that summarize their performance. As an example, using the dimensional-less coefficients $C_{a,rms}$, it is possible to highlight the role of the adopted law of motion on the root-mean-square value of the angular acceleration of the rotating heads:

$$\dot{\omega}_{L,rms} = \frac{a_{rms}}{R_T} = \frac{C_{a,rms}}{R_T}\frac{h}{T_a^2}\frac{T_a}{T} \quad (1)$$

being a_{rms} the tangential acceleration of the rotating head, calculated on the whole duration time T while the dimensionless coefficient refers only to the approaching phase T_a.

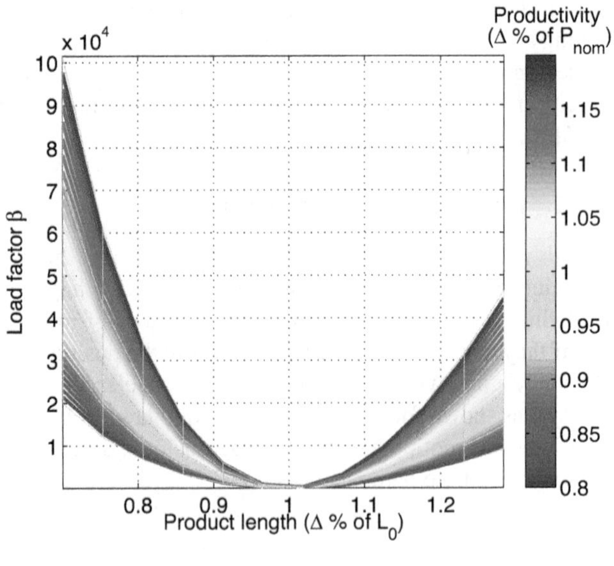

Figure 9. β as a function of product length.

4 Dynamics analysis

The sizing of the motor-reducer unit is performed under the hypothesis of pure inertial load considering that during the cutting phases, both the friction and the cutting forces are negligible. With this assumption, the only load that the motor have to face with is the rotating heads own inertial load.

4.1 α-β method

To properly size the motor-reducer unit, the α-β method is adopted (Giberti et al., 2011, 2010). This method has the advantage of highlighting and separate the terms of the power

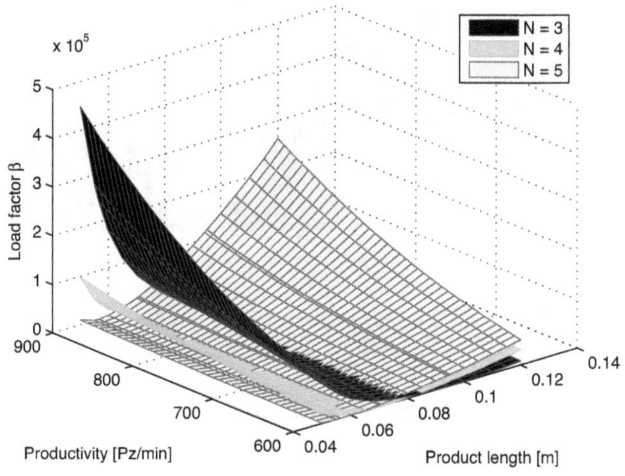

Figure 10. β surface as a function of tools number.

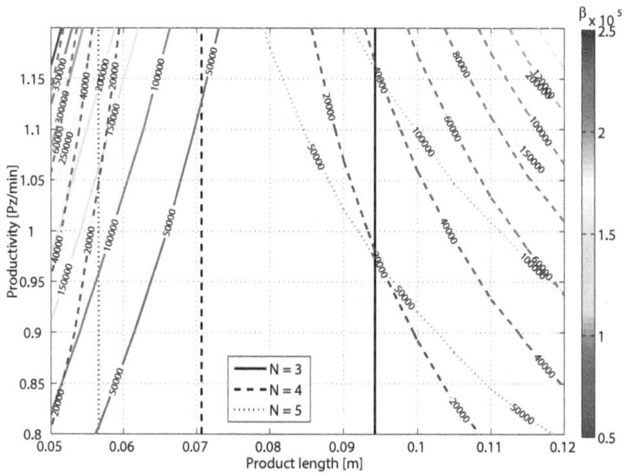

Figure 11. β as a function of input parameters.

balance that regards the motor unit and the reducer. This method allows both to avoid an iterative design procedure and to define, for each motor unit considered, the corresponding range of transmission ratios that are suitable for the analyzed application. The motor performance is described by a key-factor called accelerating factor α, defined as the ratio between the square of the nominal torque of the motor C_{m} and its own rotational inertial momentum J_{m}:

$$\alpha = \frac{C_{\mathrm{m}}^2}{J_{\mathrm{m}}} \tag{2}$$

The accelerating factor derives from the rated motor torque condition $C_{\mathrm{m,rms}} <= C_{\mathrm{m}}$ used to check the motor thermal equilibrium in which the value $C_{\mathrm{m,rms}}$ is the root mean square of the torque required by the motor to carry out the task. This is calculated by:

$$C_{\mathrm{m,rms}} = \int_0^{t_{\mathrm{a}}} \frac{1}{t_{\mathrm{a}}} \left(\tau C_{\mathrm{r}} + J_{\mathrm{m}} \frac{\dot{\omega}_{\mathrm{r}}}{\tau} \right)^2 \mathrm{d}t \tag{3}$$

where t_{a} is the cycle time, J_{m} the rotor inertia, τ the transmission ratio and C_{r} and $\dot{\omega}_{\mathrm{r}}$ the load torque and the load angular acceleration respectively. Substituting the square of $C_{\mathrm{m,rms}}$ into the rated motor torque condition it is possible to solve the inequality with respect to the accelerating factor term defined beforehand.

A more refined definition of the accelerating factor is the specific accelerating factor that is described in Giberti et al. (2014), but for the aims of this work the simpler one presented above has been considered sufficiently accurate. The load factor β contains the information regarding the root mean square load during a cycle and thus it allows to summarize in one single parameter the load the motor is subject to (using a thermal design criterion):

$$\beta = 2 \left(\dot{\omega}_{\mathrm{r,q}} C^*_{\mathrm{r,q}} + \overline{\dot{\omega}_r C^*_{\mathrm{r}}} \right) \tag{4}$$

where $\dot{\omega}_{\mathrm{r,q}}$ and $C^*_{\mathrm{r,q}}$ are, respectively, the root mean square of the angular acceleration and the resistant torque while $\overline{\dot{\omega}_r C^*_{\mathrm{r}}}$ is the mean value of their product. Having defined both α and β, the condition for the correct sizing of motor-reducer unit can be re-written (Giberti et al., 2011) as:

$$\alpha \geq \beta + f(\tau) \tag{5}$$

being τ the transmission ratio defined as:

$$\tau = \frac{\omega_{\mathrm{r}}}{\omega_{\mathrm{m}}}. \tag{6}$$

The load factor β is directly linked to the flexibility of the machine. It is equal to zero only if the product length L_{P} is equal to the design length L_0, while, as shown in Sect. 3, it grows if the rotating head needs to be accelerated or decelerated to process a greater or a product length smaller than the base one.

4.2 Load factor

As a consequence of pure inertial load assumption, the load factor β defined in Eq. (4) becomes:

$$\beta = 4 J_L \dot{\omega}^2_{\mathrm{r,rms}} \tag{7}$$

being J_L the momentum of inertia of the couple of rotating heads ($J_L(R_t) = 2J_T(R_t)$). It is important to highlight

Table 2. Test case value.

Parameter	Value	unit
R_t	$= 0.045$	[m]
L_T	$= 0.0236$	[m]
P	$= 750$	$[\mathrm{pz\,min}^{-1}]$
L_0	$= 0.0942$	[m]
J_T	$= 0.0126$	$[\mathrm{kg\,m}^2]$
$C_{\mathrm{a,rms}}$	$= 3.67$	[–]
C_{v}	$= 1.5$	[–]

that the best operative condition corresponds to β equal to zero, that implies no inertial loads, obtainable only with a null $\dot{\omega}_{\mathrm{r,rms}}$.

Combining Eqs. (7) and (1), it is possible to highlight the design terms:

$$\beta = 4J_L\left[\frac{C_{\mathrm{a,rms}}}{R_T}\frac{h}{t_{\mathrm{a}}^2}\frac{t_{\mathrm{a}}}{T}\right]^2 = 4J_L\left[\frac{C_{\mathrm{a,rms}}}{R_T}L_{\mathrm{P}}\frac{\frac{2\pi R_t}{N}-L_{\mathrm{P}}}{L_{\mathrm{P}}-L_T}\left(\frac{P}{60}\right)^2\right]^2 \quad (8)$$

This Eq. (8) is particularly important because it allows to describe the load factor as a function of the input parameters.

4.3 Gearbox

The change in transmission ratio range, defined as $\Delta\tau = \tau_{\max} - \tau_{\min}$, can also be expressed as a function of the input parameters:

$$\Delta\tau = \sqrt{J_{\mathrm{m}}}\frac{\sqrt{\alpha}\pm\sqrt{\alpha-4J_{\mathrm{r}}\dot{\omega}_{\mathrm{r,rms}}}}{2J_{\mathrm{r}}\dot{\omega}_{\mathrm{r,rms}}} = \tau_{\mathrm{opt}}\left[\sqrt{\frac{\alpha}{\beta}}\pm\sqrt{\frac{\alpha}{\beta}-1}\right] \quad (9)$$

being the ratio $\sqrt{J_{\mathrm{m}}/J_{\mathrm{r}}}$ the optimal transmission ratio τ_{opt}. It is worth noting that this equation is function only of the load factor and not of the single input parameters of which it is function. Thus it represents a general result for every input parameters combinations which produces the same value of β.

Finally, the last check on the available τ serves the purpose of ensuring that the maximum angular velocity required by the law of motion can be provided. Using the introduced formulation:

$$\omega_{\max} = \omega_{\mathrm{cost}} + \Delta\omega = \frac{v}{R_t} + \frac{v_{\max}}{R_t} = \frac{P\cdot 2\pi}{60N} + \frac{C_{\mathrm{v}}}{R_t}\frac{h}{t_{\mathrm{a}}} = \frac{P\cdot 2\pi}{60N}\left[1 + C_{\mathrm{v}}\frac{L_o - L_p}{L_{\mathrm{P}} - L_t}\right] \quad (10)$$

where C_{v} is the dimension-less velocity coefficient depending to the specific law of motion (Table 1) adopted and referring only to the approaching phase.

5 Numerical analysis and results

In the previous section, the load factor β and the admissible range for the transmission ratio $\Delta\tau = \tau_{\max} - \tau_{\min}$ were expressed as a function of the input parameters. Numerical results are obtained in this section for a real wrapping machine with the parameters shown in Table 2.

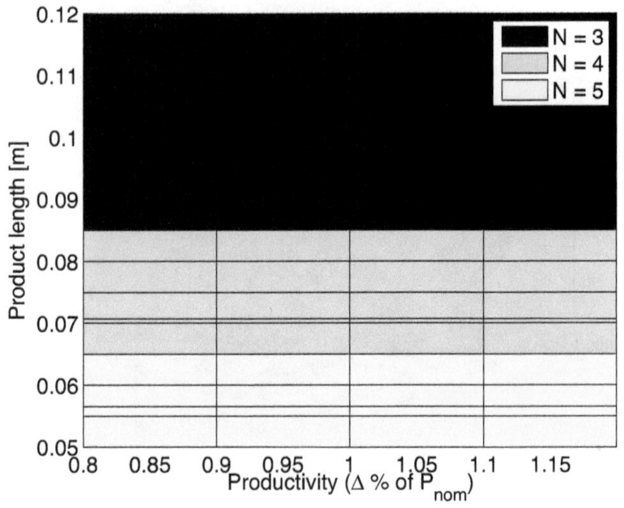

Figure 12. β surface as a function of tools number – bottom view.

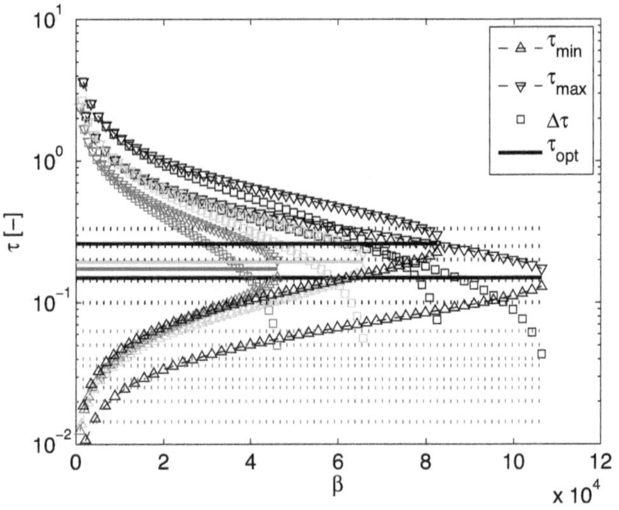

Figure 13. τ as a function of load factor β for four different motors.

It is worth underling that the acceleration and the deceleration of the driving law of motion are equal and constant (motion law labeled "Acc const symm" in Table 1). No refinements have been made on the law of motion because the aim of this work is to investigate its role on the flexibility of the machine and not comparing different adoptable solutions. Nonetheless, it is important to highlight that in some cases the selected law of motion is not able to perform the desired operation. In fact, due to a product length larger than the design one, an erroneous cut could be done if the tool happens to move backward too much as reported in Fig. 7. This fact results in a maximum product length processable using the adopted law of motion. This constraint can be avoided using a different law of motion that implies the block of the tool. In the presented results this condition was reached in order to avoid introducing a complication not useful to the aim of the present work.

As an example, Fig. 8a reports the surface that graphically describes the load factor as a function of both the variation in the productivity P and the product length L_P. For sake of clarity, Fig. 8 reports the top view of the 3-D surface.

It could be seen that if the product length is equal to the design one the load factor is still equal to zero irrespective of the assigned productivity. It is worth noting that the growing of the load factor as a consequence of the change of the product length is not symmetrical. In Fig. 9, is shown that β is more sensitive to a decrease of the product size instead of its increase.

Finally, the load factor is more sensitive to a growth in productivity than in a change of product length. Furthermore, the productivity of high-speed automated lines is defined as the one of the so called bottleneck workstation that is the station with the lowest nominal production rate Liberopoulos and Tsarouhas, 2005. As a consequence, the effect of the product length on the productivity not only affect the single machine but involves the whole automated line productivity and should be carefully taken into account by designers.

A refinement of the analysis consists in considering the effect of the number of cutting tools N. The results is a group of surfaces (Fig. 10) that represent the functions:

$$\beta = f(L_P, P, N) \tag{11}$$

with different values of input parameters. A top view is reported in Fig. 11.

This kind of comparative plot can be used to properly design the machine using the a simple procedure:

- identify the number of tools N in order to obtain the smallest load factor to package with a certain length and with an assigned productivity (Fig. 12).

- design the most flexible flying-cutting machine minimizing the curvature of the surface corresponding to a certain number of tools. In the presented application, the smoothest surface is obtained with N equal to 4 tools. The solution corresponding to N equal to 3 shows unsuitable high value corresponding to the combination of high productivity and short product length.

- define the maximum productivity allowed with a prescribed set of input parameters. The load factor grows as the third power of P and, as a consequence, high values of β could be quickly reached. Changing the number of tools allows to obtain a larger productivity than the one reachable without changing the rotating heads setup.

Finally, a collection of $\beta - \tau$ plots is presented in Fig. 13. It is important to underline that this kind of plot depends only on the selected motor unit (α) and on the load factor (β) but not to the single input parameters resulting in a more general point of view of the problem. This last plot gives two advices to the designer. The first is that, the bigger the load factor becomes, the smaller the range of admissible transmission ratio $\Delta\tau$ is. In the worst condition, corresponding to $\alpha = \beta$, the only useful transmission ratio is the optimum one (τ_{opt}). It also allows to identify the maximum load factor the motor can withstand.

6 Conclusions

This paper deals with the flexibility-oriented design of a flying-cutting machine. A general framework is provided, allowing the designer to assess different possible motor-reducer solutions and design parameter combinations, taking into account the various advantages or limitations in terms of flexibility. This new approach satisfies two requirements. The first one can verify, theoretically the cutting flexibility in an existing cutting machine. The second can design a new cutting machine capable of reaching a much higher production flexibility level.

A specific wrapping machine is used as example to describe the methodology but this choice does not limit the extendibility of the method to other flying machine layouts.

By means of the α-β sizing motor method it has been possible to obtain an expression that highlights the influence on the motor load factor with respect to the machine parameters such as the number of cutting tools installed, the motion law adopted and the size of the product required to be wrapped. Thus it is possible to compare motor-reducer solutions and to select one so as to ensure, on the one hand, larger productivity and, on the other hand, a larger range of product size.

References

Bebic, M., Rasic, N., Statkic, S., Ristic, L., Jevtic, D., Mihailovic, I., and Jeftenic, B.: Drives and control system for paper-board cross cutter, 15th International Power Electronics and Motion Control Conference and Exposition, EPE-PEMC 2012 ECCE Europe, art. no. 6397495, LS6c.31–LS6c.38, 2012.

Diekmann, A. and Luchtefeld, K.: Drive Solutions, Mechatronics for Production and Logistics, in: Intermittent drives for cross cutters and flying saws, edited by: Kiel, E., 378–389, Springer-Verlag, Berlin, Heidelberg, 2008.

Giberti, H., Cinquemani, S., and Legnani, G.: A practical approach to the selection of the motor-reducer unit in electric drive system, Mech. Based Des. Struc., 39, 303–319, 2011.

Giberti, H., Cinquemani, S., and Legnani, G.: Effects of Transmission Mechanical Characteristics on the Choice of a Motor-Reducer, Mechatronics, 20, 604–610, 2010.

Giberti, H., Clerici, A., and Cinquemani, S.: Specific Accelerating Factor: One More Tool in Motor Sizing Projects, Mechatronics, 24, 898–905, 2014.

Hansen, D., Holtz, J., and Kennel, R.: Cutter distance sensors for an adaptive position/torque control in cross cutters, IEEE Ind. Appl. Mag., 9, 33–39, 2003.

Hooper, J. H.: Confectionery Packaging Equipment, Springer, Gaithersburg, Maryland, 1999.

Liberopoulos, G. and Tsarouhas, P.: Reliability analysis of an automated pizza production line, J. Food Eng., 69, 79–96, 2005.

Matthews, J., Singh, B., Mullineux, G., and Medland, T.: Constraint-based approach to investigate the process flexibility of food processing equipment, Comput. Ind. Eng., 51, 809–820, 2006.

Peric, N. and Petrovic, I.: Flying Shear Control System, IEEE T. Ind. Appl., 26, 1049–1056, 1990.

Sethi, A. K. and Sethi, S. P.: Flexibility in manufacturing a survey, Int. J. Flex. manuf. Sys., 2, 289–328, 1990.

Shao, W., Chi, R., and Yu, L.: Synchronizing servo motion and iterative learning control for automatic high speed horizontal flow wrapper, Proceedings of the 2012 24th Chinese Control and Decision Conference, CCDC, art. no. 6244470, 2994–2998, 2012.

Shepherd, R.: A Computer-Controlled Flying Shear, Students' Quarterly Journal, 34, 143–148, doi:10.1049/sqj.1964.0006, 1964.

Shewchuk, J. P. and Moodie, C. L.: Definition and classification of manufacturing flexibility types and measure, Int. J. Flex. manuf. Sys., 10, 325–349, 1998.

Strada, R., Zappa, B., and Giberti, H.: An unified design procedure for flying machining operations, ASME 2012 11th Biennial Conference on Engineering Systems Design and Analysis, ESDA 2012, 2, 333–342, 2012.

Varvatsoulakis, M. N.: Design and implementation issues of a control system for rotary saw cutting, Control Eng. Pract., 17, 198–202, 2009.

Visvambharan, B. B.: On-line digital control system for a flying saw cutting machine in tube mills, Industrial Electronics Society, IECON '88. Proceedings, 14 Annual Conference of IEEE Industrial Electronics, vol. 2, 385–390, doi:10.1109/IECON.1988.665170, 1988.

Wu, H., Wang, C., and Zhang, C.: Design of Servo Controller for Flying Shear Machine Based on ARM and FPGA, J. Netw., = 9, 3038–3045, 2014.

Wu, Y.: China Packaging Machinery Industry is Facing Tremendous Challenges, China Food Industry, 183, 6 pp., 2010.

3

Friction Stir Welding of AA2024-T3 plate – the influence of different pin types

D. Trimble, H. Mitrogiannopoulos, G. E. O'Donnell, and S. McFadden

Department of Mechanical and Manufacturing Engineering, Trinity College Dublin, Dublin 2, Ireland

Correspondence to: S. McFadden (shaun.mcfadden@tcd.ie)

Abstract. Some aluminium alloys are difficult to join using traditional fusion (melting and solidification) welding techniques. Friction Stir Welding (FSW) is a solid-state welding technique that can join two plates of material without melting the workpiece material. This proecess uses a rotating tool to create the joint and it can be applied to alumium alloys in particular. Macrostructure, microstructure and micro hardness of friction stir welded AA2024-T3 joints were studied. The influence of tool pin profile on the microstructure and hardness of these joints was examined. Square, triflute and tapered cylinder pins were used and results from each weldment are reported. Vickers micro hardness tests and grain size measurements were taken from the transverse plane of welded samples. Distinct zones in the macrostructure were evident. The zones were identified by transitions in the microstructure and hardness of weld samples. The zones identified across the sample were the the unaffected parent metal, the Heat Affected Zone (HAZ), the Thermo-Mechanicaly Affected Zone (TMAZ), and the Nugget Zone (NZ). Measured hardness values varied through each FSW zone. The hardness in each zone was below that of the parent material. The HAZ had the lowest hardness across the weld profile for each pin type tested. The cylindrical pin consistently produced tunnel and joint-line defects. Pin profiles with flat surface features and/or flutes produced consolidated joints with no defects.

1 Introduction

The Welding Institute (TWI) developed Friction Stir Welding (FSW) in 1991 (Thomas, 1991) and it is considered to be one of the most significant developments in metal joining in recent times. It is a solid-state welding technique used for joining aluminium alloys (as shown in Fig. 1). This technique is currently being applied in the aerospace, automotive, and shipbuilding industries (Mishra and Ma, 2005) and was recently reviewed in Xiaocong et al. (2014).

Due to its solid state nature FSW has many benefits over fusion welding techniques. However, one of its main advantages is its ability to weld all series of aluminium alloys, in particular the 2xxx series alloys (Yang et al., 2008; Trimble et al., 2012). These alloys are used extensively within the aerospace industry for applications such as fuselage and wing skin panels due to their high strength-to-weight ratio. However, these alloys are mostly non-weldable using fusion welding methods due to problems with oxidisation, solidification, shrinkage, sensitivity to cracking, hydrogen solubility and the resultant porosity problems (Flores et al., 1998).

FSW involves the translation of a tool along the joint line between two plates. The tool rotates at high speeds during the process. The tool is made up of a profiled pin and a shoulder which generate frictional heat to soften the workpiece material either side of the joint line and mix the workpiece materials together. The FSW tool is not consumed in the process and is crucial to the welding process. The shoulder generates the largest component of heat in the process. The pin causes localised heating and plastic deformation of the material around the pin.

The macrostructure of a FSW joint consists of three distinct zones; the central Nugget Zone (NZ), the Thermomechanically Affected Zone (TMAZ) and the Heat Affected Zone (HAZ). The zonal microstructure of a FSW joint is shown in Figs. 1 and 2. The objective of this work was to determine the variation in hardness across a FSW joint and determine the correlation between microstructure and hard-

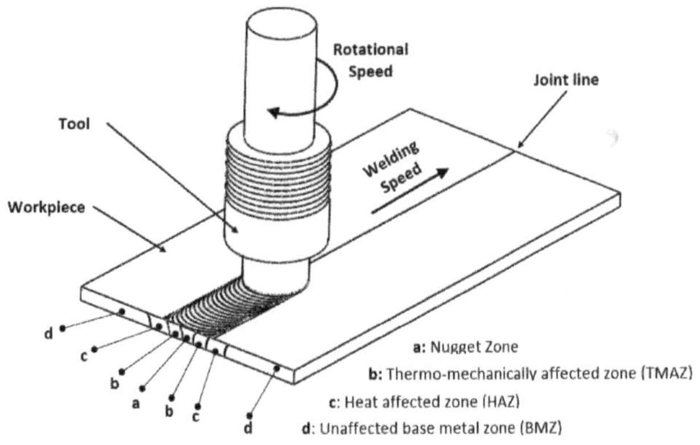

Figure 1. Friction Stir Welding (FSW) schematic.

Figure 2. Zonal macrostructure of AA2024-T3 after FSW.

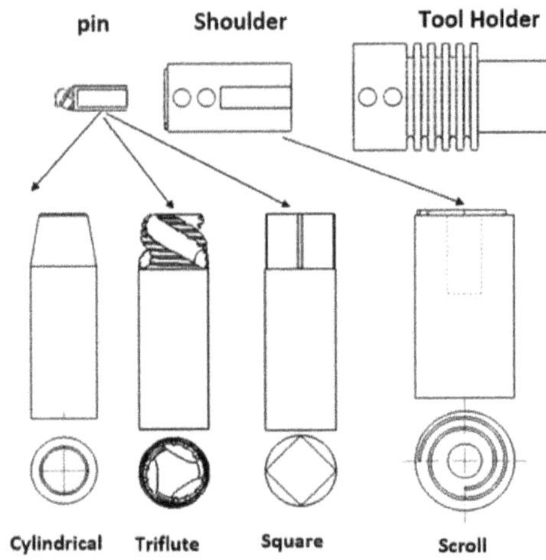

Figure 3. The tooling used including three pin profiles, the scroll shoulder, and tool holder.

ness. The effect of pin profile on the microstructure and hardness of FSW joints was examined.

2 Experiemtal procedure

A Corea F3UE vertical milling machine was modified for use in this experiment. Figure 3 gives details of the FSW tooling. Aluminium alloy AA2024-T3 plates (4.8 mm thick) were butt welded using three different pin profiles: tapered cylinder, square and triflute. All pins were 4.6 mm in length. The swept diameter of the square pin was 7 mm. The swept diameter of the tapered cylinder and triflute pins tapered from 7 mm (at the shoulder) to 2.69 mm. A scrolled shoulder was used in this investigation (no tilt angle was required). Rotational and translational speeds were 450 rpm and 180 mm min^{-1}, respectively.

Samples were cut from the cross-section of each FSW joint and prepared for post weld analysis. Samples were polished to remove all tool marks and scratches. The samples were then etched using Keller's reagent to reveal the microstructure. Samples were viewed using a Leica DM LM microscope and ImageJ software was used to measure grain size. Vickers microhardness tests were carried out using a Mitutoyo MVK-H1 micro hardness machine. Four rows and twenty five columns of indents were made across the weld samples with 1 mm horizontal spacing maintained between columns. The indent rows were positioned 1, 2, 3.25 and 4.5 mm from the base of the sample.

3 Results and discussion

Microstructural analysis revealed a complex microstructure in the FSW zone. On a macrostructural level the weld zone was found to be trapezoidal in shape for each pin. Figure 4 shows the results of macrostructural analysis for the three pin profiles. By comparing the swept diameter of each pin to the width of the nugget zone for each pin it is clear that tool geometry is directly linked to the shape of the FSW zone. Despite having the same swept diameter, it can be seen that the NZ of the triflute pin does not narrow with depth from the workpiece surface as drastically as that of the tapered cylinder pin. This was due to increased mixing action resulting from the flutes in the pin surface. The greater mixing action of the square and triflute pins resulted in a larger NZ when compared to the tapered cylinder pin.

A tunnel defect (an internal void) was observed in the welds performed by the tapered cylinder pin. This defect, shown in Fig. 5, was observed in the transition region between the NZ and the TMAZ at a depth similar to the plunge depth of the pin and on one side only. This is because the cylindrical pin produces less plastic deformation and stirring of the workpiece. Insufficient plastic flow around the pin gave rise to the conditions for the formation of the tunnel defect.

Another defect that was consistently found in the joint made with the tapered cylinder pin was a joint line remnant (Fig. 6). This feature is located at the original position of the joint interfaces at the bottom of the welded joint. Again, this feature is symptomatic of the insufficient plastic deformation and stirring of the metal when the tapered cylinder pin is used.

The HAZ, TMAZ and NZ were defined by different microstructural features. Figure 7 shows examples of the typical grain structures found in the distinct zones. Figure 7a shows the microstructure of the parent material. This microstructure

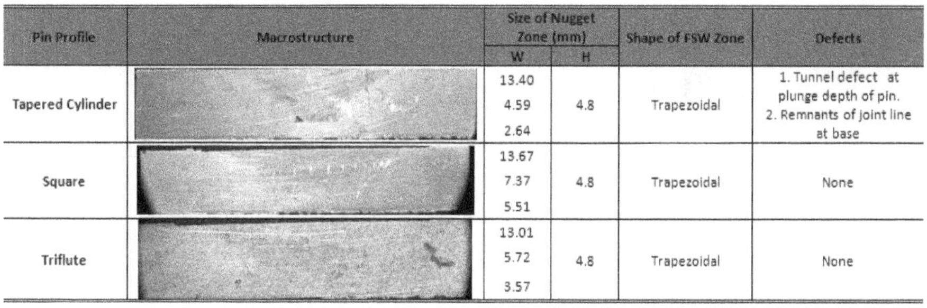

Pin Profile	Macrostructure	Size of Nugget Zone (mm) W	Size of Nugget Zone (mm) H	Shape of FSW Zone	Defects
Tapered Cylinder		13.40 4.59 2.64	4.8	Trapezoidal	1. Tunnel defect at plunge depth of pin. 2. Remnants of joint line at base
Square		13.67 7.37 5.51	4.8	Trapezoidal	None
Triflute		13.01 5.72 3.57	4.8	Trapezoidal	None

Figure 4. Macrostructural data for three pin profiles.

Table 1. Variation in grain size in the nugget zone.

Pin type	Location in Nugget zone	Grain sizes (μm) Min.	Max.	Range	Avg.
Square	Top	1.60	3.47	1.87	2.23
	Right of Centre	1.60	4.29	2.69	2.20
	Centre	1.62	3.42	1.80	2.24
	Left of Centre	1.64	3.42	1.78	2.09
	Bottom	1.60	3.50	1.90	2.16
Tapered cylinder	Top	1.61	3.66	2.05	2.17
	Right of Centre	1.61	5.00	3.39	2.46
	Centre	1.60	4.00	2.40	2.18
	Left of Centre	1.64	5.43	3.79	2.43
	Bottom	1.60	3.36	1.76	2.24
Triflute	Top	1.60	3.36	1.76	2.23
	Right of Centre	–	–	–	–
	Centre	1.60	3.46	1.86	2.17
	Left of Centre	1.60	5.01	3.41	2.35
	Bottom	1.61	3.85	2.24	2.17

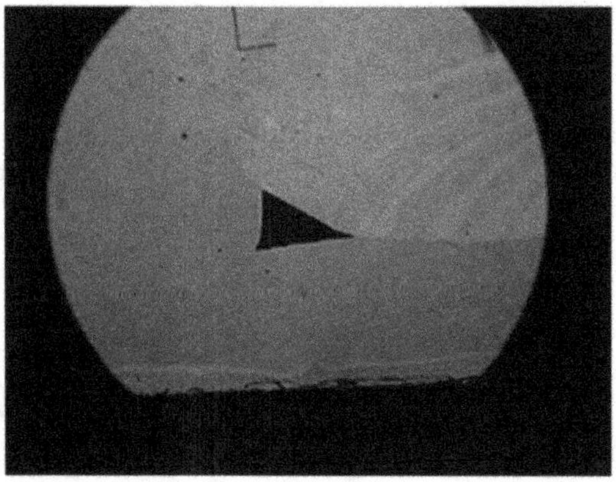

Figure 5. Tunnel defect found in the tapered cylinder FSW joint (× 50 magnification).

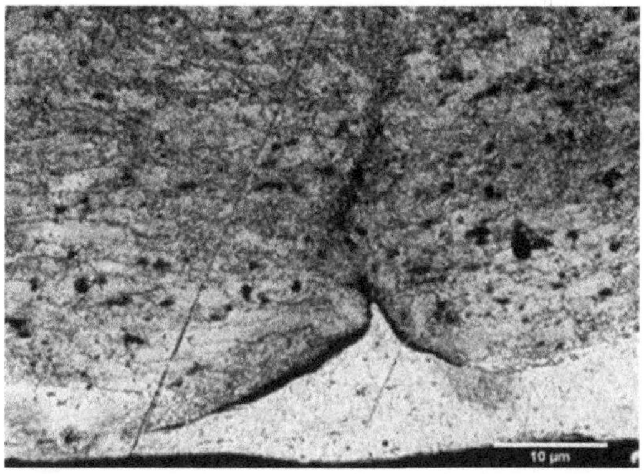

Figure 6. Remnants of joint line at the bottom of the FSW joint for a tapered cylinder pin.

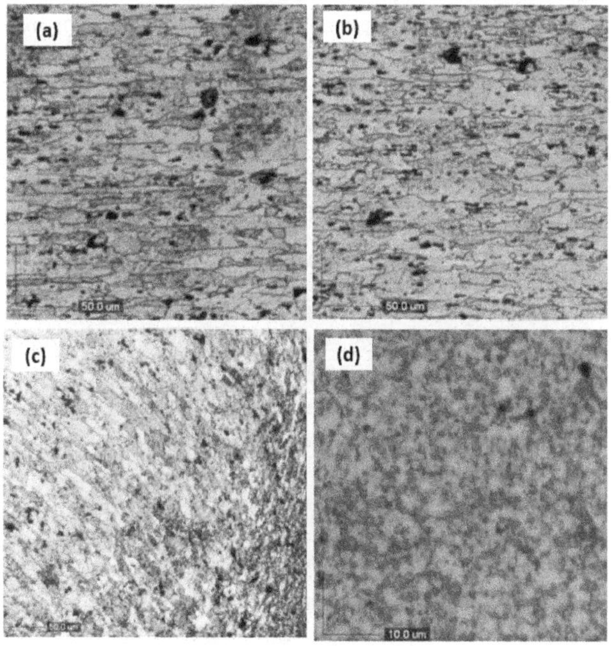

Figure 7. Grain morphology at 50 × magnification of **(a)** parent material, **(b)** HAZ, **(c)** TMAZ, and **(d)** NZ.

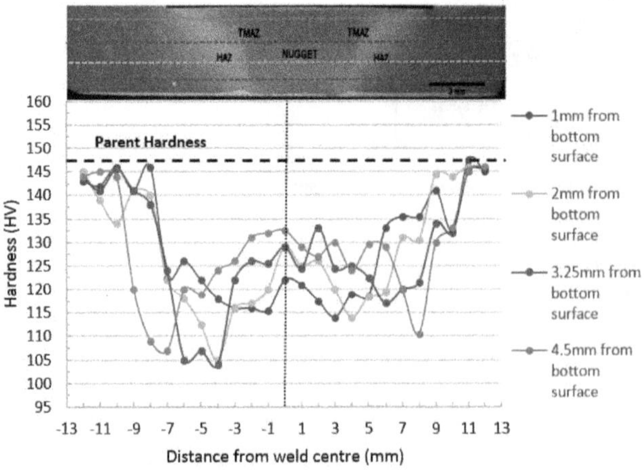

Figure 8. Tapered cylinder pin – hardness data at different heights in the weld thickness.

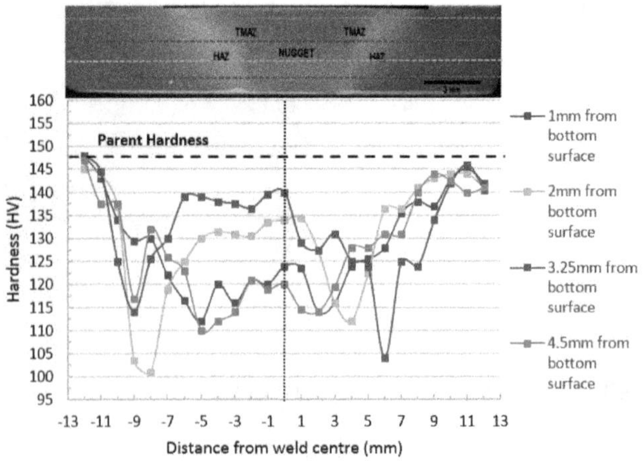

Figure 9. Square pin – hardness data at different heights in the weld thickness.

is typical of the alloy, which has temper designation T3 (solution heat treated, cold worked, and naturally aged). Figure 7b shows the microstructure of the HAZ. The HAZ and parent material were found to have similar microstructures as no deformational action from the tool is felt in this zone. There is insufficient evidence from the microstructure to suggest that recrystallization took place. Figure 7c shows the plastically strained microstructure of the TMAZ. This was due to plastic strain induced by the mechanical action of the tool. Average grain size in the TMAZ was found to be similar to the parent material however a wider range of grain size was apparent throughout the TMAZ. Figure 7c shows the fine equiaxed grain structure of the NZ.

The grain morphologies found in the various weld zones were qualitatively similar regardless of the pin used. Grain size in the NZ seemed to vary slightly as a result of pin profile (Table 1). The average grain size in the NZ for the square, tapered cylinder and triflute pins was calculated as 2.19, 2.3 and 2.23 μm, respectively. The differences in NZ grain size may have been as a result of the increased plastic deformation and stirring generated by the flutes in the triflute design and the flat faces of the square pin design.

Figures 8–10 show the measured hardness values recorded on the grid for the three tools in question. The measured hardness across each of the weld zones is below that of the parent material. This suggests that the strength of the weld will be lower than the parent material strength.

In all cases, hardness decreased from the parent material into the HAZ and increased from the HAZ, through the TMAZ, into the NZ. Hence, it is clear that the material in the HAZ experienced softening due to the process, even though changes in microstructure from the parent material to the HAZ were not apparent from the post mortem microstructure characterisation. Relatively constant hardness is found across the NZ.

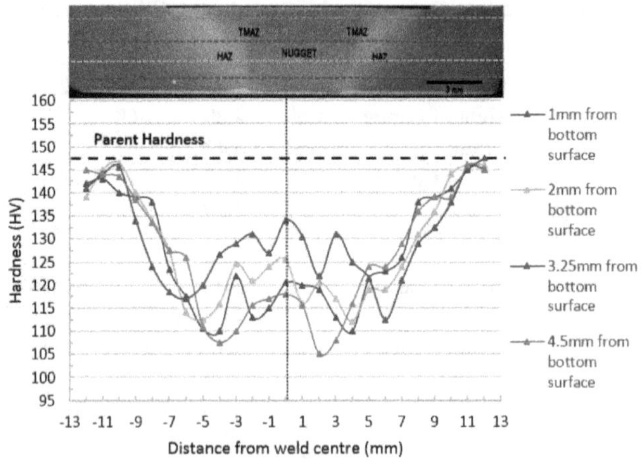

Figure 10. Triflute pin – hardness data at different heights in the weld thickness.

4 Conclusions

Flat plates of AA2024-T3 (4.8 mm thick) were butt welded using Friction Stir Welding. Three pairs of plates were welded under the same operating conditions but with differing pin profiles. The pin profiles used were tapered cylindrical, square and tri-flute. A scroll shoulder element was used for each case.

In summary, the main conclusions drawn from this work are:

1. The combined analyses of microstructure and microhardness showed a zonal transition from the unaffected parent material to a HAZ, a TMAZ, and a NZ in the centre of the weld.

2. Measured hardness varied through each FSW zone. The hardness in each zone was below that of the parent material.

3. The HAZ had the lowest hardness across the weld profile for each pin type tested. A lower hardness in this area could be due to either recovery (reduction in dislocation density) or recrystallization of the grains. Based on the evidence provided is difficult to confirm if recrystallization took place.

4. The cylindrical pin produced a tunnel defect and showed remnants of the joint line due to a lack of plastic deformation. Pin profiles with flat surface features and/or flutes produced a consolidated joint and are recommended over the tapered cylinder pin for future work.

References

Flores, O. V., Kennedy, C., Murr, L. E., Brown, D., and Pappu, S.: Microstructural issues in a Friction-Stir-Welded Aluminum Alloy, Scripta Mater., 38, 703–708, 1998.

Mishra, R. S. and Ma, Z. Y.: Friction stir welding and processing, Mater. Sci. Eng. R Rep., 50, 1–78, 2005.

Thomas, W. M.: Friction-stir butt welding, GB Patent No. 9125978.8, International patent application No. PCT/GB92/02203, 1991.

Trimble, D., Monaghan, J., and O'Donnell, G. E.: Force generation during friction stir welding of AA2024-T3, CIRP Ann. – Manufact. Technol., 61, 9–12, 2012.

Xiaocong, H., Fengshou, G., and Ball, A.: A review of numerical analysis of friction stir Welding, Prog. Mater. Sci., 65, 1–66, 2014.

Yang, Y., Kalya, P., Landers, R. G., and Krishnamurthy, K.: Automatic gap detection in friction stir butt welding operations, Int. J. Mach. Tools Manufact., 48, 1161–1169, 2008.

4

On the infinitely-stable rotational mechanism using the off-axis rotation of a bistable translational mechanism

G. Hao and J. Mullins

School of Engineering-Electrical and Electronic Engineering, University College Cork, Cork, Ireland

Correspondence to: G. Hao (g.hao@ucc.ie)

Abstract. Different from the prior art concentrating on the primary translation of bistable translational mechanisms this paper investigates the off-axis rotation behaviour of a bistable translational mechanism through displacing the guided primary translation at different positions. Moment-rotation curves obtained using the nonlinear finite element analysis (FEA) for a case study show the multiple stable positions of the rotation under each specific primary motion, suggesting that an infinitely-stable rotational mechanism can be achieved by controlling the primary motion. In addition, several critical transition points have been identified and qualitative testing has been conducted for the case study.

1 Introduction

A compliant mechanism is multistable if it has more than two stable translational/rotational positions that refer to zero force/moment points with positive stiffness (Oh, 2008; Howell et al., 2013). Multistable compliant mechanisms have a variety of successful applications such as switches, valves, relays, grasper, adaptive end effectors, sensors, energy harvesting devices and vibration isolators (Oh, 2008; Howell et al., 2013; Lassooij et al., 2012; Chen and Lan, 2012; Hansen et al., 2007; Liu et al., 2013; Shaw et al., 2013). Unlike the traditional ways of using locking mechanisms and detents, a compliant mechanism based multistable mechanism obtains multistability through the storage and release of potential energy stored in their flexible members during post-buckling.

This paper focuses on bistable translational mechanisms using fixed-clamped beams (Lassooij et al., 2012; Chen and Lan, 2012; Hansen et al., 2007; Dunning et al., 2012; Kim and Ebenstein, 2012; Holst et al., 2011; Zhang and Chen, 2013), rather than using the fixed-pinned (Qiu et al., 2004) and the pinned-pinned beams (Sonmez and Tutum, 2008) that are hard to manufacture monolithically. There are mainly two methods to design the fixed-clamped bistable translational mechanism (Dunning et al., 2012). One method is to use the fixed-guided beam via an inclined arrangement where the beam has no deformation at the initial position without the input force. The other method is to pre-stress the fixed ends of a pair of fixed-clamped non-inclined beams to introducing buckling where the beam deforms at the initial position without the input force. However, the second method is out of scope of this paper. The bistable translational mechanism studied in this paper is shown in Fig. 1a, which is composed of two fixed-clamped inclined beams connected at the middle (motion end). Under the primary force only, each beam will work like a fixed-guided beam due to the symmetrical nature. The typical force-displacement relation of this bistable mechanism in the primary direction is shown in Fig. 1c with the critical points and regions being marked.

Different from the above reported works in Lassooij et al. (2012), Chen and Lan (2012), Hansen et al. (2007), Dunning et al. (2012), Kim and Ebenstein (2012), Holst et al. (2011) and Zhang and Chen (2013) which employ the primary translation to achieve finite multi-stable statuses, this paper studies on the multistable off-axis rotation behaviour of the bistable translational mechanism in order to obtain an infinitely-stable rotational mechanism. This off-axis rotation behavior with multistable points has also potential applications in human joint rehabilitative devices, dynamic and static balancing of machines, and human mobility-assisting devices (Hou and Lan, 2013) when combining with a positive stiffness spring.

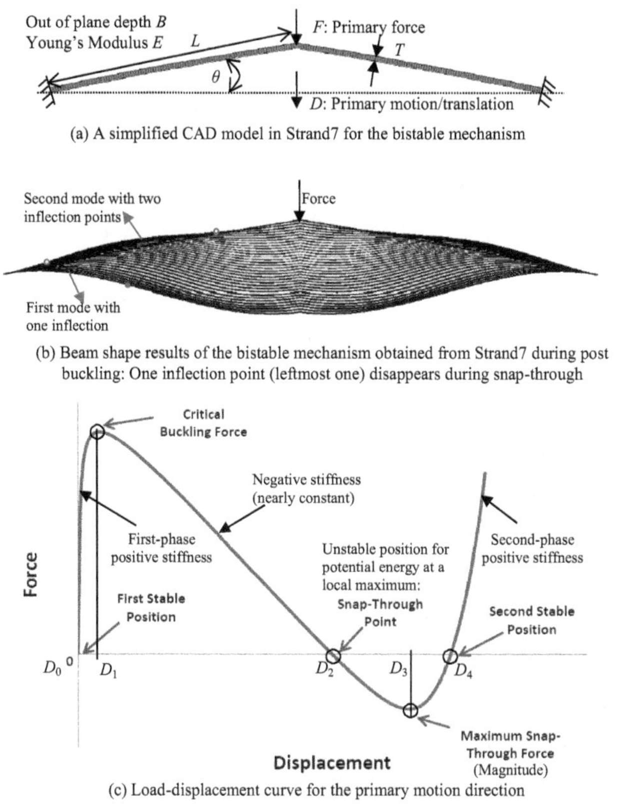

(a) A simplified CAD model in Strand7 for the bistable mechanism

(b) Beam shape results of the bistable mechanism obtained from Strand7 during post buckling: One inflection point (leftmost one) disappears during snap-through

(c) Load-displacement curve for the primary motion direction

Figure 1. Embodiment of the bistable mechanism.

This paper is organized as follows. Section 2 describes the displacement-controlled FEA (finite element analysis) simulation. The off-axis rotational analysis of a bistable translational mechanism is implemented in Sect. 3 showing the infinite stable rotational positions. A conceptual design of an infinitely-stable rotational mechanism is followed. Discussions are drawn in Sect. 4.

2 Displacement-controlled FEA simulation

Due to the strongly nonlinear characteristics of post-buckling behaviour, it is very difficult to model buckled beams correctly. There have been a large amount of works on accurately modelling the force-displacement characteristics of buckled beams as well as modelling the exact curvature and inflection points of the beam as it travels between two stable equilibrium points. These modelling techniques include the curve decomposition method (Kim and Ebenstein, 2012) and elliptic or comprehensive elliptic integral solutions (Holst et al., 2011; Zhang and Chen, 2013). Each has provided an insight into the post-buckling behaviour of compliant mechanisms, but has its own limitations. Compared to the method using the comprehensive elliptic integral in Zhang and Chen (2013), the combined method in Holst et al. (2011) or the curve decomposition method in Kim and Ebenstein (2012) is unable to solve large deflections of thin beams with multiple inflection points and subject to any kinds of load cases. The models in Kim and Ebenstein (2012) and Zhang and Chen (2013) ignoring axial elongation can also causes that the linear region that the beam undergoes before the buckling point disappears and that the snap-through region is also incorrect, while the combined model (Holst et al., 2011) incorporating the axial elongation is much closer to the actual experimental result.

Nonlinear FEA (Dunning et al., 2012), however, can accurately predict post-buckling behaviour of bistable mechanisms as well as accurate stress values without any assumptions as mentioned above, which surpasses these fast numerical methods in Kim and Ebenstein (2012), Holst et al. (2011) and Zhang and Chen (2013). In addition, nonlinear FEA can analyse off-axis characteristics that these methods in Kim and Ebenstein (2012), Holst et al. (2011) and Zhang and Chen (2013) cannot deal with. It can be easily employed by most engineers and researchers.

As the force-displacement curve for post-buckling beams goes through a negative stiffness range, simple force-controlled simulations are not suitable since the controlled force is no longer a monotonic control parameter. Therefore, the bistable mechanism analysis needs displacement-controlled FEA simulation. For the simulation cases in bistable mechanisms, there is a necessity to apply an enforced displacement and measure the load generated at the restrained node. Enforced-displacement control is what is needed to correctly model the post-buckling behaviour of a beam, especially if the off-axis analysis for example is being tested. In the off-axis analysis, the beam must be fixed (preloaded) at a specified position while other loads/displacements are applied, which cannot be achieved using force-controlled methods.

The nonlinear FEA in this paper was performed in Strand7 software, a powerful nonlinear FEA solver, using the 20-node brick meshing element. The customizability of Strand7 allows the modelling of the bistable mechanism to be easily achieved. The simulation method is to first apply an enforced displacement and then to determine the resultant reaction forces developed due to these enforced displacements. The simulations can be modified to simultaneously change restraints and loads. The Strand7 solver is also much more intuitive allowing the user to monitor the solution convergence behaviour with a convergence graph when a simulation is running for example. The FEA solver can be monitored and adjusted in much greater detail.

3 Off-axis rotational analysis

In this section, a case study is analysed for investigating the off-axis rotation behaviour. The beam length was set up to $L = 50$ mm with the in-plane thickness $T = 1$ mm, out-of-plane depth $B = 5$ mm, inclined angle $\theta = 10°$ and Young's Modulus $E = 2.4$ GPa. In order to avoid the high-order buck-

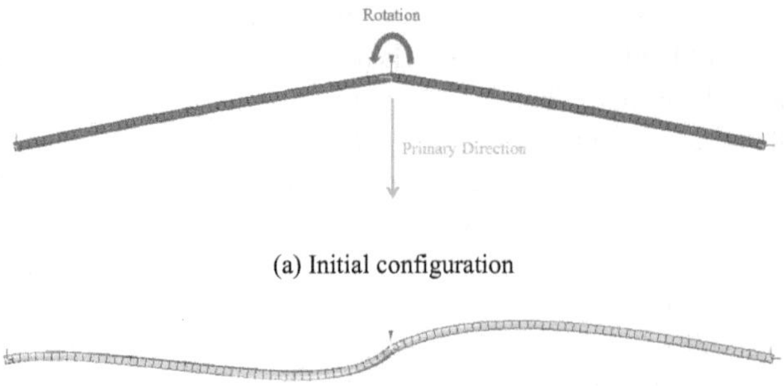

(a) Initial configuration

(b) Buckled beam being rotated

Figure 2. Off-axis rotation.

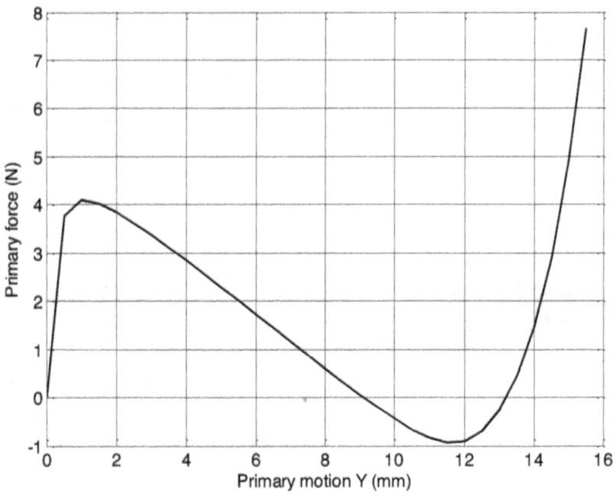

Figure 3. Load-displacement relation in the primary direction for the case study.

ling with high peek force and high stress, an optimised slight curvature ($1/1000\,\text{mm}^{-1}$) (Dunning et al., 2012) was applied to the beam, which is trivial in comparison to the beam length but is enough to perturb the beam to buckle at lower bending modes (Fig. 1b).

The FEA simulations of in-plane off-axis rotation for the bistable mechanism (Fig. 2) were performed on Strand7. Before implementing the off-axis rotational analysis, the primary translational analysis of the bistable mechanism was conducted as shown in Fig. 3. As can be seen approximately, $D_1 = 1.1$ mm, $D_2 = 9.1$ mm, $D_3 = 11.7$ mm and $D_4 = 13.2$ mm when referring to Fig. 1c.

To determine the off-axis rotational behaviour, the bistable mechanism was guided (by applied constraint) to displace only in the primary motion direction (y axis) by the displacement-controlled FEA simulation. For each guided primary displacement increment, the rotational test was performed where the bistable mechanism was displaced in the rotation direction and the reaction moment in this rotational direction was then recorded. Under the conditions that the primary motion is incremented from 0 to 14 mm with 0.5 mm each step and that the rotation is incremented from 0 to 15° with 0.15° each step, the graph of moment versus rotation is shown in Fig. 4. The stable rotational points can be observed from the FEA results when there is a zero moment point with a positive rotational stiffness. Similarly, the instable rotational points can be seen when there is a zero moment point with a negative rotational stiffness.

It is shown from Fig. 4a and b that there is no other stable rotational point except the zero point (home position) for $Y = 0$ mm or $Y > 13.0$ mm, which complies with our expectation. For $Y = 0$ the bistable mechanism also acts as a conventional compliant rotational joint with positive stiffness thereby having only one stable rotational point at the home position. For $Y > 13.0$ mm the bistable mechanism is approximately beyond the second stable point of the primary translation with positive stiffness, so it is expected to have only positive rotational stiffness over the rotation range (i.e. only one zero stable rotational point). For each primary motion Y over [0.5 mm, 12.5 mm], there are either three or two stable points of the rotation. Figure 5 shows the number of the stable rotational points with regard to each primary motion, and indicates that after about $Y = 3.0$ mm but before $Y = 12.5$ mm, the number of stable points reduces from three (two non-zero positions and one zero position) to two (two non-zero positions). The positive stable rotational position corresponding to each primary motion is shown in Fig. 6. It can be found that the stable position increases before a critical transition point at about $Y = 6.0$ mm but decreases after this critical point. It can be also noted from Fig. 4 that near the zero rotational point the rotational stiffness is approximately zero under small specific primary motion (except $Y = 0$) or under the specific primary motion between $Y = 12.5$ mm and $Y = 13.0$ mm, i.e. there is approximately constant moment.

It is interesting to be seen that with the increase of the primary motion Y, the positive rotational displacement at the maximal negative moment point increases before $Y = 3.0$ mm but decreases after $Y = 3.0$ mm, and the maximal negative moment value has an increasing trends before $Y = 7.0$ mm but a decreasing trends after $Y = 7.0$ mm.

Therefore, by controlling the primary motion from $Y = 3.5$ to 12.5 mm as shown in Figs. 5 and 6, the mechanism can automatically rotate to one stable rotational status for any primary motion since the rotation zero point is an unstable position. This suggests that an infinitely-stable mechanism can be achieved due to the fact that there are infinite stable positions of rotation corresponding to infinite primary motion positions. A conceptual design of the infinitely-stable rotational mechanism is shown in Fig. 7.

In order to qualitatively verify the multistable off-axis rotation behaviour as discussed in this section, a similar (simple) prototype made of polycarbonate was fabricated by CNC milling machining as shown in Fig. 8. By manually displacing the bistable mechanism in the primary translational direction using the sliding guide, the rotational bistable behaviors at two specific primary translation positions ($Y = 6.0$ and

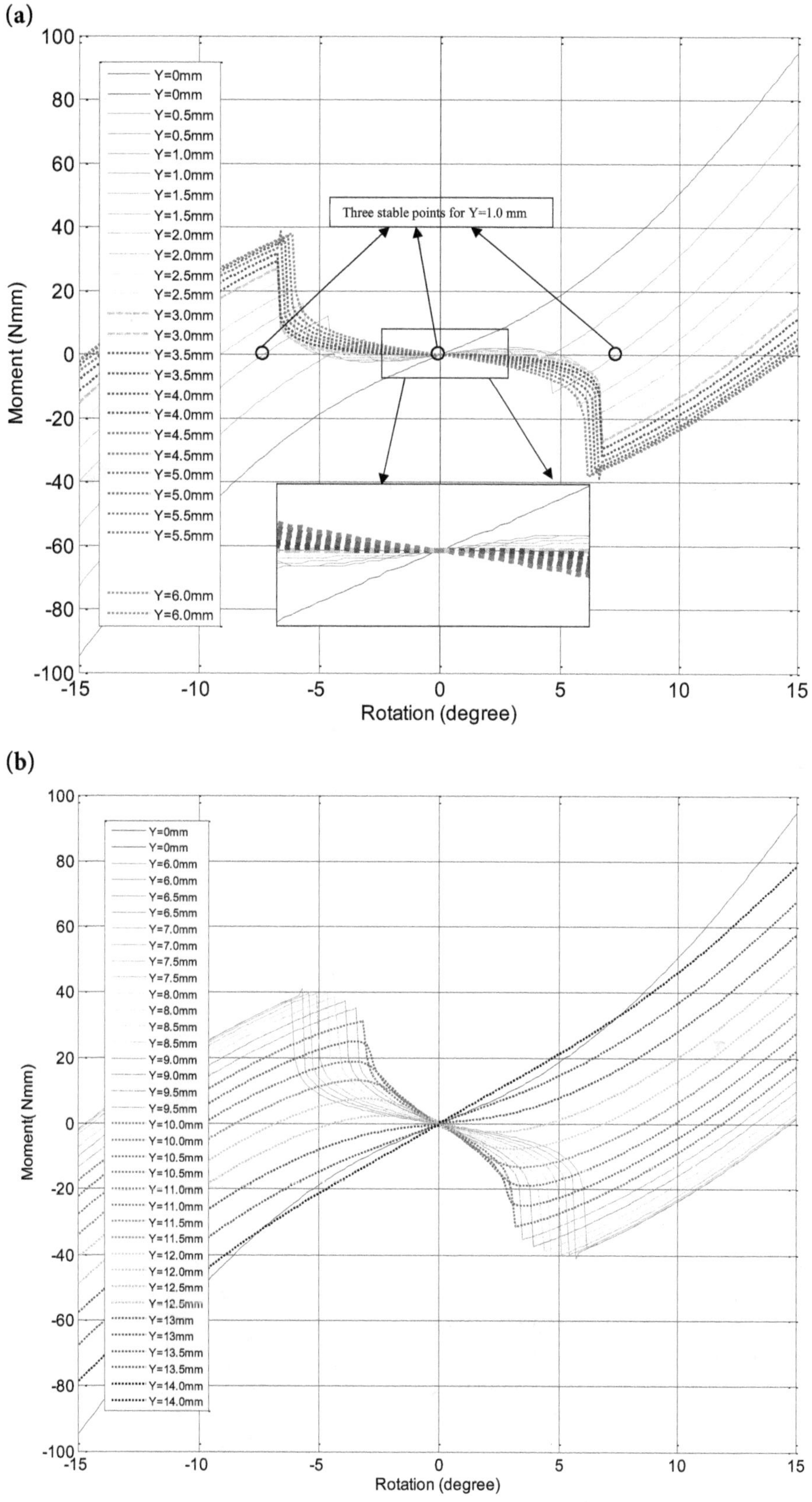

Figure 4. Moment versus rotation graph for the bistable mechanism.

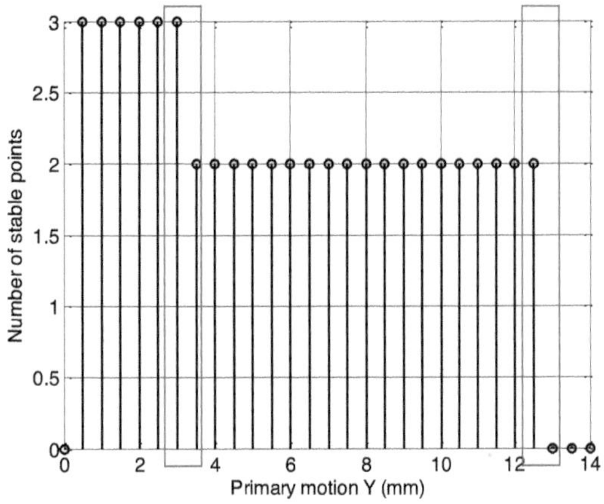

Figure 5. Number of stable rotational points under the specified primary motion.

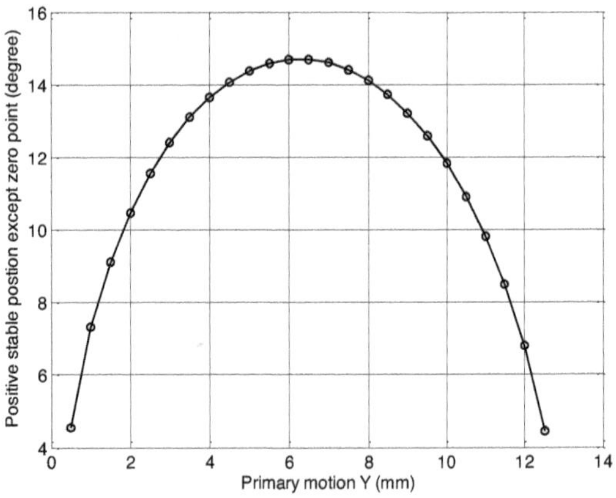

Figure 6. Positive stable rotational positions under different primary motions.

10.5 mm) have been observed (Fig. 8c–f). It is shown that the stable rotational displacement is larger under $Y = 6.0$ mm than that under $Y = 10.5$ mm, which agrees with the FEA resutls (Fig. 6).

4 Conclusions

The off-axis rotation characteristics of a bistable translational mechanism have been investigated in detail through displacing the guided primary translation/motion at different positions. For each primary motion, a moment-rotation curve has been obtained using the nonlinear FEA, which shows the multistable statues of the rotation. The idea of an infinitely-stable rotational mechanism that can be stable at any rotation position is proposed by controlling the primary motion.

Future works include fast numerical modelling, experimental testing verification and geometrical optimization for the off-axis rotation behavior.

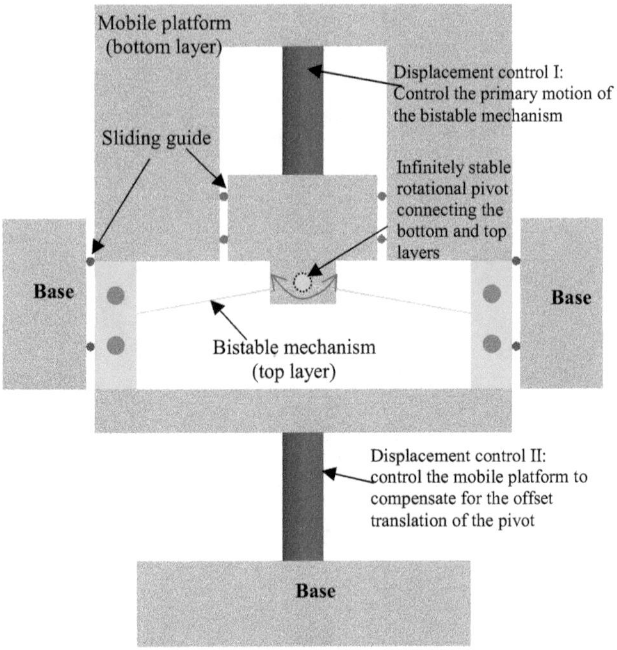

Figure 7. Schematic deisgn for the infinitely-stable rotational mechanism.

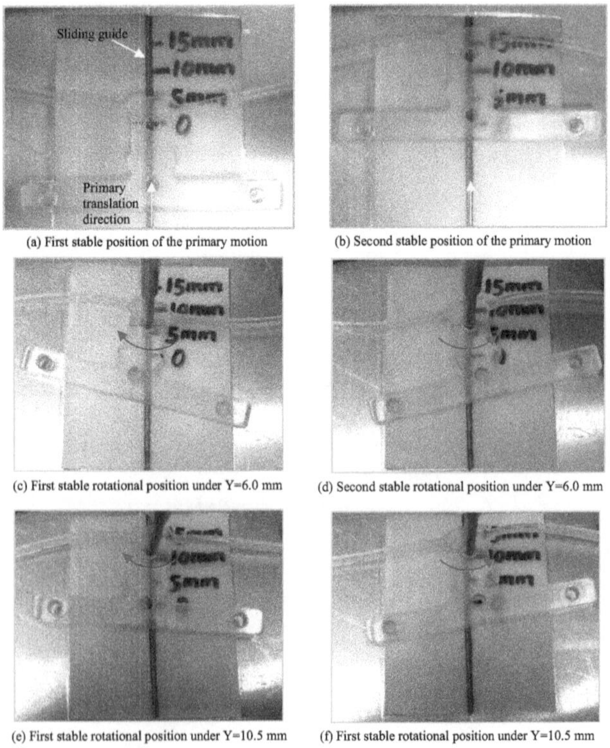

Figure 8. Fabricated prototype.

Acknowledgements. Tim Power and Michael O'Shea in University College Cork are greatly appreciated for fabricating the prototype.

References

Chen, Y. H. and Lan, C. C.: An Adjustable Constant-Force Mechanism for Adaptive End-Effector Operations, J. Mech. Design, 134, 031005, doi:10.1115/1.4005865, 2012.

Dunning, A. G., Tolou, N., and Pluimers, P. P.: Bistable Compliant Mechanisms: Corrected Finite Element Modeling for Stiffness Tuning and Preloading Incorporation, J. Mech. Design, 134, 084502, doi:10.1115/1.4006961, 2012.

Hansen, B. J., Carron, C. J., and Hawkins, A. R.: Plastic Latching Accelerometer Based on Bistable Compliant Mechanisms, Smart Mater. Struct., 16, 1967–1972, 2007.

Holst, G. L., Teichert, G. H., and Jensen, B. D.: Modeling and Experiments of Buckling Modes and Deflection of Fixed-Guided Beams in Compliant Mechanisms, J. Mech. Design, 133, 051002, doi:10.1115/1.4003922, 2011.

Hou, C.-W. and Lan, C.-C.: Functional Joint Mechanisms with Constant-torque Outputs, Mech. Mach. Theory, 62, 166–181, 2013.

Howell, L. L., Magleby, S. P., and Olsen, B. M.: Handbook of Compliant Mechanisms, Wiley, New York, 2013.

Kim, C. and Ebenstein, D.: Curve Decomposition for Large Deflection Analysis of Fixed-Guided Beams for Statically Balanced Compliant Mechanisms, J. Mech. Robot., 4, 041009, doi:10.1115/1.4007488, 2012.

Lassooij, J., Tolou, N., Tortora, G., Caccavaro, S., Menciassi, A., and Herder, J. L.: A Statically Balanced and Bi-stable Compliant End Effector Combined with a Laparoscopic 2-DoF Robotic Arm, Mech. Sci., 3, 85–93, 2012.

Liu, W. Q., Badel1, A., Formosal, F., Wu, Y. P., and Agbossou, A.: Novel Piezoelectric Bistable Oscillator Architecture for Wideband Vibration Energy Harvesting, Smart Mater. Struct., 22, 035013, doi:10.1088/0964-1726/22/3/035013, 2013.

Oh, Y.: Synthesis of Multistable Equilibrium Compliant Mechanisms, PhD thesis, University of Michigan, Michigan, USA, 2008.

Qiu, J., Lang, J., and Slocum, A.: A Curved-Beam Bistable Mechanism, J. Microelectromech. Syst., 13, 137–146, 2004.

Shaw, A. D., Neild, S. A., Wagg, D. J., Weaver, P. M., and Carrella, A.: A Nonlinear Spring Mechanism Incorporating a Bistable Composite Plate for Vibration Isolation, J. Sound Vibrat., 33, 6265–6275, 2013.

Sonmez, U. and Tutum, C. C.: A Compliant Bistable Mechanism Design Incorporating Elastica Buckling Beam Theory and Pseudo-Rigid-Body Model, J. Mech. Design, 130, 042304, doi:10.1115/1.2839009, 2008.

Zhang, A. and Chen, G.: A Comprehensive Elliptic Integral Solution to the Large Deflection Problems of Thin Beams in Compliant Mechanisms, J. Mech. Robot., 5, 021006, doi:10.1115/1.4023558, 2013.

5

Analytical model of temperature distribution in metal cutting based on Potential Theory

F. Klocke, M. Brockmann, S. Gierlings, and D. Veselovac

Laboratory for Machine Tools and Production Engineering (WZL), Aachen, Germany

Correspondence to: M. Brockmann (m.brockmann@wzl.rwth-aachen.de)

Abstract. Temperature fields evolving during metal cutting processes have also been of major interest. Temperatures in the tool influence the wear behaviour and hence costs, temperatures in the work-piece are directly responsible for later product quality. Due to the high significance of temperatures, many modelling attempts for temperature fields have been conducted, however failed to deliver satisfying results. The present paper describes a novel analytical model using complex functions based on potential theory. Relevant heat sources in metal cutting as well as changing material constants are considered. The model was validated by an orthogonal cutting process and different real machining processes.

1 Introduction

Temperatures occurring during metal cutting processes were of major interest for research since scientific investigations in this field are existing. First documented work is accredited to Thompson (1798) who examined the mechanical equivalent of heat when deephole-drilling brass. Even though his objectives for the experimental work were rather aimed at an understanding of the nature of heat itself than understanding of the metal cutting process, his work marked the starting point for consideration of thermal issues in metal cutting. The precise and quantitative determination of temperatures during the metal cutting process in terms of measurements was first conducted from Shore (1925), Gottwein (1925) and Herbert (1926) almost at the same time, who measured temperatures by means of measurement of electromotive forces, i.e. application of thermocouple method. Measurements of heat radiation in terms of pyrometers and infrared camera were first conducted by Schwerd (1933), Ueda et al. (1998) and Müller (2004). Both physical principles have their advantages and disadvantages, however in summary, measurement of temperatures in the metal cutting process are elaborate and mostly error-prone. A recent overview of temperature measurements in material removal processes is given from Davies et al. (2007).

Due to the difficulties in measurement, the need for modelling of temperatures in metal cutting was expressed by researchers several times, e.g. Shaw (2005) and Komanduri (2003). Regarding the different types of modelling, i.e. empirical, simulation and analytical, empirical and simulative models mostly fail to be transferable to other circumstances like different cutting parameters, tool and work material and cutting process. This is mainly due to the early linearization of the physical problem. Analytical models, however not applicable on complex cutting engagement situations, strengthen the basic understanding of the nature of temperature distribution and are in principle transferable to all similar physical problems. Due to this reason, the present paper describes the development of a novel approach for analytical modelling of temperatures in the metal cutting process. The derivation and parameterisation are presented in distinct sections of this paper. The model was validated by experimental trials on a fundamental cutting test rig and on a broaching machine. The aim of this paper is to assess the potential of the novel approach presented, i.e. use of the potential theory for prediction of temperature fields in metal cutting.

2 Analytical temperature models in metal cutting

In order to derive analytical models for temperature distribution in metal cutting, the governing physical equation, i.e. the partial differential equation for heat conduction has to be solved:

$$\frac{\partial^2 T(x,y)}{\partial x^2} + \frac{\partial^2 T(x,y)}{\partial y^2} = 0 \quad (1)$$

The equation is shown for two spatial dimensions x and y, stationary and constant heat conductivity, which should be the assumptions for the model derivation in this paper. Physical meaningful solutions can be found either in real-valued functions or complex-valued functions. Nowadays analytical models are exclusively using real-valued functions as will be described in the following.

2.1 Real-valued functions

Regarding the various analytical models available, the approach from Komanduri and Hou (2000) can be indicated the most advanced analytical model which was only slightly modified in recent times (e.g. Karas et al., 2013). When analysing their solution, the basic function that was used as a solution for Eq. (1) is one originally invented by Carlslaw and Jaeger (1959) and first applied by Hahn (1951):

$$T(x,y) = \frac{q_l}{2\pi\lambda} e^{\frac{-xv}{2a}} \cdot K_0\left(\frac{Rv}{2a}\right). \quad (2)$$

In physical means, the solution can be interpreted as a moving line heat source with velocity v. The solution was used by Komanduri and Hou (2000) to form a moving band heat source with an inclination angle relative to the direction of motion. The adiabatic boundaries and further boundary conditions in metal cutting were considered using mirror heat sources, exploiting the resulting symmetrical nature of the temperature distribution plot. Indeed, the vast majority of analytical models are using the Hahn (1951) solution as basic solution and differ from each other only in terms of different boundary conditions and heat source shape, like the model of Carlslaw and Jaeger (1959), which is commonly used in simulation models for temperature prediction.

Most analytical models based on Hahn solution or other similar real-valued functions do not sufficiently model the nature of the metal cutting process or are not validated at all. The main problem occurring is the deviation of the predicted distribution, which suggests that not the right mathematical function describing the nature of metal cutting was found. Furthermore, the existing analytical models are applicable only with restrictions, as the integrals of Eq. (2) are very complex and only symmetrical situations can be modelled when boundary conditions, e.g. adiabatic surfaces or heat sources, are considered.

2.2 Complex-valued functions

Besides solutions in the real-valued space, complex functions solving Eq. (1) are existing. A complex function is a function of the complex variable z, where $z = x + iy$. The two spatial dimensions are separated by the complex number i. The complex function can be expressed as:

$$F(z) = \Phi(x,y) + i \cdot \Psi(x,y). \quad (3)$$

The functions Φ and Ψ itself are real-valued functions and can be plotted in x and y graph. To find solutions of Eq. (1), there is a special group of complex functions called potential functions. These functions fulfil the following requirement:

$$\frac{\partial F}{\partial y} = i, \frac{\partial^2 F}{\partial y^2} = \frac{\mathrm{d}^2 F}{\mathrm{d}z^2} i^2. \quad (4)$$

Taking into account that $i^2 = -1$ these functions fulfil Eq. (1) if Eq. (4) is true. If the function $F(z)$ is a potential function then the real part Φ is orthogonal to the imaginary part Ψ when plotting both in an $x - y$ diagram. For potential functions furthermore the principle of superposition is true, i.e. if $F_1(z)$ and $F_2(z)$ are solutions of Eq. (1), then also $F_3(z) = F_1(z) + F_2(z)$ is also a solution.

Potential functions are firstly only mathematical solutions of Eq. (1), however they are applied successfully in other engineering fields. Most famous application can be found in fluid mechanics where solutions of potential theory are used to predict stream fields, compare e.g. Anderson (2011). The fact that exactly the same partial differential equation, i.e. Eq. (1), needs to be solved, yields the lack of available solutions for temperature distribution in metal cutting.

3 Model derivation

In order to derive new models for prediction of temperature rise distribution for metal cutting, the following method was applied:

- Identification of suitable basic complex functions from potential theory and superposition of these functions.
- Parameter study of each coefficient of the function, assessment of physical relevant numerical range.
- Application of model on metal cutting process, i.e. consideration of relevant heat sources and boundary conditions.

The three steps can be understood with iterative character, especially because the last step, i.e. the application on the metal cutting process is the most important. The results presented in this paper are concluded from a detailed iterative approach using the stated method.

3.1 Identification of function

The basic functions used in potential theory can be physically interpreted with the help of the streamline analogy. Figure 1 shows three distinct functions, a uniform flow, vortex and corner flow, when plotting the imaginary function Ψ in an $x - y$ diagram.

Figure 1. Basic functions and combined modell approach.

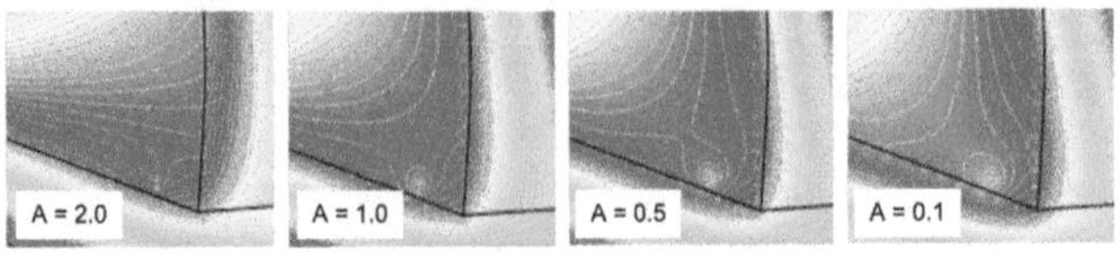

Figure 2. Variation of parameter A.

Transferring the physical interpretations to temperature lines for the metal cutting process, detailed analysis of more basic functions revealed that the superposition of these three functions yield a plot which appears similar to the temperature distribution in metal cutting. The not yet parameterised plot of Ψ of the combined function is shown at the right side of Fig. 1. The function $F(z)$ can be expressed as:

$$F(z) = F_{\text{uniform}} + F_{\text{vortex}} + F_{\text{corner}}$$
$$= Aze^{-i\alpha} + \frac{Bi}{2\pi}\ln(z + z_{\text{sum}}) + C \cdot z_{\text{rot}}^{k}. \quad (5)$$

The distinct parameters A, α, B, z_{sum}, C, z_{rot} and k are described in detail in Sect. 3.2. Taking into account the streamline analogy it appears strange at a first glance that the vortex flow and not a source or sink function is used in the combined function. As stated before this was the outcome of a detailed iterative approach yielding better results in regard to an application on the metal cutting process. The solutions were chosen due to an iterative methodology. Physical interpretations of the solutions, in terms of their influences on the run of the isotherms are known, e.g. the vortex flow bends the isothermal field.

3.2 Parameter study

The parameter study of A, α, B, z_{sum}, C, z_{rot} and k showed different kind of influences on the temperature plot and hence different kinds of possible interpretations referred to the metal cutting process. Due to the high number of unknown parameters, each parameter was varied taking into account the nature of each parameter for the basic flow analogy, e.g. parameters A and α belong to the uniform flow and influence velocity and angle of the flow. Figure 2 shows the variation of parameter A as an example. The parameter study shows that a variation of A from 0.1 to 1.0 does not change the shape of the temperature distribution field significantly, the isotherms shape does vary with rising value of A but the general behaviour stays constant.

The plot in the upper left of Fig. 2 in contrast, shows a behaviour which is not typical for temperature distribution in metal cutting, i.e. the isotherms run into the rake face of the tool, which is a discrepancy to the known literature results and own experimental trials that are presented later in this paper. In a similar way of parameter study, numerical sensible values for each parameter were determined, summarised in Table 1.

Table 1. Numerical values for $F(z)$.

Parameter	Values	correlation to ...
A	0.1–1.0	heat source strength
α	0.5–1.2	cutting speed, chip thickness
B	11.0–14.0	cutting speed
C	0.20–0.26	chip thickness
k	1.85	chip thickness
θ	$(0.025–0.05)\pi$	shear zone angle

The last row of the table gives an indication of possible correlation to cutting parameters and nature of the heat sources shear and friction. However these values are only yielded by observation and were not validated. The major conclusion of the parameter study shows that all shown parameters stay in a relative narrow numerical band when considering that apart from any experimental data only extreme situations were used to define the numerical boundaries as described in Fig. 3. In Table 1, the parameter θ is replacing the term z_{rot} from Eq. (5) considering:

$$z_{\text{rot}} = \sqrt{(x\cos\theta - y\sin\theta)^2 + (x\sin\theta + y\cos\theta)^2}. \quad (6)$$

Summarised, the parameters presented in this Section seem to be almost constant respectively only slightly changing for all temperature fields that can be predicted with this model approach. The parameter z_{sum} in Eq. (5) however was found not to be constant.

3.3 Application on metal cutting process

Considering the stream flow analogy of Eq. (5), the term z_{sum} describes number and direction of vortexes that can be placed at different locations. The term z_{sum} can be expressed as:

$$z_{\text{sum}} = z_{\text{ch}} + z_{\text{sh}} + z_{\text{con}}, \quad (7)$$

where the term z_{ch} locates the vortexes on the contact line between rake face and chip, the term z_{sh} considers the strength shear plane heat source and z_{con} the length and angle of the shear plane, compare Fig. 3.

For term z_{ch} the expression

$$z_{\text{ch}} = \sum_{-1}^{m}(x_{\text{ch},m} + i \cdot y_{\text{ch},m}) = \sum_{-1}^{m}[1 + i \cdot (2m+1)] \quad (8)$$

can be formulated. The number of vortexes along chip side hereby is m. The initial negative value under the summation is due to the chosen coordinate system. For the vortexes

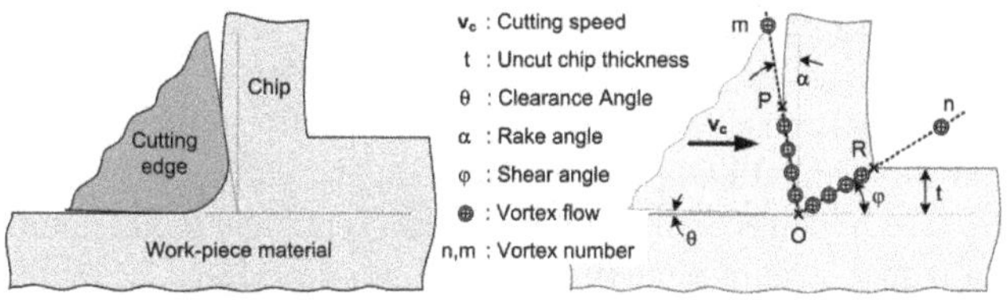

Figure 3. Vortex locations and metal cutting process.

along the shear plane heat source pertains:

$$z_{\text{sh}} = \sum_{0}^{n} x_{\text{sh},n}$$

$$+ i \cdot y_{\text{sh},n} = \sum_{0}^{n} \{(2n+1) + i \cdot [(2n+1) \cdot \tan(\phi)]\} \quad (9)$$

where n is the number of vortexes along the shear plane and φ is the shear angle. The last term z_{con} can be formulated as:

$$z_{\text{con}} = t + i \cdot 0.5 \cdot \text{OR} \cdot \cos(\phi) \quad (10)$$

where t is the uncut chip thickness, OR is the length of the shear plane heat source and φ again the shear angle. Plugging Eqs. (8), (9) and (10) into Eq. (5) yields the following expression for the function $\Psi(x, y)$:

$$\Psi_{\text{total}}(x, y) = A\sin(\alpha) - \frac{B}{2\pi}\ln\left(\sqrt{x_{\text{total}}^2 + y_{\text{total}}^2}\right) + C$$

$$\cdot [\sqrt{(x\cos\theta - y\sin\theta)^2 + (x\sin\theta + y\cos\theta)^2}]^k \sin(k\theta). \quad (11)$$

As stated in Sect. 2.2, Ψ hereby is only the imaginary part of $F(z)$, i.e. all expressions that contain the imaginary unit i. The real part of $F(z)$ is not plotted but can be considered as a graph with orthogonal lines. The terms x_{total} and y_{total} are defined as:

$$x_{\text{total}} = x + 1 + t + \sum_{0}^{n} (2n+1)$$

$$y_{\text{total}} = y + 0.5 \cdot \text{OR} \cdot \cos(\phi)$$

$$+ \sum_{0}^{n} (2n+1)\tan(\phi) + \sum_{-1}^{m} (2m+1). \quad (12)$$

Plotting the function $\Psi(x, y)$ yields a graph which need to be scaled to get temperature distribution fields, where the parameters A, α, B, C, θ and k can be considered as nearly constant as described in Sect. 3.2. The uncut chip thickness t is known, solely length of the shear plane heat source OR and number of vortexes on shear plane n and friction zone m have to be measured or modelled.

4 Model validation

The presented model in form of Eqs. (11) and (12) were parameterised using the findings from the parameter study.

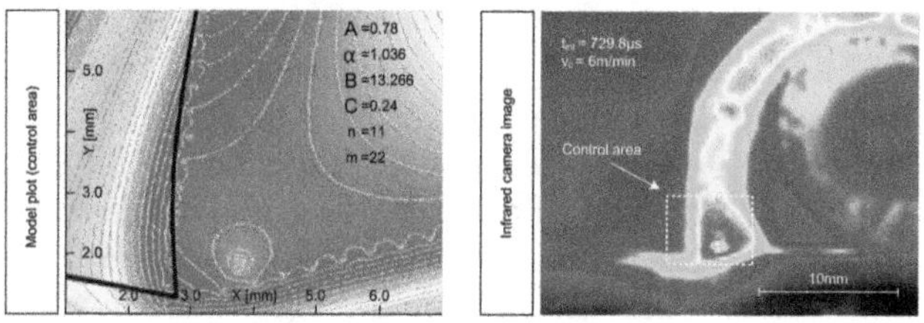

Figure 4. Comparison of model plot and infrared image (for Inconel 718, HSS tool, cutting depth 30 μm).

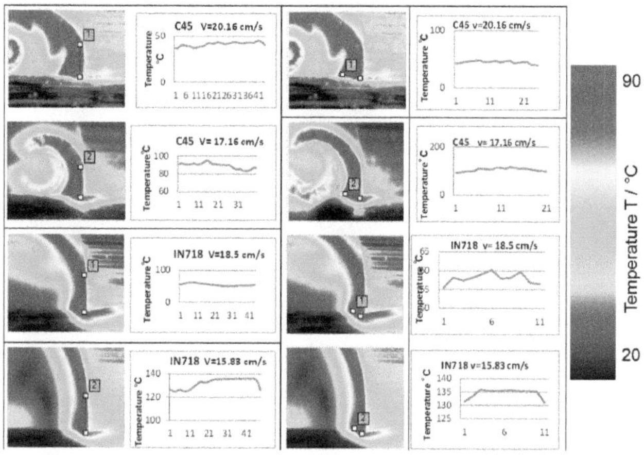

Figure 5. Model validation for Inconel 718 and C45 steel.

The mentioned data for length of shear plane heat source and shear angle were taken out of an infrared image for simplification reasons. Figure 4 shows a comparison between the parameterised model and results from experimental data. As only the shape of the isotherms is of importance no scale is provided.

The experiments were conducted on an analogy test rig for orthogonal metal cutting, the test rig is described in Klocke et al. (2011). The infrared images were taken by an online calibration measurement method combining two-colour parameter with a high-speed infrared camera, details can be found in Gierlings and Brockmann (2013).

The comparison shown in Fig. 4 shows a good accordance in the chip and near the contact zone between tool and work piece, especially in terms of temperature field distribution. The line in the model plot shows that the tool geometry can be found in the model plot. The comparison presented here should be understood as a first qualitative validation showing promising results for the potential of the derived model.

Further validation work was done for a broaching process which can also be found in Gierlings and Brockmann (2013). Figure 5 shows further validation trials for steel C45 and an Nickel-based alloy (Inconel 718).

The temperature rise distribution along the shear plane heat source and frictional heat source for chip-tool interface was found by using experimental result for steel material C45 and IN718 respectively at different cutting speeds. Therefore, the average temperature for these examples was considered to be a reference for calculate the temperature between the shear plane heat source and frictional heat for chip-tool interface. Figure 5 shows the temperature distribution for the C45 and IN718 work material along the shear plane heat source for different cutting velocity with plots the temperature along the selected profile length. The following parameters were used from Shaw (2005):

- Chip velocity: $v_{\text{ch}} = 60\,\text{cm}\,\text{s}^{-1}$
- Friction force: $F_{\text{f}} = 605\,\text{N}$
- Heat flux: $q = 605 \times 60/100/(0.0833 \times 0.025) = 68\,042.8\,\text{W}\,\text{cm}^{-2}\,°\text{C}$
- Chip thickness: $t_2 = 0.0833\,\text{cm}$
- Shear angle: $\varnothing = 16.7°$

The chosen model parameters are:

- Number of vortex flow along the shear plane line: OR = 11
- Number of vortex flow along the chip-tool interface line: OP = 22
- Parameter for uniform flow: $A = D \times (q'')//\lambda = 5 \times 10^{-5} \times 68\,042.8/0.436 = 0.78\,°\text{C}\,\text{cm}^{-1}$
- Parameter of (Alpha): $\alpha = v_c t_1/(40 \times a) = 1.036$
- Parameter for vortex flow: $B = E \times v_c/a = 8 \times 10^{-3} \times 200/0.1206 = 13.266\,\text{cm}^{-1}$
- Parameter for corner flow: $C = r/(5.5w) = 0.24\,\text{cm}^{-1}$
- Parameter for corner flow: $k = 1.8, \theta = \pi/30$.

The temperature rise along the chip-tool interface is also calculated. The average temperature rise distribution are considered, this will be also the reference temperature that can used to approximate the temperature rise in the chip-tool interface for the complex potential flow model. Using the average value for temperature rise distribution for the shear plane heat source and chip-tool interface, the combined effect for both the shear plane heat source and chip-tool interface frictional heat source can be represented by using this data. For the final plots for the temperature rise distribution further calibration was required.

5 Summary and discussion

The presented paper shows a novel approach for analytical modelling of temperature field distribution for metal cutting processes. For the derivation of the model, complex functions solving the partial differential equation for heat conduction were considered. The complex functions belong to a certain group of functions called potential functions, which already showed good results in other engineering fields e.g. fluid mechanics, where solutions of exactly the same partial differential equation are needed. The analysis revealed the combination of three basic functions to yield temperature plots that show a huge potential for application in the metal cutting process. The combined function was parameterised to suit the situation in metal cutting, mainly by taking into account length and strength of the two heat sources friction and shear zone. Beyond this first approach of using complex functions of the potential theory for modelling temperature fields in metal cutting, still a lot of parameters need to be correlated systematically to the cutting parameters. The final verification of the model is outstanding as more cutting process need to be investigated using the basic function described in this paper.

Acknowledgements. The presented modelling work is part of project A02 of the Collaborative Research Center SFB/TR 96: “Thermo-energetic design of machine tools”, funded by the German Science Foundation DFG.

References

Anderson, J. D.: Introduction to flight, McGraw-Hill Science, 7, 251–362, 2011.

Carslaw, H. S. and Jaeger, J. C.: Conduction of heat in solids. 2nd ed. Oxford [Oxfordshire], New York, Clarendon Press, Oxford University Press, 1959.

Davies, M. A., Ueda, R., M'Saoubi, R., Mullany, B., and Cooke, A. L.: On the Measurements of Temperature in Material Removal Processes, CIRP Annals Manufacturing Technology, 56, 581–604, 2007.

Gierlings, S. and Brockmann, M.: Analytical Modelling of Temperature Distribution using Potential Theory by Reference to Broaching of Nickel-Based Alloys, Adv. Mat. Res., 769, 139–146, 2013.

Gottwein, K.: Die Messung der Schneidentemperatur beim Drehen, Maschinenbau Betrieb, 4, 1129–1135, 1925.

Hahn, R. S.: On the temperature development at the shear plane in the metal cutting process, Proc. of First National Congress of Applied Mech., 661–666, 1951.

Herbert, E. G.: The measurement of cutting temperatures, Proceeding of the Institution of Mechanical Engineers, 1, 289–329, 1926.

Karas, A., Bouzit, M., and Belarbi, M.: Development of a thermal model in the metal cutting process for prediction of temperature distributions at the tool-chip-workpiece interface, J. Theor. Appl. Mech., 51, 553–567, 2013.

Klocke, F., Bergs, T., Busch, M., Rohde, L., Witty, M., and Cabral, G. F.: Integrated approach for a knowledge-based process layout for simultaneous 5-Axis milling of advanced materials, Adv. in Tribology, 2, 108–115, 2011.

Komanduri, R.: NSF workshop on research needs in thermal aspects of material removal, Oklahoma, 2003.

Komanduri, R. and Hou, Z. B.: Thermal modeling of the metal cutting process – part I, Int. J. Mech. Sci., 42, 1715–1752, 2000.

Müller, B., Renz, U., Hoppe, S., and Klocke, F.: Radiation thermometry at high-speed turning process, J. Manuf. Sci. E.-T. ASME, 126, 488–495, 2004.

Schwerd, F.: Über die Bestimmung des Temperaturfeldes beim Spanablauf, Zeitschrift des VDI, 9–77, 211–216, 1933.

Shaw, M. C.: Metal cutting principles, Oxford University Press, 2, 29–30, 2005.

Shore, H.: Thermoelectric measurement of cutting tool temperatures, J. Washington Academy of Science, 15, p. 85, 1925.

Thompson, B.: An inquiry concerning the source of heat which is excited by friction, Philos. T. R. Soc. Lond., 18, 278–287, 1798.

Ueda, T., Sato, M., and Nakayama, K.: The temperature of a single crystal diamond tool in turning, Annals of the CIRP, 47, 41–44, 1998.

6

Dimensional synthesis of mechanical linkages using artificial neural networks and Fourier descriptors

N. Khan[1], I. Ullah[2], and M. Al-Grafi[2]

[1]University of Engineering and Technology Peshawar, Pakistan
[2]College of Engineering, Taibah University, Yanbu, Saudi Arabia

Correspondence to: I. Ullah (drirfanullah61@hotmail.com)

Abstract. Dimensional synthesis of mechanisms to trace given paths is an important problem with no exact solution. In this paper, the problem is divided into representation of curve shape and learning the relation between curve shape and mechanism dimensions. Curve shape is represented by Fourier descriptors of cumulative angular deviation of the curve, which do not depend on the position or scale of the curve. An artificial neural network (ANN) is trained to learn the (unknown) relation between the Fourier descriptors of a planar curve and the dimensions of the mechanism tracing that curve. Presented with any simple, closed, planar curve, the ANN suggests the dimensions of a four-bar whose coupler curve is similar in shape. A local optimization procedure further refines the results. Examples presented indicate the method is successful as long as the curve shape is such that the mechanism is able to trace it.

1 Introduction

Mechanical linkages are used extensively for transforming motion and transmitting power in industrial machinery, agriculture, construction, automobiles and household gadgets. The design of linkages involves determining dimensions of constituting links so that the linkage moves in a manner necessary to carry out the required task (Sandor and Erdman, 1988).

Figure 1 shows a common four-bar linkage whose tracer point P generates a path known as the coupler curve when the input link rotates. The problem of dimensional synthesis for path generation requires calculation of suitable dimensions for the linkage so that point P traces a desired continuous path. An exact solution for this problem is not possible because of the limited number of dimensions available, but various techniques have been used for approximate solutions. The most common techniques used include conventional optimization methods (Tomas, 1968; Sancibrian et al., 2004; Diab and Smaili, 2008), using atlases of mechanisms (Zhang et al., 1984), simulated annealing (Ullah and Kota, 1996), and genetic algorithms or evolutionary algorithms (Cabrera et al., 2002; Laribi et al., 2004; Starosta, 2008; Lin, 2010).

In this research, normalized Fourier descriptors are used to represent the shape of coupler curves and artificial neural networks (ANNs) are used for approximating the inverse relationship between curve shape and linkage dimensions. A brief introduction to Fourier descriptors is presented below, followed by review of the relevant research.

Fourier descriptors are functions used for description of shapes of object boundaries (or of any two-dimensional, simple closed curve). The boundary of a shape is represented by a periodic function that, when expanded in a Fourier series, yields a set of coefficients containing the shape information. Different types of Fourier descriptors have been proposed in the literature, including Fourier descriptors of the boundary (McGarva and Mullineux, 1993; Persoon and Fu, 1986) and Fourier descriptors of the angular orientation of the curve (Zahn and Roskies, 1972; Buskiewicz et al., 2009). The latter method is used here, which describes shape using the net amount of angular bend of the curve relative to the starting angle as a function of arc length $\phi(l) = \theta(l) - \theta(0)$, as shown in Fig. 2. The main advantage of using this formulation is invariance to translation, rotation and scaling of the curve so that the *shape* of a coupler curve is represented, independently of the position and size of the linkage producing it. The Fourier descriptors can also be normalized to

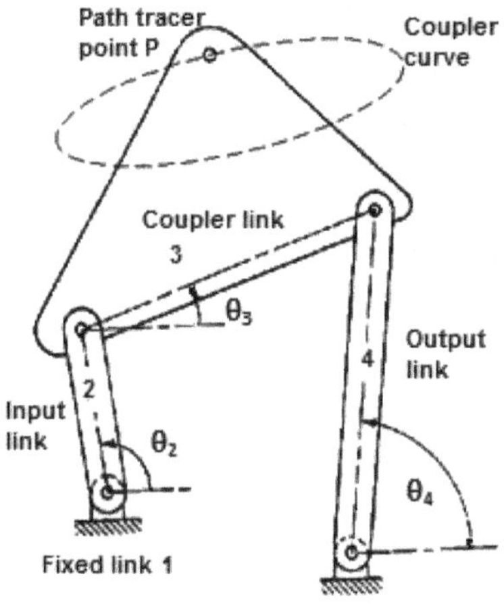

Figure 1. A four-bar linkage.

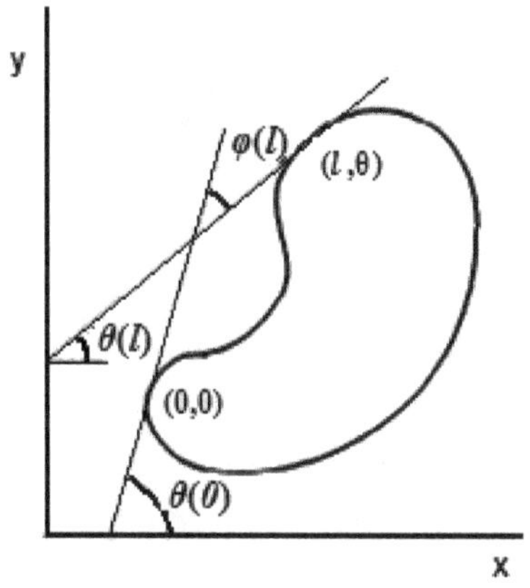

Figure 2. Net angular bend of curve boundary as a function of curve length.

remove dependence on starting point of the curve, direction of traversing the curve and to remove the difference between mirror image curves; these properties are irrelevant to synthesis of mechanisms.

After a change of variable to remove length dependence, $t = \frac{2\pi l}{L}$, and addition of a linear part to remove discontinuities, the function to be expanded in a Fourier series is given by Eq. (1).

$$\varphi^*(t) = \varphi\left(\frac{Lt}{2\pi}\right) + t \quad t \in [0, 2\pi], \tag{1}$$

where L is the total length of the curve. This periodic function is expanded in a Fourier series and the coefficients of the series are termed the Fourier descriptors.

Work on linkage synthesis using ANNs is not extensive. Hoskins and Kramer (1993) used a radial basis function (RBF) ANN in combination with optimization techniques to construct an approximate inverse model of four-bar linkage. In that work, coupler curve shape is represented by power spectrum Fourier transform of curvature versus length. The RBF. ANN used consists of 50 input nodes, 50 to 750 nodes in the single hidden layer and 5 nodes in the output layer. A constrained local optimization technique is used to refine linkage parameters. Vasiliu and Yannou (2001) use normalized Fourier descriptors of the boundary to represent curve shape and then use an ANN with 17 input neurons, 2 hidden layers of 22 neurons each and 5 output neurons to synthesize four-bar mechanisms for path generation. Xie and Chen (2007) discuss motion synthesis of a crank-rocker based on neural networks. The angle of the coupler link and the position of the coupler point are first mapped onto a curve in image space, which is then described by a two-dimensional Fourier transform. A separate three-layered ANN is designed for relating each link dimension to the curve shape. It is not clear, however, how a network can model the relation of curve shape to one link dimension, without reference to other links. Galan-Marin et al. (2009) use wavelet descriptors and neural networks for linkage synthesis.

Peñuñuri et al. (2011) propose combined synthesis for function, path, and motion generation by combining the objective functions of all three. Kim and Yoo (2012) demonstrate simultaneous number and dimension synthesis for function generation problems. Ebrahimi and Payvandy (2015) use the structural error objective function but add a penalty for violation of workspace limits. They demonstrate the use of the imperial competitive algorithm, a variation of genetic algorithms, to solve the problem.

In the next section, the methodology of the work is outlined, including the equations used. Section 3 presents the design of the ANN, including the structure of network and details of training and testing, followed by some example problems. Finally, a discussion of the results is presented and conclusions drawn in Sect. 4.

2 Methodology

The methodology used in this work comprises the following steps:

1. Dimensions for a large number of four-bar mechanisms were generated at random, keeping link length ratios within a reasonable limit and ensuring all mechanisms are crank-rockers so that the coupler curves produced are closed curves.

2. Coupler curves of the generated mechanisms were obtained, each as a set of points in the plane.

3. Each coupler curve was transformed into its Fourier descriptors and descriptors were normalized.

4. An ANN was designed to take Fourier descriptors of the generated curves as the input and corresponding mechanism dimensions as the output or target values. The ANN was trained using the data generated.

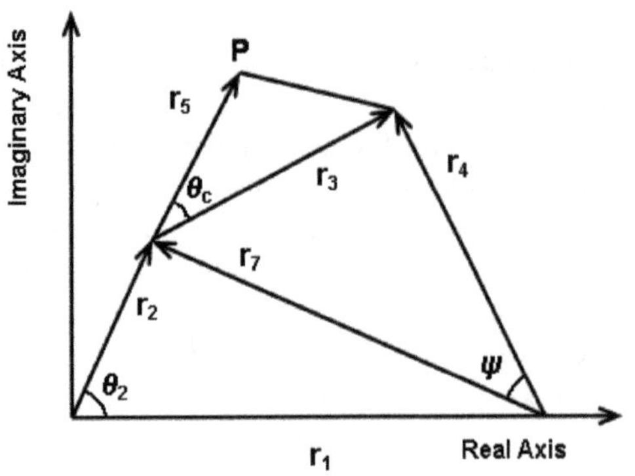

Figure 3. Four-bar representation for analysis.

5. The trained network was tested by presenting new data to it. Results were confirmed by numerical as well as visual comparison of desired and proposed curves.

6. To determine the four-bar mechanism that will reproduce any desired curve, the curve is first converted to its Fourier descriptors, which are then presented as input to the trained ANN. The mechanism dimensions produced by the ANN are further refined by using a local optimization procedure.

Each step is explained below in some detail.

2.1 Generation of crank-rockers

With reference to Fig. 3, a four-bar linkage can be represented by the lengths of its four links plus two variables for locating the coupler point P, the path tracing point, on link 3. For scale independence, the length of link 2 is taken as unity and limits are imposed on the lengths of other links to obtain a practical linkage. The five variables selected to represent a four-bar are as given in Eq. (2).

$$1 \leq \boldsymbol{r}_i \leq 5 \quad i = 1,3,4,5 \quad \text{and} \quad 0 \leq \theta_c \leq 2\pi \tag{2}$$

Link lengths are generated at random but still meeting the Grashof conditions, as follows:

- Choose $1 \leq p \leq 5$ at random.
- Choose l at random such that $p \leq l \leq 5$.
- Choose q at random such that $s + l \leq p + q \leq p + l$, where $s = \boldsymbol{r}_2$ to ensure a crank-rocker.
- The three lengths p, q, and l are arbitrarily assigned to $\boldsymbol{r}_1$, $\boldsymbol{r}_3$, and $\boldsymbol{r}_4$.

The other two variables are generated at random between chosen limits. Five thousand random crank-rocker linkages were generated using this procedure.

2.2 Calculation of coupler curves

As shown in Fig. 3, the positions of various links are represented as variables in the complex plane. All link lengths are known, as well as angles of $\boldsymbol{r}_1$ (taken as zero) and $\boldsymbol{r}_2$. The following calculations (Eqs. 3–8) determine the position of the coupler point P.

$$\boldsymbol{r}_7 = \boldsymbol{r}_2 - \boldsymbol{r}_1 \tag{3}$$

$$\psi = \cos^{-1}\left[\frac{r_4^2 + r_7^2 - r_3^2}{2 r_4 r_7}\right] \tag{4}$$

$$\arg(\boldsymbol{r}_4) = \arg(\boldsymbol{r}_7) - \psi \tag{5}$$

$$\boldsymbol{r}_3 = \boldsymbol{r}_4 - \boldsymbol{r}_7 \tag{6}$$

$$\arg(\boldsymbol{r}_5) = \arg(\boldsymbol{r}_3) + \theta_c \tag{7}$$

$$\boldsymbol{r}_{\mathrm{P}} = \boldsymbol{r}_2 + \boldsymbol{r}_5 \tag{8}$$

Repeating the calculations by varying the crank angle in the range $\leq \theta_2 \leq 2\pi$, a set of complex numbers, $\boldsymbol{V}_i$, are obtained, which are the vertices of the coupler curve.

2.3 Calculation of Fourier descriptors

The curve is converted from vertices to length–angle representation, using Eqs. (9)–(11).

$$\Delta \boldsymbol{l}_i = \boldsymbol{V}_{i+1} - \boldsymbol{V}_i \tag{9}$$

$$\boldsymbol{l}_i = \boldsymbol{l}_{i-1} + |\Delta \boldsymbol{l}_i| \tag{10}$$

$$\Delta\varphi_i = \arg(\Delta \boldsymbol{l}_{i+1}) - \arg(\Delta \boldsymbol{l}_i), \tag{11}$$

where $\Delta \boldsymbol{l}_i$ is the ith side of the polygonal curve, $\boldsymbol{l}_i$ is the length from the first to the ith vertex, and $\Delta\varnothing_i$ is the change in angle at the ith vertex.

The Fourier descriptors of the curve, $2m + 1$ in number, are then calculated from Eqs. (12)–(16) (Zahn and Roskies, 1972). The polar form of Fourier descriptors was selected as it is easier to normalize.

$$\mu_0 = -\pi - \frac{1}{L}\sum_{i=1}^{n} l_i \,\Delta\varphi_i \tag{12}$$

$$a_k = -\frac{1}{k\pi}\sum_{i=1}^{n} \Delta\varphi_i \sin\frac{2\pi k l_i}{L} \quad k = 1, \ldots, m \tag{13}$$

$$b_k = -\frac{1}{k\pi}\sum_{i=1}^{n} \Delta\varphi_i \cos\frac{2\pi k l_i}{L} \quad k = 1, \ldots, m \tag{14}$$

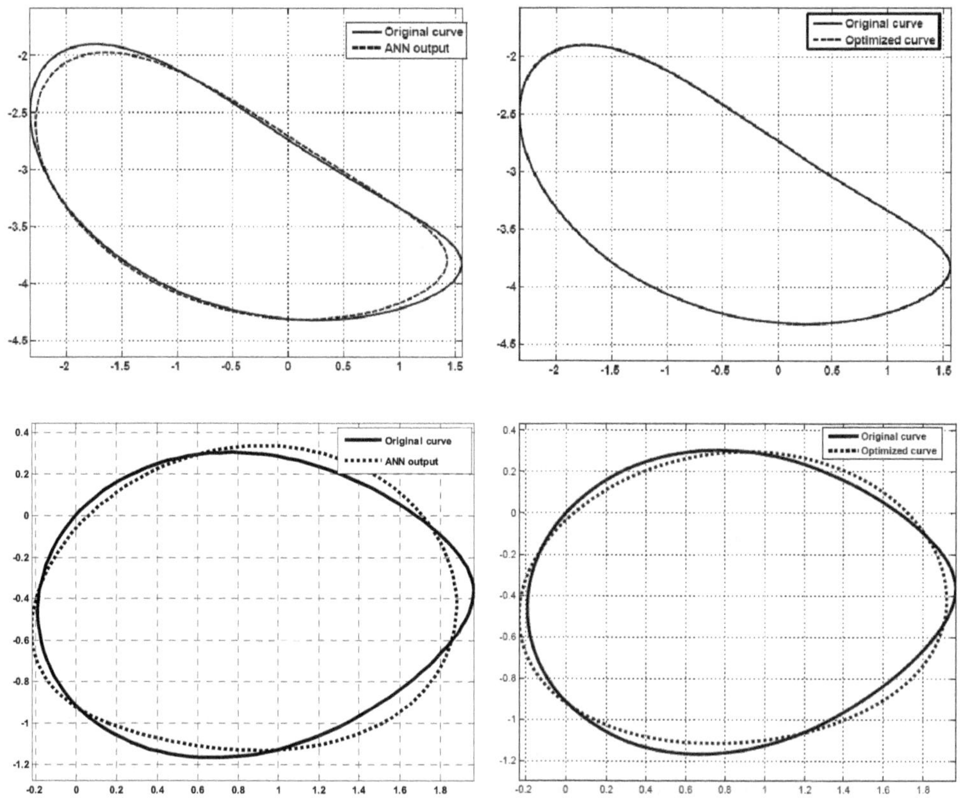

Figure 4. Example 1 (top row) and 2 (bottom row). On the left, the ANN output and on the right the optimized curve, both superposed on the original curve.

$$A_k = \sqrt{a_k^2 + b_k^2} \quad k = 1, \ldots, m \tag{15}$$

$$\alpha_k = \tan^{-1}\frac{a_k}{b_k} \quad k = 1, \ldots, m \tag{16}$$

The amplitudes A_k are already normalized, while the phase angles α_k are normalized as follows to remove dependence on starting point, direction of curve traversal and mirroring (Buskiewicz et al., 2009). $\Delta\alpha = \frac{\alpha_j}{j}$, where j is the index of the first non-zero harmonic.

For $k = 1, \ldots, m$,

$$\alpha_k = (\alpha_k - k\Delta\alpha) \bmod 2\pi. \tag{17}$$

$$\text{If } \alpha_k > \pi, \alpha_k = (\alpha_k - \pi). \tag{18}$$

$$\text{If } \alpha_k > \frac{\pi}{2}, \alpha_k = (\pi - \alpha_k). \tag{19}$$

3 Design of artificial neural network

An artificial neural network (ANN) is a mathematical structure that takes a vector of inputs (22 normalized Fourier descriptors of a curve) and produces a set of outputs (the five dimensions of a four-bar mechanism). To train the network, it is provided with a large set of inputs and corresponding outputs so that it "learns" the relationship between inputs and outputs. After the network has been trained and tested, it can be used by providing an input (Fourier descriptors of a desired curve) and obtaining the output (dimensions of the mechanism that will produce a curve shaped like the desired curve).

A multilayer feed-forward neural network was selected and trained using back propagation. After some experimentation, $m = 11$ was selected in the above equations so that a total of 22 Fourier descriptors were generated for each curve. Accordingly, the neural network had 22 neurons in the input layer. The output layer had 5 neurons, representing linkage dimensions. The number of hidden layers and the number of neurons in each layer were varied until satisfactory performance was obtained. Specifications of the final network architecture are shown in Table 1. The functions named in that table are Matlab Neural Network Toolbox™ functions.

It is useful to scale the inputs and outputs before training the neural network so that they fall within a specific range. This allows for efficient training but requires keeping track of scaled and unscaled values. Matlab mapping function mapminmax was used to provide linear scaling in the range $[-1, 1]$. Training was carried out using the Broyden–

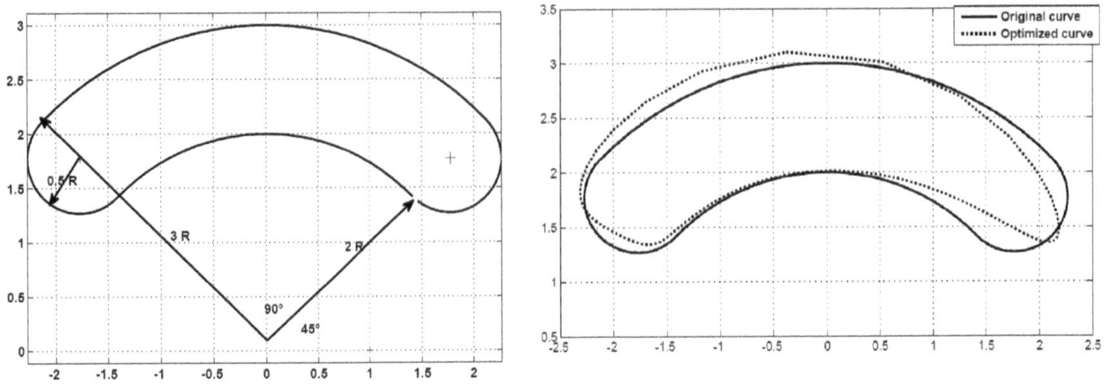

Figure 5. Example 3. (Left panel) The original curve, and (right panel) the original and the final curves compared.

Fletcher–Goldfarb–Shanno algorithm to reduce the performance function msereg, which is the average squared error between actual output and the desired output of the network.

3.1 Simulation and testing

After designing and training the neural network using 5000 cases, it was tested for a new set of 50 coupler curves not used in training. The dimensions of 50 four-bar linkages output by the network were then used to construct coupler curves. Checking of results was carried out by comparing dimensions of linkages and values of Fourier descriptors as well as visual comparison of the shapes of curves.

3.2 Example problems

In the first two example problems, a four-bar was chosen at random, the Fourier descriptors of its coupler curve calculated and supplied as input to the trained ANN. The ANN output dimensions of a four-bar were then optimized to minimize the difference between Fourier descriptors of input and output curves. Table 2 shows the comparison of linkage dimensions for the original four-bar, the output of ANN, and the final, optimized four-bar for each example. Figure 4 shows the comparison of the curve shapes in each case. For comparison, the output curves were superposed on the original curve as follows (Buskiewicz et al., 2009): the output curve is translated so that its centroid is positioned over the centroid of the original curve. The curve is then rotated about the centroid so that the orientations of the principal axes of the two curves match and finally the curve is scaled so that the lengths of the two curves match.

In the third example, a general desired curve was chosen rather than a coupler curve, as shown in Fig. 5 (left panel). The four-bar mechanism obtained has the coupler curve shown in Fig. 5 (right panel) and the following dimensions:

$$r_1 = 1.1711, r_3 = 13.8151, r_4 = 13.8736,$$
$$r_5 = 2.6779, \theta_c = -0.8907 \text{ rad.} \quad (20)$$

Table 1. Structure of ANN used.

No. of input neurons	22
No. of hidden layers	03
No. of neurons in hidden layers	[22,5,5]
Transfer function of hidden layers	Tan-sigmoid
No. of output neurons	5
Transfer function of output neurons	Purelin
Training function	Trainbfg
Network performance function	mseregs

4 Discussion and conclusions

In the first two examples presented, the desired curve was a coupler curve obtained from a known four-bar mechanism. The algorithm was able to find the same four-bar mechanism in the first case but converged to a different four-bar having similar coupler curve. This was observed in other example cases also. Sometimes the curve obtained is a mirror image of the original curve; however, that problem is easily corrected by "flipping" the four-bar.

In the third example, a general desired curve (not being a coupler curve) was used to determine the dimensions of a four-bar that will produce a coupler curve similar in shape. The results show that although the algorithm gets close, a significant difference remains in curve shape. This is most likely due to the limitations of available coupler curves that a four-bar can generate, rather than problems in the algorithm.

In conclusion, a new method for approximate dimensional synthesis of planar linkages has been proposed and examples of application presented. The method is seen to be successful provided the desired path is close to the type of curves the mechanism is able to create. Although four-bar coupler curves have wide-ranging shapes, it should be noted that the method is limited to simple, closed curves; open curves (such as those produced by double-rocker mechanisms) and curves with crunodes are excluded.

Table 2. Results for examples 1 and 2.

Four-bar dimensions	r_1	r_3	r_4	r_5	θ_c (rad)
			Example 1		
Input four-bar	2.9587	3.4723	3.5771	3.3454	3.3771
ANN output four-bar	3.1876	3.4862	3.6964	3.1337	3.4049
Optimized four-bar	2.9662	3.5503	3.6504	3.3455	3.3798
			Example 2		
Input four-bar	3	3	2.5	1	5
ANN output four-bar	1.9640	2.1175	2.0717	0.7483	2.4803
Optimized four-bar	3.5368	2.7720	2.2270	1.0612	1.0333

References

Buskiewicz, J., Starosta, R., and Walczak, T.: On the application of the curve curvature in path synthesis, Mech. Mach. Theory, 44, 1223–1239, 2009.

Cabrera, J. A., Simon, A., and Prado, M.: Optimal synthesis of mechanisms with genetic algorithms, Mech. Mach. Theory, 37, 1165–1177, 2002.

Diab, N. and Smaili, A.: Optimum exact/approximate point synthesis of planar mechanisms, Mech. Mach. Theory, 43, 1610–1624, 2008.

Ebrahimi, S. and Payvandy, P.: Efficient constrained synthesis of path generating four-bar mechanisms based on the heuristic optimization algorithms, Mech. Mach. Theory, 85, 189–204, 2015.

Galan-Marin, G., Alonso, F., and Del Castillo, J.: Shape optimization for path synthesis of crank-rocker mechanisms using a wavelet-based neural network, Mech. Mach. Theory, 44, 1132–1143, 2009.

Hoskins, J. C. and Kramer, G. A.: Synthesis of mechanical linkages using artificial neural networks and optimization, Neural Networks, IEEE International Conference, 2, 822-j–822-n, 1993.

Kim, B. S. and Yoo, H. H.: Unified synthesis of a planar four-bar mechanism for function generation using a spring-connected arbitrarily sized block model, Mech. Mach. Theory, 49, 141–156, 2012.

Laribi, M. A., Mlika, A., Romdhane, L., and Zeghloul, S.: A combined genetic algorithm–fuzzy logic method (GA–FL) in mechanisms synthesis, Mech. Mach. Theory, 39, 717–735, 2004.

Lin, W. Y.: A GA–DE hybrid evolutionary algorithm for path synthesis of four-bar linkage, Mech. Mach. Theory, 45, 1096–1107, 2010.

McGarva, J. and Mullineux, G.: Harmonic representation of closed curves, Appl. Math. Model., 17, 213–218, 1993.

Peñuñuri, F., Peón-Escalante, R., Villanueva, C., and Pech-Oy, D.: Synthesis of mechanisms for single and hybrid tasks using differential evolution, Mech. Mach. Theory, 46, 1335–1349, 2011.

Persoon, E. and Fu, K.: Shape discrimination using Fourier descriptors, IEEE Trans. Pattern Analysis and Machine Intelligence, PAMI-8, 388–397, 1986.

Sancibrian, R., Viadero, F., Garcıa, P., and Fernandez, A.: Gradient-based optimization of path synthesis problems in planar mechanisms, Mech. Mach. Theory, 39, 839–856, 2004.

Sandor, G. and Erdman, A.: Advanced Mechanism Design: Analysis and Synthesis, Prentice-Hall, 1988.

Starosta, R.: Application of genetic algorithm and Fourier coefficients (GA–FC) in mechanism synthesis, J. Theor. Appl. Mech., 46, 395–411, 2008.

Tomas, J.: The synthesis of mechanisms as a nonlinear programming problem, J. Mechanisms, 3, 119–130, 1968.

Ullah, I. and Kota, S.: Globally-optimal synthesis of mechanisms for path generation using simulated annealing and Powell's method, Proceedings of the 1996 ASME Design Engineering Technical Conferences, 96-DETC/MECH-1225, Irvine, California, USA, 1996.

Vasiliu, A. and Yannou, B.: Dimensional synthesis of planar mechanisms using neural networks: Application to path generator linkages, Mech. Mach. Theory, 36, 299–310, 2001.

Xie, J. and Chen, Y.: Application of back-propagation neural network to synthesis of whole cycle motion generation mechanism, Twelfth IFToMM World Congress, Besancon, France, 2007.

Zahn, C. T. and Roskies, R. Z.: Fourier descriptors for plane closed curves, IEEE T. Pattern Anal., C-21, 269–281, 1972.

Zhang, C., Norton, R., and Hammond, T.: Optimization of parameters for specified path generation using an atlas of coupler curves of geared five bar linkages, Mech. Mach. Theory, 19, 459–466, 1984.

7

Hybrid Position-Force Control of a Cable-Driven Parallel Robot with Experimental Evaluation

W. Kraus, P. Miermeister, V. Schmidt, and A. Pott

Fraunhofer Institute for Manufacturing Engineering and Automation IPA in Stuttgart, Germany

Correspondence to: W. Kraus (wek@ipa.fhg.de)

Abstract. For cable-driven parallel robots elastic cables are used to manipulate a mobile platform in the workspace. In this paper, we present a hybrid position-force control, which allows for applying defined forces on the environment and simultaneous movement along the surface. We propose a synchronous control of the cable forces to ensure the stability of the platform during movement. The performance of the controller is experimentally investigated regarding contact establishment and dynamic behavior during a motion on the cable robot IPAnema 3.

1 Introduction

Due to their huge workspace, high dynamics and lightweight structure, cable-driven parallel robots, in the following referred to as cable robots, received high interest in the past.

Processes like grinding and polishing demand an accurately controlled contact force to give reliable results. One application example is the surface grinding in the airplane maintenance, which is exemplary setup using oil barrels shown in Fig. 1. Force control also allows compensating for deviations in the work piece geometry by keeping the tool constantly in contact with the surface. There exist solutions like air controlled cylinders or stiffness actuators, which can be installed at the robot's platform to enable for the control of process forces. Osypiuk and Kröger (2011) The drawback is that the system adds additional weight to the platform and the direction of the force is typically fixed w.r.t. the platform frame. Furthermore, these systems make a power and signal supply on the platform necessary.

The aim of the hybrid position-force control is to apply process forces in a programmable direction to a surface while the movement perpendicular to the force vector is position controlled.

The basic approach for the hybrid position-force control was already presented in the early eighties by Raibert and Craig (1981) and its basic structure is shown in Fig. 3. The control problem is divided into parallel control loops for position and force control with their set-points $\boldsymbol{x}_{\text{set}}$ and $\boldsymbol{F}_{\text{set}}$,

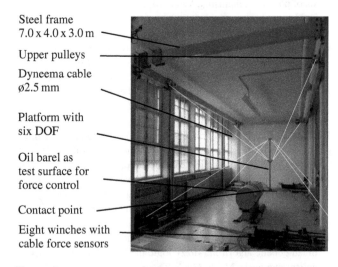

Figure 1. Cable-driven parallel robot IPAnema 3: Robot test set-up.

respectively. With the selection matrix $\mathbf{S}$, the degrees-of-freedom (DOF) are assigned to be either force or position controlled. The hybrid control law was implemented already on parallel manipulators in Madani and Moallem (2011) and Deng et al. (2010). For the cable robot, several adaptions are made. The major difference to the classic hybrid control law is the additional cable force control loop to control the internal cable tensions. For a serial robot, the concept of an additional null-space control loop is described in Park (2006).

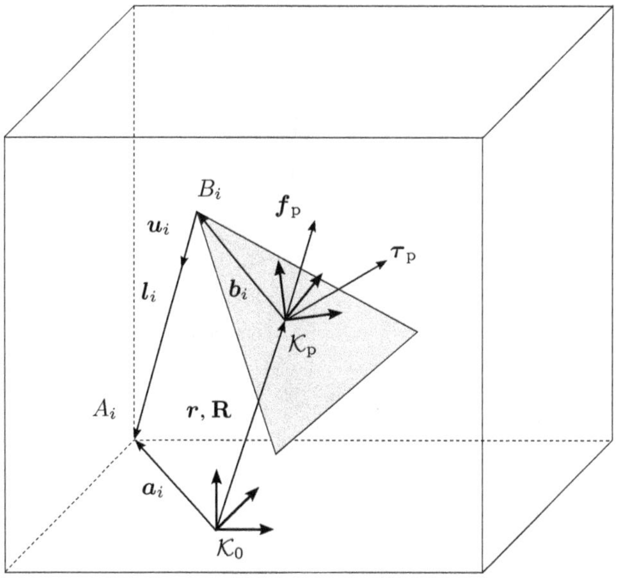

Figure 2. General kinematic parameters.

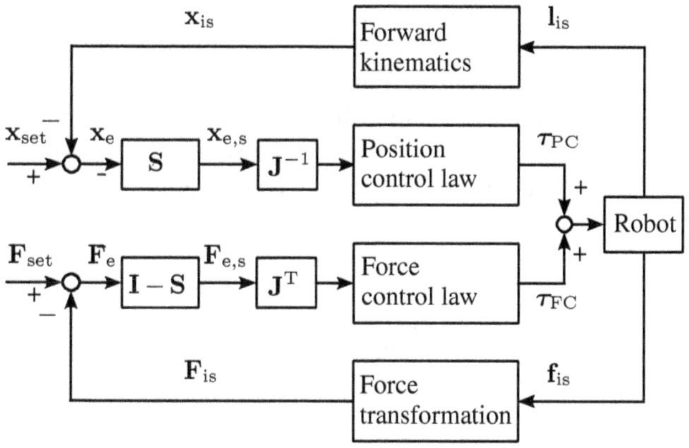

Figure 3. Hybrid position and force control proposed by Raibert and Craig (1981).

Another difference is the use of joint space force measurement and transformation into operational space instead of a force-torque sensor at the platform. The cable force sensors are already integrated for cable force control. By using them for wrench measurement, additional costs for sensors can be avoided and no additional sensor has to be added to the platform. We assume a cable robot using closed-loop position control in the joint space, instead of closed-loop position control in operational space. Therefore, no forward kinematics is used which is computationally expensive for a cable robot. The position control is encapsulated in the motion generator. Therefore, the selection matrix **S** cannot be realized for position control and the decoupling is realized in the force control implementation.

Beside the hybrid position-force control, the contact establishment is challenging. The major concern regarding stability are the transient system properties when the platform comes in contact with the environment. The contact problem was investigated for both serial (Albu-Schaeffer et al., 2003; Raibert and Craig, 1981), and classical parallel robots Tang et al. (2012). In Assuncao and Schumacher (2003) this problem is solved by a state machine based approach using acceleration and brake controllers for bringing the robot into contact. An alternative approach for eliminating the switching strategy is presented in Almeida et al. (1999) using an adaptive gain of the impedance controller. In Reisinger et al. (2011) Moreau's sweeping process is applied for the transition during contact establishment.

In this paper, we present the design and experimental evaluation of a operational space force control for a cable robot.

In addition to the known contact problem, we have to ensure that the cables remain under tension.

The state machine based approach presented in Assuncao and Schumacher (2003) is adapted for the cable robot. To keep the cables under tension, we introduce a decoupled cable force controller to tension the cables during the contact force control. For this, the cable force control presented in Kraus et al. (2014) is modified regarding the set-point determination. A task coordinate system is established to realize the hybrid position-force control. For motion control, the trajectory planning and interpolation is done in operational space and transformed with the inverse kinematics to position set-points in joint space. This structure is typically used in commercial CNC controllers. We chose position-controlled drives and not a torque-based approach to make use of state-of-the art industrial servo drives and the robustness of the decentralized position control. To program the desired force vector with standard industrial machine interfaces, new commands based on standardized G-Code are established.

This paper is organized as follows: The robot model including kinematic and the determination of cable force distributions are summarized in Sect. 2. The design and implementation of the operational space force controller is described in Sect. 3. Experimental results are presented and discussed in Sect. 4. Finally, conclusions and an outlook on future work are given in Sect. 5.

2 Kinematic and static model

For completeness, we briefly review the kinematic model for a spatial robot with $n = 6$ DOF. Verhoeven (2004) The geometry of the robot is described by the proximal anchor points on the robot base $\boldsymbol{a}_i$ and the distal anchor points on the platform $\boldsymbol{b}_i$ as shown in Fig. 2. The index i denotes the cable number and m is the absolute number of cables. By applying a vector loop, the cable vector $\boldsymbol{l}_i$ follows as

$$\boldsymbol{l}_i = \boldsymbol{a}_i - \boldsymbol{r} - \mathbf{R}\boldsymbol{b}_i, \quad (1)$$

while $\boldsymbol{r}$ is the platform position vector and rotation matrix $\mathbf{R}$ describes the orientation of the mobile platform frame $\mathcal{K}_\mathrm{p}$ w.r.t. the base coordinate frame $\mathcal{K}_0$.

The structure equation with $\mathcal{K}_0$ the structure matrix $\mathbf{A}^\mathrm{T}$ describes the force and torque equilibrium at the platform for a given cable force distribution $\boldsymbol{f}$ and can be written as

$$\underbrace{\begin{bmatrix} \boldsymbol{u}_1 & \cdots & \boldsymbol{u}_m \\ \boldsymbol{b}_1 \times \boldsymbol{u}_1 & \cdots & \boldsymbol{b}_m \times \mathbf{u}_m \end{bmatrix}}_{\mathbf{A}^\mathrm{T}(\boldsymbol{r},\mathbf{R})} \underbrace{\begin{bmatrix} f_1 \\ \vdots \\ f_m \end{bmatrix}}_{\boldsymbol{f}} = -\underbrace{\begin{bmatrix} \boldsymbol{f}_\mathrm{p} \\ \boldsymbol{\tau}_\mathrm{p} \end{bmatrix}}_{\boldsymbol{w}}, \qquad (2)$$

with $\boldsymbol{u}_i = \frac{\boldsymbol{l}_i}{\|\boldsymbol{l}_i\|}$ being the unit vector of the cables directing from the mobile platform to the base. The applied wrench $\boldsymbol{w}$ consists of external forces $\boldsymbol{f}_\mathrm{p}$ and torques $\boldsymbol{\tau}_\mathrm{p}$ and also includes the weight m_p of the platform.

For the cable force control, a force distribution $\boldsymbol{f}$ has to be calculated, which solves the structure Eq. (2) and is continuous along a trajectory. There exist several approaches which fulfill these requirements like Lamaury and Gouttefarde (2012) or Mikelsons et al. (2008). In this paper, the advanced closed-form solution presented in Pott (2013) is applied. It is based on the closed-form solution for the structure matrix presented in Pott et al. (2009). In this approach, a reference cable force $\boldsymbol{f}_\mathrm{ref} = (\boldsymbol{f}_\mathrm{min} + \boldsymbol{f}_\mathrm{max})/2$ is introduced which is based on the minimum cable force $\boldsymbol{f}_\mathrm{min}$ and maximum cable force $\boldsymbol{f}_\mathrm{max}$, respectively. With the Moore-Penrose matrix inverse $\mathbf{A}^{+\mathrm{T}}$, the least square optimal force distribution is obtained by

$$\boldsymbol{f} = \boldsymbol{f}_\mathrm{ref} - \mathbf{A}^{+\mathrm{T}}(\boldsymbol{w} + \mathbf{A}^\mathrm{T}\boldsymbol{f}_\mathrm{ref}). \qquad (3)$$

The redundancy $r = m - n$ defines the rank of the null-space of the structure matrix. For the IPAnema cable robot with $r = 2$, the workspace derived with the closed-form solution amounts only 57.5 % of the possible workspace. This motivated for the further development and ended in the advanced closed-form method. The algorithm and its application to workspace analysis of cable robots was firstly presented in Pott (2013).

The advanced closed-form algorithm consists of the evaluation of the closed-form solution and an order reduction of the linear equation system in the case, the cable bounds are violated. The largest violation of the cable bounds is detected and the cable force is set to the corresponding limit value. Next, this cable is excluded from structure matrix and the exerted force and torque of this cable is added to the wrench. This procedure is repeated until a feasible solution is reached or the structure matrix is quadratic and no further reduction is possible. The computational effort is strictly bounded. In the worst case, r iterations have to be performed. The description of the robot model is hereby completed and next the controller design for hybrid position-force control is presented.

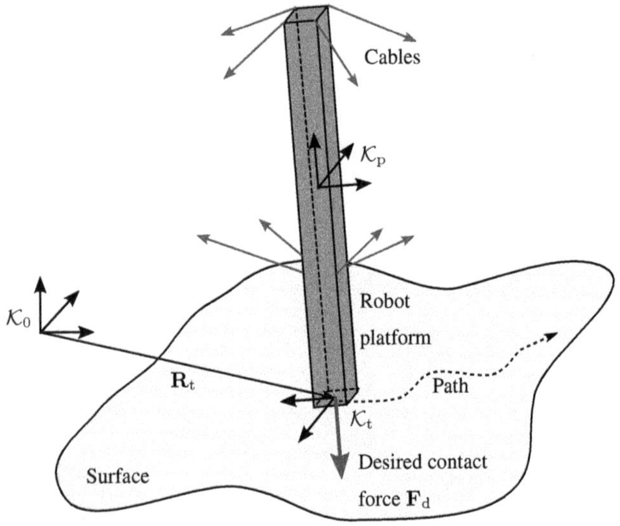

Figure 4. Principle set-up at the mobile platform and coordinate systems for hybrid position-force control.

3 Controller design and implementation

3.1 Controller design

The operational space force control aims at the control of the wrench $\boldsymbol{w}$ in a given direction e.g. perpendicular to the surface of the environment. While using a cable robot for applying the contact force, additionally the cables have to be kept under tension. Especially for the simultaneous position controlled movement in the tangential plane this is important, as friction forces have to be overcome. To realize the hybrid position-force control for a cable robot, the control problem is divided into three parts:

- The operational space position control gives a desired position in operational space $\boldsymbol{x}_\mathrm{set,PC}$ to move the platform along a path.
- The operational space force control generates an offset in the operational space position $\boldsymbol{x}_\mathrm{set,OFC}$ allowing to control the external platform wrench i.e. contact force using joint space position control.
- The cable force control for keeping the cables under tension. For this, a feasible force distribution is calculated for the measured wrench. The controller output is the modification $\boldsymbol{l}_\mathrm{CFC}$ of the cable length set-point.

With the inverse kinematics (IK) the output of the three controllers are summed up by

$$\boldsymbol{l}_\mathrm{set} = \mathrm{IK}\{\boldsymbol{x}_\mathrm{set,PC} + \boldsymbol{x}_\mathrm{set,OFC}\} + \boldsymbol{l}_\mathrm{CFC}, \qquad (4)$$

to the cable length set-point $\boldsymbol{l}_\mathrm{set}$ which is commanded to the servo drives. The control structure is visualized in Fig. 5. In the following, the operational space and cable force control are described in detail.

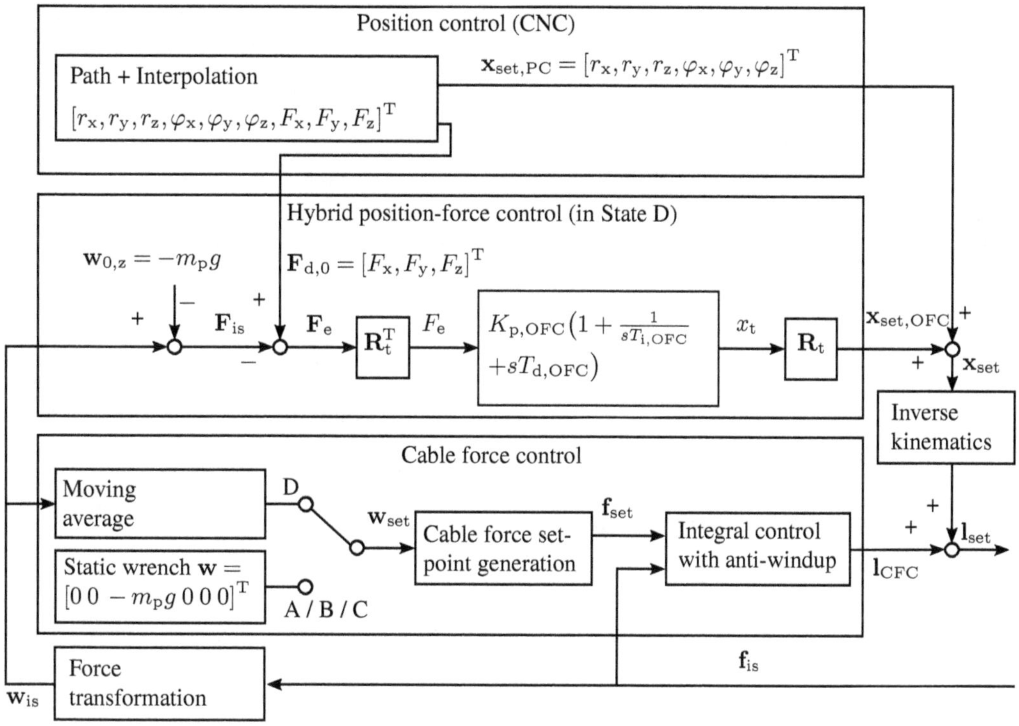

Figure 5. Proposed implementation of the hybrid position-force control with cable force control.

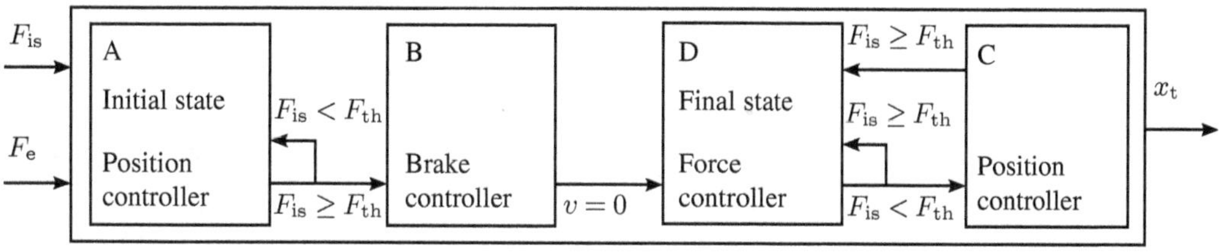

Figure 6. State machine with transition conditions for the contact establishment.

3.2 Operational Space Force Control

The main challenge of the operational space force control is the contact establishment. When the platform comes in contact, the system properties change abruptly. We adapt the approach presented by Assuncao and Schumacher (2003) which uses a state machine. Basically, the contact is established in position control (States A–C) and closed-loop force control is performed when the platform is in contact. The contact establishment is structured into four states, as illustrated in Fig. 6. The inputs of the state machine are the actual contact force F_{is} and the deviation from the desired contact force $F_e = F_{set} - F_{is}$. The output is the position offset x_t in the direction of the desired force vector.

In state A, the platform is moved in the direction of the surface with a maximal velocity v_{max}. During the movement, the wrench is observed. The contact between the platform and the surface is detected when the threshold force F_{th} is exceeded. As the measured wrench is disturbed by measurement errors, we also observe the derivative of the force to identify the contact. When the contact is detected, the platform is slowed down till stand still in state B. Next in state D, the closed-loop force control is activated as shown in Fig. 5. A PID controller with the transfer function

$$G_c(s) = \frac{\mathcal{L}\{x_t(t)\}}{\mathcal{L}\{F_e(t)\}} = K_{p,OFC}\left(1 + \frac{1}{sT_{i,OFC}} + sT_{d,OFC}\right), \quad (5)$$

is applied, where $K_{p,OFC}$ denotes the gain, $T_{i,OFC}$ the integrator reset time and $T_{d,OFC}$ the derivative time. The control parameters are determined experimentally and presented in Table 1. When the robot loses the contact to the surface, this means the force is less than the threshold force, the state machine switches to state C and a position controller brings the platform again in contact.

The operational space force control is based on an indirect force measurement using the cable force sensors. For the transformation from joint to operational space, the structure

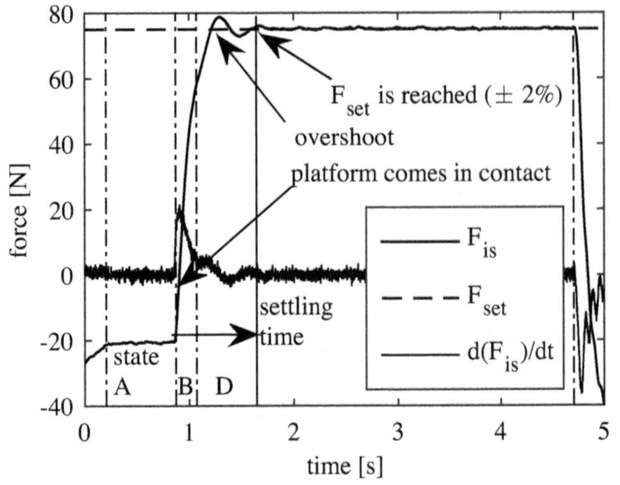

Figure 7. Progression of the force during contact establishment, force control and release.

matrix $\mathbf{A}^{\mathrm{T}}$, describing the actual cable directions, is used. The estimated wrench $\boldsymbol{w}_{\mathrm{is}}$ is derived by

$$\boldsymbol{w}_{\mathrm{is}} = -\mathbf{A}^{\mathrm{T}} \boldsymbol{f}_{\mathrm{is}}, \tag{6}$$

where $\boldsymbol{f}_{\mathrm{is}}$ denotes the measured cable forces. The actual force vector w.r.t. the world frame $\boldsymbol{F}_{\mathrm{is}}$ is derived from the actual wrench under consideration of the own-weight of the platform $m_{\mathrm{P}}g$.

We introduce a task coordinate system $\mathcal{K}_{\mathrm{t}}$ as visualized in Fig. 4, which lies in the contact point and the z axis coincides with the desired force direction. The transformation from world $\mathcal{K}_0$ to task coordinate system $\mathcal{K}_{\mathrm{t}}$ is described by the rotation matrix $\mathbf{R}_{\mathrm{t}}$. The desired force vector $\boldsymbol{F}_{\mathrm{set},0}$ is transformed by

$$\boldsymbol{F}_{\mathrm{set},t} = \mathbf{R}_{\mathrm{t}}^{\mathrm{T}} \boldsymbol{F}_{\mathrm{set},0}, \text{ with } \boldsymbol{F}_{\mathrm{set},0} = [0 \; 0 \; \boldsymbol{F}_{\mathrm{set}}]^{\mathrm{T}}, \tag{7}$$

to the task coordinate system, where the z component represents the scalar desired force $\boldsymbol{F}_{\mathrm{set}}$. With the coordinate transformation, the force control can be designed as a single input single output controller.

3.3 Cable force control

Redundant cable robots have the property that the cables can be tensed against each other without exerting an external force. In the present control architecture, the external forces are controlled by the operational space force control. Therefore, the cable force control aims at the control of the internal tensions and has to be decoupled from the operational space force control. As the wrench is controlled by the operational space force control, we propose to use the measured wrench $\boldsymbol{w}_{\mathrm{is}}$ for the determination of the set-point of the cable forces $\boldsymbol{f}_{\mathrm{set}}$ according to (Eq. 3). This ensures that the controlled cable force is compatible to the actual external forces and does

Table 1. Control parameters of hybrid position-force control (controller is experimentally tuned).

parameter	definition	value	unit
T_{MA}	time constant of moving average for wrench filtering	2.0	s
$T_{\mathrm{i,OFC}}$	integrator reset time	2.6	s
$T_{\mathrm{d,OFC}}$	derivative time constant	0.03	s^{-1}
$K_{\mathrm{p,OFC}}$	proportional gain (compliance)	$22\,000^{-1}$	$\mathrm{m\,N}^{-1}$
$f_{\min}$	minimum cable force	100	N
$f_{\max}$	maximum cable force	1000	N
F_{th}	threshold force	25	N
$v_{\max}$	maximum approaching velocity	0.05	$\mathrm{m\,s}^{-1}$

not influence the motion of the platform. To avoid instabilities due to the synchronous control of the external wrench and the internal cable forces, a moving average is applied on the wrench to smooth the measurement signal (low pass filter) and, thus, increase the stability margins of the control loop. The control error in cable force $\boldsymbol{f}_{\mathrm{e}}$ is established by

$$\boldsymbol{f}_{\mathrm{e}} = \boldsymbol{f}_{\mathrm{set}} - \boldsymbol{f}_{\mathrm{is}}. \tag{8}$$

The integral control with anti-windup is realized according to Kraus et al. (2014).

During the contact establishment, the system properties and also the wrench change abruptly. This leads to the destabilization of the contact controller. In this transient phase, the cable force controller also influences the external forces and additionally disturbs the contact controller. To avoid this problem, the cable force control assumes the gravitational force $m_{\mathrm{P}}g$ as wrench during contact establishment which corresponds to the states A to C of the state machine. With this approach, we can eliminate additional sources of instabilities due to the internal cable force control. After contact establishment, the cable force control is enabled again.

3.4 Implementation

The CNC executes the position control for a path programmed in G-code. Three additional axes were added to the CNC, which are interpreted as desired force vector. These axes were handled as movement axes and were interpolated in the interpolation cycle. In this way, the force vector changes smoothly and is consistent with the movement commands along the surface.

4 Experimental evaluation

4.1 Test platform

For the experimental evaluation the control algorithms are implemented on the cable robot IPAnema 3 using eight ca-

Table 2. Experimental evaluation of the contact establishment at different positions and resulting overshoot and settling time.

position	overshoot		settling time	
	mean [%]	σ [%]	mean [s]	σ [s]
1	9.357	1.171	0.781	0.191
2	8.698	0.840	1.378	0.242
3	5.692	0.351	0.981	0.093
4	4.299	0.927	0.866	0.138
5	8.573	0.486	0.977	0.081
6	8.329	0.461	1.035	0.019
7	18.106	2.227	0.804	0.156
8	15.024	2.857	0.742	0.209
9	5.100	0.353	0.889	0.010
10	3.910	0.067	3.517	0.413
average	8.709	–	1.197	–

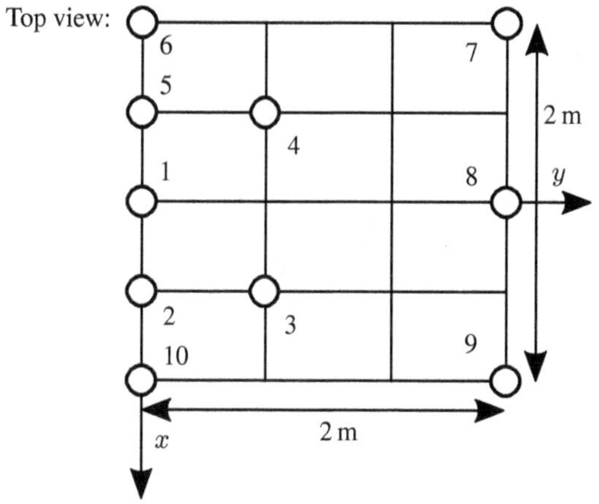

Figure 8. Experimental evaluation of the contact establishment at different positions.

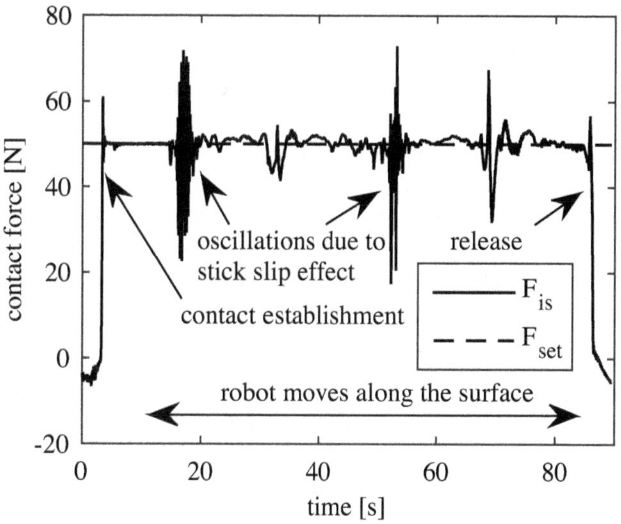

Figure 9. Progression of the force normal to the surface on the oil barrel.

bles and a straight line platform as shown in Fig. 1. The robot frame measures 7 m by 4 m and is 3 m high.

The robot control is realized on an industrial PC with Beckhoff TwinCAT 3.1 CNC at a cycle time of 1 ms. The field bus protocol is EtherCAT. For the force measurement each winch is equipped with a cable force sensor of type Tecsis F2301 with a measurement range up to 4000 N. The analog output signal of the force sensors is digitized in A/D-converters and sent via the field bus to the control. The cable type is LIROS D-Pro 01505-0250 based on Dyneema SK 75 fiber (Polyethylene) with a diameter of 2.50 mm.

4.2 Contact establishment

For the investigation of the contact establishment, a vertical force on the floor is applied. The resulting progression of the force during force control is shown in Fig. 7. The time derivative of the contact force $d(F_{is})/dt$ is noisy but clearly signals when the platform comes in contact.

One interesting point is the contact establishment under the condition of changing robot stiffness in the contact. For this, the operational space force control was applied on the floor at ten different positions which are visualized in Fig. 8. Position 1 lies in the center of the workspace, whereas 7 and 9 are on the edge of robot's footprint.

The evaluation criteria for the contact establishment are the overshoot and settling time. For the experiment the approaching velocity is $50\,\mathrm{mm\,s^{-1}}$ and the desired force amounts to 75 N. The experimental results are presented in Table 2. We can observe a maximum overshoot of 18.2 % whereas the average lies at 8.7 %. The settling time is defined as the time from the first contact to the surface until the control error is smaller than 2 %. The mean settling time is in average 1.2 s.

The results show a compromise between an acceptable overshoot and a settling time. By changing the approaching velocity, this compromise can be influenced in favor of one evaluation criteria.

The settling time lies in a range of experimental results which can be found in literature. In Kroeger et al. (2004) the settling time for a serial industrial manipulator lies between 0.6 and 1.7 s. In another scenario shown in Lange et al. (2013) a settling time of 0.8 s for an industrial robot was achieved.

4.3 Following behavior

The following behavior of the force control during a simultaneous motion of the platform was investigated on oil barrels.

The test set-up is shown in Fig. 1. The test scenario is exemplary for a spherical surface which can be found at wings of airplanes or blades of wind turbines. The path along the oil barrel is programmed offline. In addition, we also programmed the desired force vector which is aligned to the normal vector of the surface.

The quantitative behavior of the contact force along the surface is shown in Fig. 9. The desired force is 50 N. As the contact point of the platform to the oil barrel is eccentric, the stiffness is low and stick slip effects in the contact generate oscillations. The oscillation is relatively high, when the robot moves position controlled into a direction where it is not very stiff. Since the cables are tensed, these critical positions were overcome. For the position and orientation measurement during the trajectory no measurement equipment was available. Qualitatively, the platform followed well the trajectory.

5 Conclusions

In this paper, we proposed an adaption of the well-known hybrid position-force control for a cable robot. By using the measured wrench as input for the set-point determination of cable forces we were able to tense the cables during the contact control. The experimental evaluations on the IPAnema 3 are promising. Further investigation could analyze a specific process by attaching a tool and running the process and analyzing how the controller performs when including additional noise associated with running the process. Another interesting point is the online identification of the stiffness in the contact and how such knowledge can improve the control quality.

Acknowledgements. This work was supported by the FhG Internal Programs under Grant no. WISA 823 244.

References

Albu-Schaeffer, A., Ott, C., Frese, U., and Hirzinger, G.: Cartesian Impedance Control of Redundant Robots: Recent Results with the DLR-Light-Weight-Arms, IEEE International Conference on Robotics and Automation, 2003.

Almeida, F., Lopes, A., and Abreu, P.: Force-Impedance Control a new control strategy of robotic manipulators, Recent Advances in Mechatronics, 1999.

Assuncao, V. and Schumacher, W.: Hybrid Force Control for Parallel Manipulators, 11th Mediterranean Conference on Control and Automation (MED), Rhodos, Griechenland, 2003.

Deng, W., Lee, H. J., and Jeh-Won, L.: Dynamic Hybrid Position-Force Control for Parallel Robot Manipulators, ROMANSY 18 Robot Design, Dynamics and Control, 2010.

Kraus, W., Schmidt, V., Rajendra, P., and Pott, A.: System Identification and Cable Force Control for a Cable-Driven Parallel Robot with Industrial Servo Drives, IEEE International Conference on Robotics and Automation, 2014.

Kroeger, T., Finkemeyer, B., Heuck, M., and Wahl, F. M.: Adaptive implicit hybrid force/pose control of industrial manipulators: compliant motion experiments, IEEE/RSJ International Conference on Intelligent Robots and Systems (IROS), 2004.

Lamaury, J. and Gouttefarde, M.: A Tension Distribution Method with Improved Computational Efficiency, in: Cable-Driven Parallel Robots, Springer, 71–85, 2012.

Lange, F., Bertleff, W., and Suppa, M.: Force and trajectory control of industrial robots in stiff contact, IEEE International Conference on Robotics and Automation, 2013.

Madani, M. and Moallem, M.: Hybrid position/force control of a flexible parallel manipulator, Journal of the Franklin Institute, 999–1012, 2011.

Mikelsons, L., Bruckmann, T., Schramm, D., and Hiller, M.: A Real-Time Capable Force Calculation Algorithm for Redundant Tendon-Based Parallel Manipulators, in: ICRA, Pasadena, 2008.

Osypiuk, R. and Kröger, T.: Parallel Stiffness Actuators with Six Degrees of Freedom for Efficient Force/Torque Control Applications, in: Robotic Systems for Handling and Assembly, Springer, vol. 67, 2011.

Park, J.: Control strategies for robots in contact, Ph.D. thesis, Stanford University, 2006.

Pott, A.: An improved Force Distribution Algorithm for Over-Constrained Cable-Driven Parallel Robots, in: Computational Kinematics, Springer, 2013.

Pott, A., Bruckmann, T., and Mikelsons, L.: Closed-form Force Distribution for Parallel Wire Robots, in: Computational Kinematics, Springer, 25–34, 2009.

Raibert, M. H. and Craig, J. J.: Hybrid Position/Force Control of Manipulators, in: Journal of Dynamic Systems, Measurement, and Control, vol. 103, 1981.

Reisinger, T., Wobbe, F., Kolbus, M., and Schumacher, W.: Integrated Force and Motion Control of Parallel Robots – Part 2: Constrained Space, in: Robotic Systems for Handling and Assembly, Springer, 2011.

Tang, H., Yao, J., Chengz, L., and Zhao, Y.: Hybrid Position Force Control Investigation of parallel machine tool with redundant action, Applied Mechanics and Materials, 2012.

Verhoeven, R.: Analysis of the Workspace of Tendon-based Stewart Platforms, Ph.D. thesis, University of Duisburg-Essen, Duisburg, 2004.

Influence of gear loss factor on the power loss prediction

C. M. C. G. Fernandes[1], P. M. T. Marques[1], R. C. Martins[1], and J. H. O. Seabra[2]
[1]INEGI, Universidade do Porto, Campus FEUP, Rua Dr. Roberto Frias 400, 4200-465 Porto, Portugal
[2]FEUP, Universidade do Porto, Rua Dr. Roberto Frias s/n, 4200-465 Porto, Portugal

Correspondence to: C. M. C. G. Fernandes (cfernandes@inegi.up.pt)

Abstract. In order to accurately predict the power loss generated by a meshing gear pair the gear loss factor must be properly evaluated. Several gear loss factor formulations were compared, including the author's approach.

A gear loss factor calculated considering the load distribution along the path of contact was implemented.

The importance of the gear loss factor in the power loss predictions was put in evidence comparing the predictions with experimental results. It was concluded that the gear loss factor is a decisive factor to accurately predict the power loss. Different formulations proposed in the literature were compared and it was shown that only few were able to yield satisfactory correlations with experimental results. The method suggested by the authors was the one that promoted the most accurate predictions.

1 Introduction

According to Kragelsky et al. (1982) tribology is an important field in engineering which can contribute to develop more reliable and efficient mechanisms like gearboxes.

According to Höhn et al. (2009) the power loss in a gearbox consists of gear, bearing, seals and auxiliary losses. Gear and bearing losses can be separated in no-load and load losses. No-load losses occur with the rotation of mechanical components, even without torque transmission. No-load losses are mainly related to lubricant viscosity and density as well as immersion depth of the components on a sump lubricated gearbox, but it also depends on operating conditions and internal design of the gearbox casing. Rolling bearing no-load losses depend on type and size, arrangement, lubricant viscosity and immersion depth.

Load dependent losses occur in the contact of the power transmitting components. Load losses depended on the transmitted torque, coefficient of friction and sliding velocity in the contact areas of the components. Load dependent rolling bearing losses also depend on type and size, rolling and sliding conditions and lubricant type (SKF, November 2005).

At nominal loads the power loss generated in a gearbox is mainly dependent of the gears load power losses, which puts in evidence the importance of the evaluation of the gear loss factor.

This work shows the influence of the gear loss factor formulation (considering different gear geometries) in the prediction of the power loss. The gear loss factor formulations will be compared with experimental results previously published by Fernandes et al. (2015).

2 Load dependent power loss in meshing gears

Ohlendorf (1958) introduced an approach for prediction of the load dependent losses on spur gears. The power loss generated between gear tooth contact can be calculated according to Eq. (1),

$$P_{VZP} = P_{\mathrm{IN}} \cdot H_V \cdot \mu_{mZ}. \tag{1}$$

H_V is the gear loss factor.

Originally Eq. (1) was obtained assuming a constant coefficient of friction (μ_{mZ}). This was a simplification of the problem.

Equation (1) can be used to calculate the average power loss between gear teeth, given the correct gear loss factor H_V. Despite considering β_b the Eq. (2) initially proposed by Ohlendorf (1958) is mostly valid for spur gears (Wimmer, 2006).

$$H_V^{\mathrm{Ohl}} = (1+u)\cdot\frac{\pi}{z_1}\cdot\frac{1}{\cos\beta_\mathrm{b}}\cdot\left(1-\epsilon_\alpha+\epsilon_1^2+\epsilon_2^2\right) \tag{2}$$

The classical formulas for gear loss factor (Eqs. 3 and 4) consider a rigid load distribution, and a constant coefficient of friction, but tooth profile modifications are disregarded. In depth details about these formulas can be found in the classical works of Niemann and Winter (1989) and Buckingham (1949).

Niemann and Winter (1989) proposed the gear loss factor that is shown in Eq. (3).

$$H_V^{\mathrm{Nie}} = (1+u)\cdot\frac{\pi}{z_1}\cdot\frac{1}{\cos\beta_\mathrm{b}}\cdot\epsilon_\alpha \cdot\left(\frac{1}{\epsilon_\alpha}-1+\left(2k_0^2+2k_0+1\right)\cdot\epsilon_\alpha\right) \tag{3}$$

Buckingham (1949) also introduced a Eq. (4) for the gear loss factor of a meshing gear pair.

$$H_V^{\mathrm{Buc}} = (1+u)\cdot\frac{\pi}{z_1}\cdot\frac{1}{\cos\beta_\mathrm{b}}\cdot\epsilon_\alpha\cdot\left(2k_0^2-2k_0+1\right) \tag{4}$$

where

$$k_0 = \frac{z_1}{2\pi\cdot\epsilon_\alpha\cdot u}\cdot\left(\left(\left(\frac{r_{a2}}{r_{p2}}\right)^2\cdot\frac{1}{\cos\alpha_t^2}-1\right)^{\frac{1}{2}}-\tan\alpha_\mathrm{t}\right) \tag{5}$$

The more recent approach of Velex and Ville (2009) includes the effects of profile modifications, keeps the constant coefficient of friction assumption, but no a priori assumptions about the load distribution are made.

Velex and Ville (2009) which did no a priori assumption on tooth load distribution by using generalized displacements, in order to calculate the efficiency of a meshing gear pair, obtained a closed form solution for the efficiency of a meshing gear pair (constant coefficient of friction was assumed) as presented in Eq. (6). It turns out that Eq. (4) suggested by Buckingham is an approximation of the one suggested by Velex and Ville (2009) when $\mu \ll 1$.

$$\rho = 1-\mu\cdot(1+u)\cdot\frac{\pi}{z_1}\cdot\frac{1}{\cos\beta_\mathrm{b}}\cdot\epsilon_\alpha\cdot\Lambda(\mu) \tag{6}$$

where $\Lambda(\mu)$ is the loss factor described in Eq. (7).

$$\Lambda(\mu) = \frac{2k_0^2-2k_0+1}{1-\mu\cdot\left(\frac{\tan\alpha_\mathrm{t}\cdot(2k_0-1)-\frac{\pi}{z_1}\cdot\epsilon_\alpha\cdot\left(2k_0^2-2k_0+1\right)}{\cos\beta_\mathrm{b}}\right)} \tag{7}$$

The load distribution (force per unit of length along the path of contact) disregarding elastic effects can be calculated dividing the total normal force $F_\mathrm{n} = \frac{M_i}{r_{\mathrm{b}i}}$ by the total length of the lines of contact along the path of contact.

The total length of the lines of contact along the path of contact can be calculated with the algorithm presented in Appendix A. The load distribution per unit of length along the path of contact can then be calculated according to Eq. (8). An example of the load distribution in a helical gear is presented (Fig. 1).

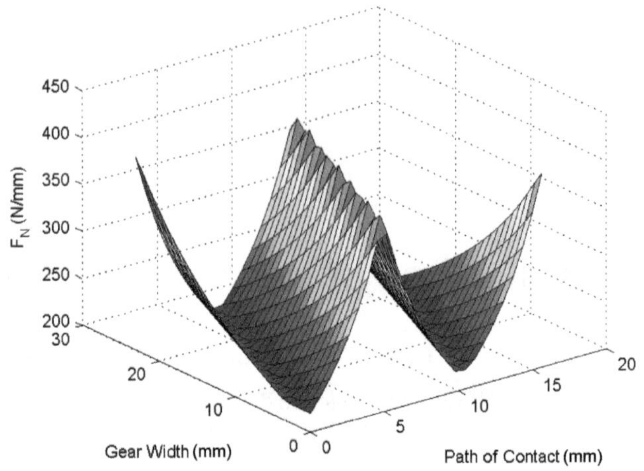

Figure 1. Load distribution of a helical gear with an applied torque of 320 Nm.

$$F_\mathrm{N}(x,y) = \frac{F_{b\mathrm{n}}}{L(x,y)} \tag{8}$$

The gear loss factor can now be calculated according to Eq. (9) proposed by Wimmer (2006)

$$H_V^{\mathrm{num}} = \frac{1}{p_\mathrm{b}}\int_0^b\int_A^E \frac{F_\mathrm{N}(x,y)}{F_b}\cdot\frac{V_g(x,y)}{V_\mathrm{b}}\mathrm{d}x\mathrm{d}y. \tag{9}$$

To solve Eqs. (8) and (9) the total length of contacting lines should be known at each point along the path of contact. To perform this task, an algorithm was developed and implemented (Appendix A).

3 Average coefficient of friction

Several authors (Ohlendorf, 1958; Eiselt, 1966; Naruse et al., 1986; Michaelis, 1987; Schlenk, 1994; Doleschel, 2002) have introduced different formulas to calculate the average coefficient of friction between gear teeth for different gear geometries. Due to the complexity of the problem, these equations are usually based in experimental results, and naturally, the results yielded by these models vary for the same operating conditions. In this work, instead of calculating the coefficient of friction yielded by these formulations, a value is calculated from the experimental procedure used in a previous work (Fernandes et al., 2013) and then compared to the models.

Assuming that P_{VZ0}, P_{VL} and P_{VD} are correctly calculated the power loss generated by the meshing gears can be obtained according to Eq. (10). The rolling bearing, seals

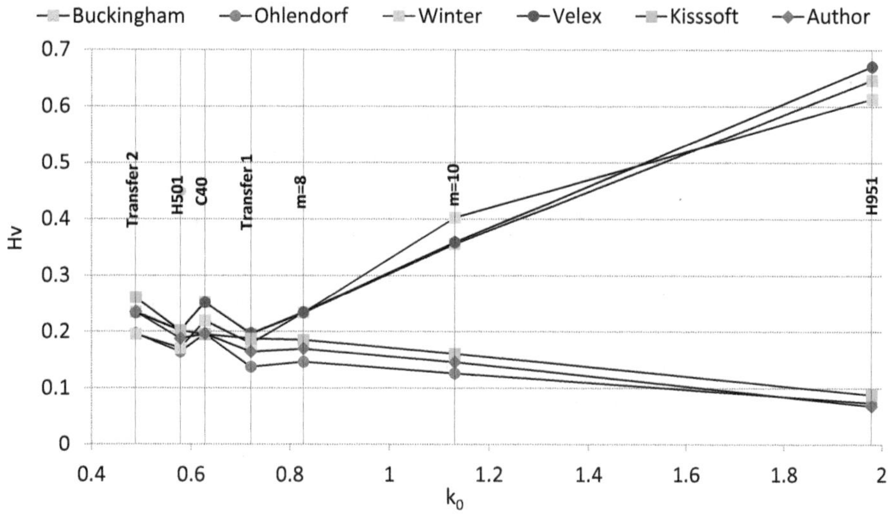

Figure 2. Gear loss factor comparisson with different formulas.

and load independent gear losses were discussed in previous works of Fernandes et al. (2013, 2015).

$$P_{VZP}^{\text{exp}} = P_V^{\text{exp}} - (P_{VZ0} + P_{VL} + P_{VD}) \tag{10}$$

Considering the power loss generated by the gears in the gearbox (Eq. 10) an average coefficient of friction (μ_{mZ}^{exp}) can be calculated. It can be calculated according to different approaches:

1. From Ohlendof's approach (Eq. 11).

$$\mu_{mZ}^{\text{exp}} = \frac{P_{VZP}^{\text{exp}}}{P_{\text{IN}} \cdot H_V^i} \tag{11}$$

H_V^i is the gear loss factor which can assume various forms, depending on the formulation that is used. Four H_V were defined according to Eq. (2) H_V^{Ohl}, Eq. (9) H_V^{num}, Eq. 3 H_V^{Nie}, Eq. 4 H_V^{Buc}.

2. Considering the average power loss generated between gear teeth along the path of contact according to Velex and Ville (2009), μ_{mZ}^{exp} can be obtained solving Eq. (12) to find μ_{mZ}^{exp}.

$$P_{VZP} = P_{\text{IN}} \cdot \mu_{mZ}^{\text{exp}} \cdot (1+u) \cdot \frac{\pi}{z_1 \cdot \cos\beta_b} \cdot \epsilon_\alpha \cdot \Lambda\left(\mu_{mZ}^{\text{exp}}\right) \tag{12}$$

The coefficient of friction extracted from the gear mesh power loss obtained with Eq. (10) will be dependent of the formulation that is used to calculate the gear loss factor. In order to decide which gear loss factor formulation is better suited for the authors study, this factor was calculated for seven different gear geometries, in which, spur, helical and low loss gears are included (Table 1) (Fernandes et al., 2015).

The gear loss factor was also calculated based on the results obtained with the commercial software *KissSoft* which accounts for elastic effects.

Figure 2 shows the comparison between the different gear geometries as a function of the k_0 (Eq. 5) parameter. There are clearly two groups of results that diverge at a certain point. A deviation is found in the solutions proposed by Buckingham (1949), Niemann and Winter (1989) and Velex and Ville (2009) because Eq. (5) is expected to yield values between 0 and 0.5. which means that it is not suitable for gears with profile shift.

The H501 and H951 geometries were previously tested for power loss in an FZG test rig (Fernandes et al., 2015). The results presented were collected for FZG load stages with a lever arm of 0.35 m, i.e. K5 = 105, K7 = 199 and K9 = 323 Nm applied on wheel. Changing from H501 to H951 resulted in a dramatic power loss reduction (Fig. 3), which was attributed to the H951 gear geometry (everything but the gear geometry was kept the same). These experimental results suggest that the gear loss factor of the H951 must be lower than that of the H501. The trends shown by the gear loss factors obtained with *KissSoft*, the author's method and Ohlendorf are in agreement with the experimental observations of Fernandes et al. (2015). The gear loss factors obtained with Eq. (9) are close to those obtained with the ones derived from the *KissSoft* computations. Aiming for simplicity and fast computing the gear loss factor was calculated using Eq. (9).

Following Fig. 2 it becomes clear that Buckingham, Velex and Winter's approaches are not suitable for all gear geometries and can only be applied over a limited range of the k_0 parameter.

Table 1. Geometrical parameters of the gears.

Gears		Parameters										
		z [/]	m [mm]	a [mm]	α [°]	β [°]	b [mm]	x_z [/]	d_a [mm]	ϵ_α [/]	ϵ_β [/]	Ra [μm]
C40	Pinion Gear	16 24	4.5	91.5	20	0	40	+0.1817 +0.1715	82.64 115.54	1.44	0	0.7
H501	Pinion Gear	20 30	3.5	91.5	20	15	23	+0.1381 +0.1319	80.37 116.57	1.45	0.54	0.3
H951	Pinion Gear	38 57	1.75	91.5	20	15	23	+1.6915 +2.0003	76.23 111.73	0.93	1.08	0.3
Transfer 1	Pinion Gear	32 23	3.5	105.0	20	20	35	+0.3810 +0.4150	128.45 95.17	1.32	1.09	0.4
Transfer 2	Pinion Gear	28 17	4	95.0	20	20	33.5	−0.2400 +0.0510	125.22 80.73	1.49	0.91	0.4
$m = 8$	Pinion Gear	17 69	8	355	20	9	124	+0.4965 +0.3985	160.74 580.36	1.40	0.77	–
$m = 10$	Pinion Gear	19 77	10	500	20	9	175	+0.6500 +0.8877	222.65 814.63	1.32	0.87	–

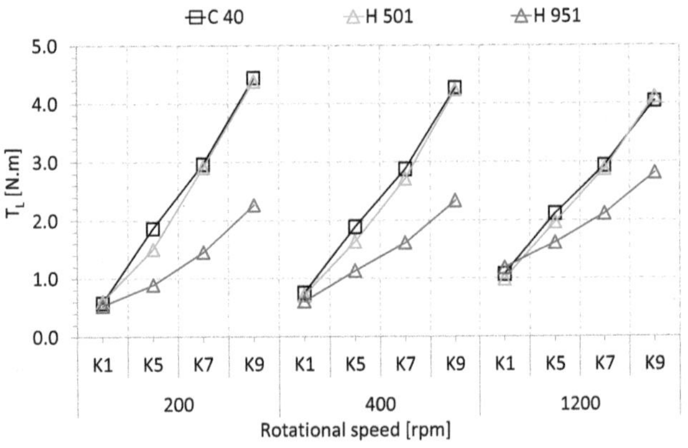

Figure 3. Torque loss for different gear geometries lubricated with a mineral wind turbine gear oil (Fernandes et al., 2015).

4 Validation with experimental results

In order to validate the gear loss factor that was proposed, Schlenk's (Schlenk, 1994) coefficient of friction was used (Eq. 13). The lubricant parameter (X_L) was previously determined with a spur gear geometry (C40) for different wind turbine gear oil formulations (Fernandes et al., 2013). Alternatively, experimental results obtained with H501 and H951 gear geometries were presented in Fig. 3 (Fernandes et al., 2015). The gear loss factors calculated according to different approaches for the C40, H501 and H951 gear geometries are presented in Table 2.

Table 2. Gear loss factor calculated according to different approaches.

H_V	C 40	H 501	H 951
Ohlendorf	0.1959	0.1639	0.0739
Author	0.1959	0.1873	0.0684
KissSoft	0.2039	0.2011	0.0882

$$\mu_{mZ}^{\text{Schlenk}} = 0.048 \cdot \left(\frac{F_{bt}/b}{v_{\Sigma C} \cdot \rho_{\text{redC}}} \right)^{0.2} \cdot \eta^{-0.05} \cdot \text{Ra}^{0.25} \cdot X_L \quad (13)$$

In Fig. 4 the absolute error of the power loss model prediction using the KissSoft, Ohlendorf and Author gear loss factors is presented. The results suggest that the gear loss factor presented by the authors in Eq. (9), considering the rigid load distribution, present a much lower absolute error for the prediction of a mineral wind turbine gear oil power loss for with helical gears, previously published by Fernandes et al. (2015).

Schlenk's Equation should be valid for both helical and spur gear geometries, also H_V^{Ohl} is mostly valid for spur gears. This means that using the lubricant parameter X_L extracted from experimental results with spur gears and applying it to helical gears resulted in excellent correlations between numerical and experimental data when using H_V^{num}.

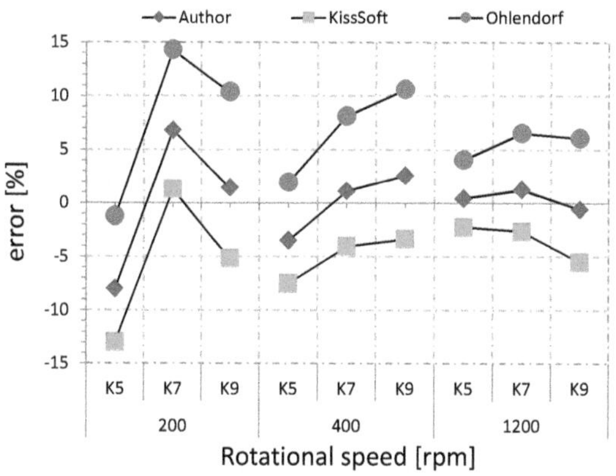

(a) H501

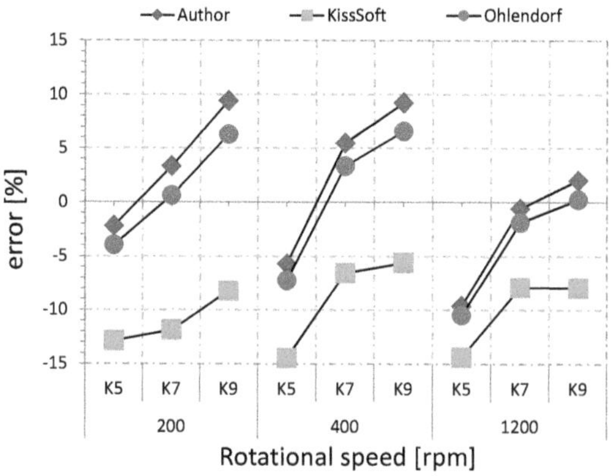

(b) H951

Figure 4. Correlation between the experimental power loss measured and the predicted with Author, Ohlendorf or KissSoft gear loss factors.

5 Conclusions

In this work several gear loss factors were compared. The gear loss factor results were indirectly compared with experimental gear power loss measurements in order to assess the validity of each one of the formulations.

An alternative formulation based on the numerical integration of the rigid load distribution is suggested. The method presented by the authors to solve the gear loss factor formula proposed by Wimmer (2006) disregards the elastic effects of the gears but proved to be reliable to predict the average power loss of helical and spur gears as proven with experimental results.

The results suggest that the classical formulas are accurate only in very specific scenarios. The comparison with the experimental results indicates that the approach suggested by the authors works quite well.

This study has shown the importance of a correct evaluation of the gear loss factor in the prediction of the power loss generated in meshing gears.

Appendix A: Load distribution along the path of contact

Before enter the contact zone of a gear, or the path of contact which value is given by Eq. (A1), a teeth contact line has the representation of Fig. A1a.

$$\mathrm{AE} = \epsilon_\alpha \cdot p_{\mathrm{bt}} \tag{A1}$$

When the contact starts, the length of the contacting line increases proportionally to the coordinate of the path of contact, represented by the first condition of Eq. (A2) (Fig. A1a and b). The contact then continues to increase up to the situation of a full line of contact, that occur at the coordinate $x = \epsilon_\beta \cdot p_{bt} = b \cdot \tan \beta_b$ up to the end of contact at $x = \epsilon_\alpha \cdot p_{bt}$ which is given by second condition of Eq. (A2) (Fig. A1c). Then, the teeth start to go out from the contact and the line length starts to decrease as shown in the third condition of Eq. (A2) and Fig. A1d.

$$l(x) = \begin{cases} \frac{x}{\sin \beta_\mathrm{b}} & 0 < x < \epsilon_\beta \cdot p_{\mathrm{bt}} \\ \frac{b}{\cos \beta_\mathrm{b}} & \epsilon_\beta \cdot p_{\mathrm{bt}} < x < \epsilon_\alpha \cdot p_{\mathrm{bt}} \\ \frac{b}{\cos \beta_\mathrm{b}} - \frac{x - \epsilon_\alpha \cdot p_{\mathrm{bt}}}{\sin \beta_\mathrm{b}} & \epsilon_\alpha \cdot p_{\mathrm{bt}} < x < \left(\epsilon_\alpha + \epsilon_\beta\right) \cdot p_{\mathrm{bt}} \end{cases} \tag{A2}$$

Equation (A2) previously presented is valid for the length of a single line along the path of contact. The other teeth have the same behaviour of the single line yet presented but at the distance of a transverse pitch (p_{bt}), which is the distance between the teeth along the path of contact as represented in Fig. A1.

The same equations deduced for a single line can be used, but the coordinates should be transformed according to Eq. (A3). The value i of the Eq. (A3) is calculated with Eq. (A4) that represents the lines screened from the single line with value $i = 0$, from behind and behead in integer steps.

$$x^*(x) = x + i \cdot p_{\mathrm{bt}} \tag{A3}$$

$$i = -\mathrm{ceil}\left(\epsilon_\alpha + \epsilon_\beta\right) : 1 : \mathrm{ceil}\left(\epsilon_\alpha + \epsilon_\beta\right) \tag{A4}$$

Ceil is a function that rounds the value for the highest close integer.

It is also possible to do a 3-D representation of the line length as function of x and y. To do that, the y coordinate representing the tooth width that changes from 0 up to b. Since the tooth line of contact of a helical gear has a helix angle the y coordinate is function of the x coordinate which can be expressed with Eq. (A5).

$$x^*_{(x,y)} = x + i \cdot p_{\mathrm{bt}} + y \tan(\beta_\mathrm{b}) \tag{A5}$$

Applying the coordinate transformation of Eq. (A5) and the formulas of Eq. (A2), the line length of each tooth screened from the teeth i is presented in Eq. (A6).

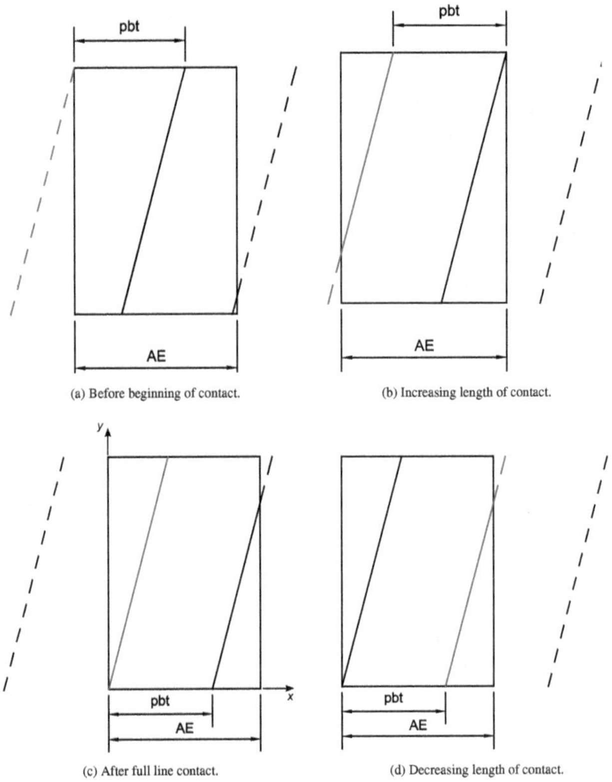

Figure A1. Evolution of a single line along the path of contact.

$$l_i(x,y) = \begin{cases} \frac{x^*}{\sin \beta_\mathrm{b}} & 0 < x^* < \epsilon_\beta \cdot p_{\mathrm{bt}} \\ \frac{b}{\cos \beta_\mathrm{b}} & \epsilon_\beta \cdot p_{\mathrm{bt}} < x^* < \epsilon_\alpha \cdot p_{\mathrm{bt}} \\ \frac{b}{\cos \beta_\mathrm{b}} - \frac{x^* - \epsilon_\alpha \cdot p_{\mathrm{bt}}}{\sin \beta_\mathrm{b}} & \epsilon_\alpha \cdot p_{\mathrm{bt}} < x^* < \left(\epsilon_\alpha + \epsilon_\beta\right) \cdot p_{\mathrm{bt}} \end{cases} \tag{A6}$$

The formulation presented is valid for gears with a contact ratio $\epsilon_\alpha > \epsilon_\beta$.

For the case that one complete line is not in contact, the cycle of meshing is slightly different and the path of contact is smaller than the transverse pitch. In such cases usually the overlap contact ratio is $\epsilon_\beta > \epsilon_\alpha$.

The equation is slightly different from that presented before because the domains change in a different way as presented in Eq. (A7).

$$l_i(x,y) = \begin{cases} \frac{x^*}{\sin \beta_\mathrm{b}} & 0 < x^* < \epsilon_\alpha \cdot p_{\mathrm{bt}} \\ \frac{\epsilon_\alpha \cdot p_{\mathrm{bt}}}{\sin \beta_\mathrm{b}} & \epsilon_\alpha \cdot p_{\mathrm{bt}} < x^* < \epsilon_\beta \cdot p_{\mathrm{bt}} \\ \frac{\epsilon_\alpha \cdot p_{\mathrm{bt}}}{\sin \beta_\mathrm{b}} - \frac{x^* - \epsilon_\beta \cdot p_{\mathrm{bt}}}{\sin \beta_\mathrm{b}} & \epsilon_\beta \cdot p_{\mathrm{bt}} < x^* < \left(\epsilon_\alpha + \epsilon_\beta\right) \cdot p_{\mathrm{bt}} \end{cases} \tag{A7}$$

The total sum of lines can be easily done with Eq. (A8). It is important to note that the algorithm also calculate the line contact length of spur gears using only the second row of Eq. (A2).

$$L(x,y)=\sum_{-\text{ceil}(\epsilon_\alpha+\epsilon_\beta)}^{\text{ceil}(\epsilon_\alpha+\epsilon_\beta)} l_i(x,y) \quad \text{(A8)}$$

Stepwise functions

The algorithm previously presented is based on the identification of different domains in the meshing cycle of helical gears. However, the different domains can be combined using stepwise functions like Heaviside (Eq. A9) or hyperbolic tangent (Eq. A10).

$$\xi=\frac{1}{1+e^{-2k(x-a)}} \quad \text{(A9)}$$

$$\xi=\frac{1}{2}\cdot(\tanh(k\cdot(x-a))+1) \quad \text{(A10)}$$

The coordinate a is the point when the step is desired.

Using the hyperbolic tangent equation, the three domains can be expressed in Eq. (A11) for the beginning of contact ($a=0$), Eq. (A12) for a complete line ($a=\epsilon_\beta\cdot p_{\text{bt}}$) and Eq. (A13) for a line going out from the contact ($a=\epsilon_\alpha\cdot p_{\text{bt}}$). The constant k changes the precision of the algorithm. For the case it was considered $k=1000$.

$$\xi_1=\frac{1}{2}\cdot\left(\tanh(k\cdot x)-\tanh\left(k\cdot\left(x-\left(\epsilon_\alpha+\epsilon_\beta\right)\cdot p_{\text{bt}}\right)\right)\right) \quad \text{(A11)}$$

$$\xi_2=\frac{1}{2}\cdot\left(\tanh\left(k\cdot\left(x-\epsilon_\beta\cdot p_{\text{bt}}\right)\right)+1\right) \quad \text{(A12)}$$

$$\xi_3=\frac{1}{2}\cdot(\tanh(k\cdot(x-\epsilon_\alpha\cdot p_{\text{bt}}))+1) \quad \text{(A13)}$$

For each single line the length along the path of contact is given by Eq. (A14).

$$l(x)=\frac{1}{\sin\beta_{\text{b}}}\cdot\xi_1\cdot\left(x-\xi_2\cdot\left(x-\epsilon_\beta\cdot p_{\text{bt}}\right)-\xi_3\cdot(x-\epsilon_\alpha\cdot p_{\text{bt}})\right) \quad \text{(A14)}$$

For spur gears the length for each line is given by Eq. (A15).

$$l(x)=b\cdot\xi_1 \quad \text{(A15)}$$

For the lines screened from the one considered the length is computed with Eq. (A3) previously explained which results in Eq. (A16).

$$l_i(x,y)=l\left(x^*_{(x,y)}\right) \quad \text{(A16)}$$

The total sum of lines is then given by Eq. (A8).

Using such type of function or other stepwise function is great to get a continuous function. However, the computational time can increase due to the expense of computing the step function. The algorithm with step function works for all the type of gear geometries and the transverse and overlap contact ratios (ϵ_α and ϵ_β) do not need to follow any rule.

Table A1. Notation and units.

a	Centre distance, [mm]
AE	Path of contact, [mm]
b	Face width, [mm]
d_{a}	Addendum diameter, [mm]
H_V	Gear loss factor from, [–]
F_b	Tooth normal force on transverse plane, [N]
F_{bn}	Tooth normal force, [N]
$F_{\text{N}}(x,y)$	Normal force per length, [N mm^{-1}]
k_0	Gear geometry factor , [–]
$l(x)$	Length of a single line of contact, [mm]
$L(x,y)$	Sum of the lengths of the lines of contact, [mm]
m	Module, [mm]
n	Rotational speed, [rpm]
M_i	Torque in gear i, [Nm]
p_{b}	Base pitch, [mm]
p_{bt}	Transverse base pitch, [mm]
P_{IN}	Input power, [W]
P_V	Total power loss, [W]
P_{VD}	Seals power loss, [W]
P_{VL}	Rolling bearing power loss, [W]
P_{VZ0}	Gears no-load loss, [W]
P_{VZP}	Gears load loss, [W]
r_{ai}	Tip radius, [m]
r_{bi}	Base radius of gear i, [mm]
r_{pi}	Pitch radius of gear i, [mm]
Ra	Average roughness, [μm]
u	Gear ratio, [–]
$V_g(x,y)$	Sliding velocity, [m s^{-1}]
V_{b}	Base cylinder transverse tangential speed, [m s^{-1}]
$v_{\Sigma C}$	Sum velocity at pitch point, [m s^{-1}]
x_z	Profile shift coefficient, [–]
x	Coordinate along the tooth path of contact, [mm]
X_{L}	Lubricant parameter, [–]
y	Coordinate along the tooth width, [mm]
z_i	Number of teeth of gear i, [–]
α	Pressure angle, [rad]
α_{t}	Transverse pressure angle, [rad]
β	Helix angle, [rad]
β_{b}	Base helix angle, [rad]
ϵ_1	Addendum contact ratio, [–]
ϵ_2	Deddendum contact ratio, [–]
ϵ_α	Transverse contact ratio, [–]
ϵ_β	Overlap ratio, [–]
$\Lambda(\mu)$	Efficiency parameter, [–]
η	Dynamic viscosity, [mPa s^{-1}]
μ_{mZ}	Average coefficient of friction, [–]
ρ	Efficiency of a gear pair, [–]
ρ_{redC}	Equivalent contact radius at pitch point, [mm]
ξ	Step function, [–]

Acknowledgements. The authors gratefully acknowledge the funding supported by:

- National Funds through Fundação para a Ciência e a Tecnologia (FCT), under the project EXCL/SEM-PRO/0103/2012;
- COMPETE and National Funds through Fundação para a Ciência e a Tecnologia (FCT), under the project Incentivo/EME/LA0022/2014;
- Quadro de Referência Estratégico Nacional (QREN), through Fundo Europeu de Desenvolvimento Regional (FEDER), under the project NORTE-07-0124-FEDER-000009 – Applied Mechanics and Product Development;

without whom this work would not be possible.

References

Buckingham, E.: Analytical mechanics of gears, Dover Books for Engineers, republished by Dover Publication, 1963, McGraw-Hill Book Co., New York, 1949.

Doleschel, A.: Wirkungsgradtest, Vergleichende Beurteilung des Einflusses von Schmierstoffen auf den Wirkungsgrad bei Zahnradgetrieben, FVA Forschungsvorhaben Nr. 345, FVA Forschungsheft Nr. 664, FVA, Germany, 2002.

Eiselt, H.: Beitrag zur experimentellen und rechnerischen Bestimmung der Fresstragfähigkeit von Zahnradgetrieben unter Berücksichtigung der Zahnflankenreibung, PhD thesis, Dissertation TH Dresden, Dresden, 1966.

Fernandes, C. M., Martins, R. C., and Seabra, J. H.: Torque loss of type C40 FZG gears lubricated with wind turbine gear oils, Tribol. Int., 70, 83–93, doi:10.1016/j.triboint.2013.10.003, 2013.

Fernandes, C. M., Marques, P. M., Martins, R. C., and Seabra, J. H.: Gearbox power loss, Part II: Friction losses in gears, Tribol. Int., 88, 309–316, doi:10.1016/j.triboint.2014.12.004, 2015.

Höhn, B.-R., Michaelis, K., and Hinterstoißer, M.: Optimization of gearbox efficiency, goriva i maziva, 48, 462–480, 2009.

Kragelsky, I., Dobychin, M., and Kombalov, V.: Friction and Wear: Calculation Methods, Pergamon Press, Oxford, 1982.

Michaelis, K.: Die Integraltemperatur zur Beurteilung der Fresstragfähigkeit von Stirnrädern, PhD thesis, Dissertation TU München, München, 1987.

Naruse, C., Haizuka, S., Nemoto, R., and Kurokawa, K.: Studies on Frictional Loss, Temperature Rise and Limiting Load for Scoring of Spur Gear, Bull. Jpn. Soc. Mech. Eng., 29, 600–608, doi:10.1299/jsme1958.29.600, 1986.

Niemann, G. and Winter, H.: Maschinenelemente: Band 2: Getriebe allgemein, Zahnradgetriebe – Grundlagen, Stirnradgetriebe, Maschinenelemente/Gustav Niemann, Springer, München, Germany, 1989.

Ohlendorf, H.: Verlustleistung und Erwärmung von Stirnrädern, PhD thesis, Dissertation TU München, München, 1958.

Schlenk, L.: Unterscuchungen zur Fresstragfähigkeit von Grozahnrädern, PhD thesis, Dissertation TU München, München, 1994.

SKF: SKF General Catalogue 6000 EN, SKF, November 2005.

Velex, P. and Ville, F.: An analytical approach to tooth friction losses in spur and helical gears-influence of profile modifications, J. Mech. Design, 131, 1–10, 2009.

Wimmer, A. J.: Lastverluste von Stirnradverzahnungen, PhD thesis, Fakultät für Maschinenwesen der Technischen Universität München, München, 2006.

9

Realisation of model reference compliance control of a humanoid robot arm via integral sliding mode control

S. G. Khan[1] and J. Jalani[2]

[1]Department of Mechanical Engineering, College of Engineering Yanbu, Taibah University, Al Madinah, Saudi Arabia

[2]Department of Electrical Engineering Technology, University Tun Hussein Onn Malaysia, Batu Pahat, Malaysia

Correspondence to: S. G. Khan (engr_ghani@hotmail.com)

Abstract. Human safety becomes critical when robot enters the human environment. Compliant control can be used to address some safety issues in human-robot physical interaction. This paper proposes an integral sliding mode controller (ISMC) based compliance control scheme for the Bristol Robotics Laboratory's humanoid BERT II robot arm. Apart from introducing a model reference compliance controller, the ISMC scheme is aimed to deal with the robot arm dynamic model's inaccuracies and un-modelled nonlinearities. The control scheme consists of a feedback linearization (FL) and an ISMC part. In addition, a posture controller has been incorporated to employ the redundant DOF and generate human like motion. The desired level of compliance can be tuned by selecting the stiffness and damping parameters in the sliding mode variable (compliance reference model). The results show that the compliant control is feasible at different levels for BERT II in simulation and experiment. The positioning control has been satisfactorily achieved and nonlinearities and un-modelled dynamics have been successfully overcome.

1 Introduction

As robots are stepping into social environment, the need for more suitable and safe (compliant) control strategies are even more important (Khan et al., 2010a, b, 2014a, b; Rezoug and Hamerlain, 2014; Islam et al., 2014; Iqbal et al., 2014; Ding and Fang, 2013). Both human and robot safety is still one of the major issues in social robotics. Compliance (low stiffness) control can help addressing some safety issues (Khan et al., 2014a). There are many ways to realise compliance control schemes for human robot physical interaction. Most of the compliance control schemes are fully or partially dynamic model based. However, dynamic models are prone to inaccuracies and nonlinearities. In such cases integral sliding mode controller (ISMC) can provide a robust control solution. In addition, model reference compliance scheme can be implemented via the ISMC (Herrmann et al., 2014) more easily. Both conventional sliding mode controller (SMC) and ISMC have the following design steps (see Fig. 1):

1. reaching phase: the system state is driven from any initial state to reach the switching manifolds (the anticipated sliding modes) in finite time;

2. sliding-mode phase: the system is forced into sliding motion, $r = 0$, on the switching manifolds, i.e. the switching manifold becomes an attractor.

The above two phases correspond to the following two main design steps.

1. Switching manifold selection: a set of switching manifolds are selected with prescribed desirable dynamical characteristics. Common candidates are linear hyperplanes.

2. Discontinuous control design: a discontinuous control strategy is formed to ensure the finite time reachability of the switching manifolds. The controller may be either local or global, depending upon specific control requirements.

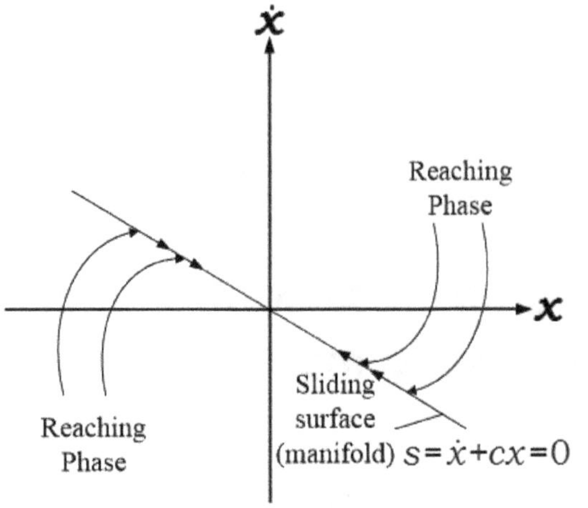

Figure 1. Sliding surface design consists of sliding phase and reaching phase.

For an ISMC, there are a few advantages outlined by Shi et al. (2008) as compared to the SMC counterpart: Sliding motion occurs right from the start of the control action, i.e. robustness (i.e. sliding mode) is guaranteed. Thus, the reaching phase is eliminated. This is not the case for the SMC. Secondly, the position control error can have large overshoots and large deviation with respect to the designed error dynamics when a conventional SMC is used in contrast to an ISMC. Lastly, the SMC control signals often have large amplitude due to chattering. Shi et al. (2008) showed these facts for a robotic system simulation. Moreover, ISMC controller design based on a perturbation estimator has significantly reduced the amplitude of chattering. Castaños and Fridman (2006) have exploited the advantages of ISMC by combining an ISMC and H_∞-control to bring robustness to the controller against unmatched disturbances.

One of the main advantages of the ISMC and the main reason to employ it in this work is its suitability to be used as a model reference compliance controller. This will be discussed later in the paper to modify the sliding mode variable to include force feedback for compliance control.

ISMC is a well investigated control method, although practical applications are still limited, particularly in the robotics context. Some of the reasons of this may be the jittery nature and aggressiveness in the control action of sliding mode control. These type of control schemes may pose a risk of damaging the robotic devices if not implemented with care. However, due to better understanding of these schemes and availability of good control hardware these methods are getting popularity. Shi et al. (2008) have compared traditional Sliding Mode Control and ISMC through simulation for a two-link rigid manipulator. Makoto et al. (2010) have applied the ISMC to a power-assisted manipulator with a single degree of freedom through simulation and experiment. Eker and Akinal (2008) have suggested a similar scheme using an integral sliding surface in combination with the conventional sliding mode control. They have tested the scheme on a rather simple electro-mechanical system. In our previous work (Jalani et al., 2010a, b), an ISMC scheme for underactuated robotic fingers was simulated and experimentally tested . In this paper, experimental results of the Cartesian (x and y) ISMC control in combination with a feedback linearization (FL) and a posture controller for the four degrees of freedom of the humanoid BERT II arm (see Fig. 2) are presented. ISMC has been employed here to deal with uncertainties and unmodelled nonlinearities. In addition, external force (the interaction force between robot and its environment (including human) has been included in the ISMC which results into a model reference compliance controller (Jalani et al., 2013). The dynamic model of the BERT II robot arm for the FL is obtained via MapleSim (a package of Maple) (Maplesoft, 2015).

The redundant DOF have been used to generate movement in the most human-like posture (Spiers et al., 2009a, b; Khan et al., 2010a, b; De Sapio et al., 2005). Spiers et al. (2009a, b) have implemented a robust version of the technique for a two degrees of freedom of a humanoid arm manipulator in real time. In their work, they were able to overcome a kinematic redundancy (i.e. multiple solutions produced by the robot arm manipulator) and human like movement of the arm manipulator was realized. This posture control is obtained by minimizing a gravity dependent cost function using inertia matrix of the 4 DOF dynamic model of the BERT II arm.

As mentioned before, compliance control in robotics is increasingly getting attention due to its relevance to safety in human robot interaction (HRI). Many researchers are looking into passive and active approaches to address part of safety issues in HRI (Braun et al., 2013; Potkonjak et al., 2011; Kim et al., 2012; Ficuciello and Villani, 2012; Di Natali and Valdastri, 2012; Ajwad et al., 2014, 2015). In this paper, we employ ISMC to produce compliant behaviour. A Cartesian (X, Y and Z) model reference type of ISMC compliant scheme is simulated for 4 DOF of the humanoid BERT II robot arm[1]. Different compliance levels are produced by choosing spring constant and damping coefficients in the reference model (Jalani et al., 2013; Herrmann et al., 2014). The scheme will be used to address safety issues in human-robot interaction.

The contributions of the paper can be summarised as follows:

- introduction of an active compliance scheme via integral sliding mode control for a humanoid robot;
- combination of bio-inspired posture controller with ISMC scheme employing the redundancy;
- successful experimental testing of the ISMC tracking scheme on the Humanoid robot prototype.

[1]The mechanical design and manufacturing for the BERT II torso including hand and arm has been conducted by Elumotion (www.elumotion.com).

Figure 2. Bristol Robotics Laboratory BERT II robot.

2 Integral sliding mode controller

Integral slide mode (ISMC) control law (Shi et al., 2008) is presented here. The ISMC based task space (Cartesian) controller is briefly explained here and experimentally tested on the Bristol Robotics Laboratory's robot arm. The integral sliding mode variable r is defined as follows (Herrmann et al., 2014):

$$r = \dot{e} + K_r e + K_i \int_0^t e(\xi) d\xi - \dot{e}(0) - K_r e(0). \tag{1}$$

The position error vector is given by $\boldsymbol{e} = X_\mathrm{d} - X$, where $X_\mathrm{d} = [x_\mathrm{d}, y_\mathrm{d}]^T$ the Cartesian demand position (as we are controlling x and y only). The vectors $\dot{\boldsymbol{e}}(0)$ and $\boldsymbol{e}(0)$ are the initial values of the Cartesian velocity and position error respectively. It should be noted that the values of K_r and K_i define the desired behaviour of the control scheme once sliding motion $r = 0$, is achieved. Hence, it acts like a reference model for the control scheme. The matrices $\mathbf{K}_r$ and $\mathbf{K}_i$ in this case are diagonal (2×2) and having positive scalar values.

Table 1. Physical Properties of BERT II arm.

BERT II arm for 4 DOF, Physical Properties			
Link	Mass (Kg)	Length (m)	Radius (m)
1	$M_1 = 4$	$L_1 = 0.21$	$R_1 = 0.03$
2	$M_2 = 1.5$	$L_2 = 0.23$	$R_2 = 0.02$
3	$M_3 = 0.85$	$L_3 = 0.23$	$R_3 = 0.02$
4	$M_4 = 1.8$	$L_4 = 0.23$	$R_4 = 0.02$

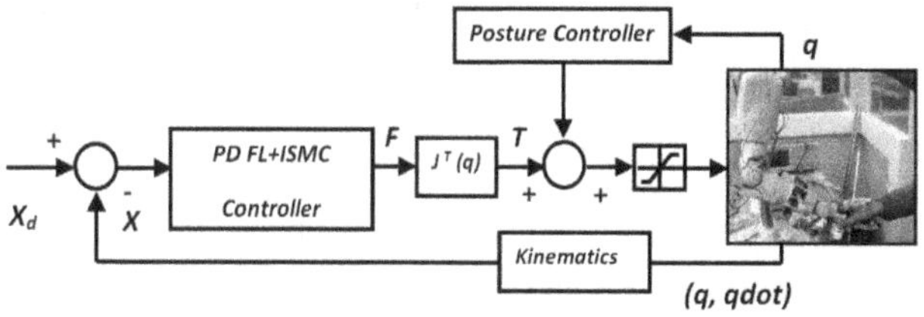

Figure 3. ISMC scheme.

As mentioned earlier, ISMC approach has an advantage in relation to other SMC methods (Shi et al., 2008; Utkin and Shi, 1996). From Eq. (1), it is easily seen that if the correct initial values, i.e. $\dot{\boldsymbol{e}}(0)$ and $\boldsymbol{e}(0)$, are used then $r(t = 0) = 0$. This is important, as sliding motion is secured from the start and aggressive controller action is kept minimal.

The general structure of the robot dynamics is given by:

$$\Lambda(q)\ddot{q} + \mu(q, \dot{q}) + \nu(q) = \tau \tag{2}$$

where $\boldsymbol{\Lambda} \in Re^{n \times n}$ is the inertia matrix, $\boldsymbol{\mu} \in Re^{n \times 1}$ is the coriolis/centripetal vector. $\boldsymbol{\nu} \in Re^{n \times 1}$ is the gravity vector. τ is the input torque. The $\Lambda(q)$, $\mu(q, \dot{q})$ and $\nu(q)$ of the dynamic model (Lewis et al., 2003) for the BERT II arm has been obtained with the help of MapleSim (Khan, 2012; Maplesoft, 2015; Khan et al., 2010b) (an associated software package of Maple). Physical parameters of the robot are listed in Table 1.

The Cartesian space dynamics are now given as follows: Instead of joint torques, the dynamics consider the forces acting on the end effector:

$$A(q)\ddot{X} + \mu_{cc}(q, \dot{q}) + g(q) = f \tag{3}$$

where $A = (J \Lambda^{-1} J^T)^{-1}$, $\mu_{cc} = \overline{J}^T \mu - A \dot{J} \dot{q}$, $g = \overline{J}^T \nu$, $f = \overline{J}^T \tau$ and X is the robot end-effector Cartesian position ($X = [x, y]^T$), and J is the Jacobian ($J = \frac{\delta X}{\delta q}$). Hence, the Cartesian velocities are defined as $\dot{X} = J\dot{q}$. The matrix $\overline{\mathbf{J}}$ is the inertia weighted pseudo Jacobian inverse (Khatib, 1987; Nemec and Zlajpah, 2000):

$$\bar{\mathbf{J}} = \lambda^{-1} J^T \left(J \lambda^{-1} J^\tau \right)^{-1}. \tag{4}$$

The Cartesian position (x and y) of the end effector is used here to define the motion task of the robot. The combination of feedback linearization and ISMC Cartesian/task control law is:

$$F = A\ddot{X}_d + AK_p e + AK_d \dot{e} + g(q) + \mu_{cc}(q, \dot{q}) + \alpha \frac{r}{(||r|| + \zeta)}. \tag{5}$$

The scalars K_p and K_d are the proportional and the derivative gains respectively. The scalar $\alpha > 0$ is normally selected large enough to reduce the effect of uncertainty and achieve robustness. The scalar ζ, is introduced to minimize chattering effect (Jalani et al., 2010a). It should be noted that we are controlling only x and y. Hence, we have two redundant degrees of freedom. These redundant degrees of freedom are used to produce human like motion employing the posture controller presented in the next section.

2.1 Posture torque controller

As mentioned in the previous section, the ISMC scheme is applied to a multi-redundant system (i.e. 4 DOF, shoulder flexion, shoulder abduction, humeral rotation and elbow flexion are used, while, only 2 DOF (x and y) of the end-effector are controlled. Hence, the motion is under-constrained. Hence, a posture torque controller is employed which deals with the redundant motion, to produce human like posture. This posture controller also gives some sense of safety to the human interacting with the robot by generating human-like motion. This human like motion is achieved by minimizing an effort function (based on gravity) during reaching a particular point in the work space of the robot arm. The method here is adopted from the previous work by Spiers et al. (2009a, b) (see also De Sapio et al., 2005). The detailed description of the "posture" controller scheme along with experimental results can be found in Spiers et al. (2009a, b) and Khan et al. (2010a, b). The "posture" controller τ_p is in the null space of the main controller given by Eq. (5) (Cartesian controller i.e. FL + ISMC); hence, it does not affect the main controller:

$$\tau = J^T F + N^T \tau_p, \; N^T = \left(\mathbf{I} - J^T \bar{J}^T \right) \tag{6}$$

where $\mathbf{I}$ is the identity matrix. $\bar{J}$ is the the inertia weighted pseudo Jacobian inverse:

$$\bar{J} = \boldsymbol{\lambda}^{-1} J^T \left(J \boldsymbol{\lambda}^{-1} J^T \right)^{-1} \tag{7}$$

where $\boldsymbol{\lambda}$ is the inertia matrix given by Eq. (2). The posture torque, τ_p is defined as:

$$\tau_p = -K_{pp} \frac{\delta U}{\delta q} - K_{dp} \dot{q} \tag{8}$$

where, K_{pp} and K_{dp} are proportional and derivative gains respectively. U is the effort function defined as:

$$U = \boldsymbol{\nu}^T (K_\epsilon)^{-1} \boldsymbol{\nu} \tag{9}$$

and $\boldsymbol{\nu}$ is the gravitational vector term from Eq. (2), and K_ϵ has $n \times n$ dimensions in our case, actuator activation diagonal matrix (having positive diagonal elements). The diagonal values $K_{\epsilon i}$, define the relative preferential weighting of each actuator ($i = 1 \ldots n$).

To get acceptable posture control performance, a good model of the inertia matrix $\boldsymbol{\lambda}$ is required (in order to calculate Jacobian inverse given by Eq. 7).

The Cartesian control law (ISMC + FL) Eq. (5) is required for a better tracking accuracy. The control scheme is shown in Fig. 3 and the applied torques to the robot joints are given by $\tau = J^T F + (I - J^T \bar{J}^T)\tau_p$

2.2 ISMC Cartesian controller results discussion

In this section experimental results are included for ISMC based tracking controller for 4 DOF robot arm. Also, simulation is included for the model reference compliance control scheme in Sect. "Model reference compliance control simulation using ISMC".

For the real time implementation of the control scheme, a dSPACE DS1106 embedded system is employed. A sampling time of 1 ms is used. The BERT II robot arm uses optical encoders for position and velocity measurements (see Fig. 4). As mentioned before, four degrees of freedom namely, shoulder flexion, shoulder abduction, humeral rotation and elbow flexion of the BERT II robot arm (Fig. 2) are used. The end-effector position is specified with respect to the fixed coordinate frame in the shoulder of the BERT II arm (see Fig. 4).

ISMC controller tuning is simple. Choosing suitable value for $\alpha > 0$ large enough to reduce the effect of uncertainty and achieve robustness. The scalar ζ, is selected to minimize the chattering effect (Jalani et al., 2010a; Herrmann et al., 2014).

Figures 5–6 show experimental results for ISMC based task space controller. In the real robot experiment 1 (Fig. 5), a multi-step demand is applied in the x direction while a constant demand is applied in the y direction. Tracking error is also shown which is very small. The control tuning parameters for experiment 1 are: $K_p = 2000$, $K_d = 10$, $K_{r1} = K_{r2} = 8$, $K_{i1} = K_{i2} = 16$, $\alpha = 50$ and $\zeta = 5$.

In another real experiment (Fig. 6), a sine wave demand is applied in x and a constant demand is applied in the y direction. The tracking error shown in Fig. 6 for experiment 2. Efficient real-time tracking results demonstrate the effectiveness of the proposed control scheme for our BERT II robot arm. The following control tuning parameters are used for experiment 2: $K_p = 2500$, $K_d = 10$, $K_{r1} = K_{r2} = 8$, $K_{i1} = K_{i2} = 16$, $\alpha = 50$ and $\zeta = 5$.

The scheme is effectively dealing with the deficiencies and uncertainties in the dynamic model. The ISMC has almost

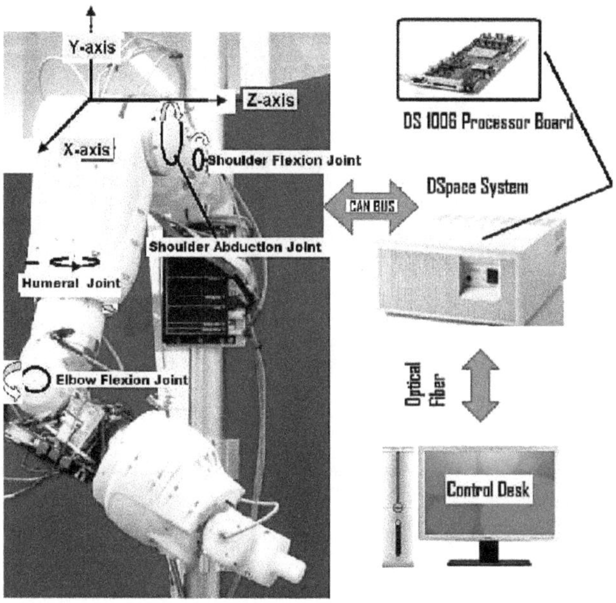

Figure 4. Experimental setup using dSPACE.

eliminated the chattering effect which normally would occur with traditional sliding mode schemes. The real advantage of this scheme here is to enhance the feedback linearization controller (using the 4 DOF dynamic model of the BERT II robot arm) with the ISMC to overcome the nonlinearities and unmodelled dynamics.

3 Model reference compliance control using ISMC

In this section, an ISMC based model reference compliance controller scheme is presented. This compliance scheme is similar in nature to the model reference adaptive compliance presented in (Khan et al., 2010b, 2014b), Khan and Herrmann (2014) and Colbaugh et al. (1996) However, here this scheme is much simpler and solely based on ISMC. The reference model is inside the sliding mode variable r. While in Khan et al. (2010b, 2014b), Khan and Herrmann (2014) and Colbaugh et al. (1996) the reference model is "external" where, tracking demand is modified (using force feedback) to compensate the external forces. The sliding mode element r given by Eq. (1) is modified to introduce the compliance effect based on the external sensed force via joint torque sensors (Jalani et al., 2013; Herrmann et al., 2014):

$$r = \dot{e} + K_r e + K_i \int_0^t e(\xi)d\xi - \int_0^t G_f H - \dot{e}(0) - K_s e(0) \quad (10)$$

where, G_f is a positive scalar and H is the externally sensed force via joint torque sensors. The K_r is the damping coefficient here and K_i is the spring constant. The values of spring constant and damping coefficient can be used to get suitable level of compliance.

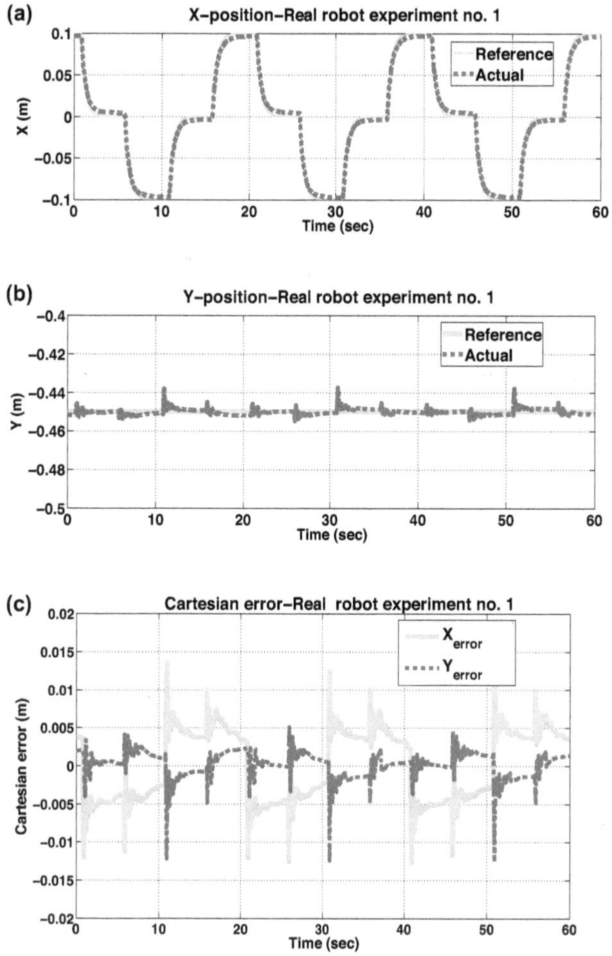

Figure 5. Position x (**a**) and y (**b**) and Position errors (**c**) experiment 1).

If we consider the derivative of sliding mode element s in Eq. (10) (Jalani et al., 2013):

$$\dot{r} = \ddot{e} + K_r\dot{e} + K_i e - G_f H. \quad (11)$$

When the sliding mode is achieved i.e. $r = 0$ and $\dot{r} = 0$, then Eq. (11) becomes:

$$\ddot{e} + K_r\dot{e} + K_i e = G_f H. \quad (12)$$

Hence, the Cartesian error dynamics reduces to a second order mass-spring-damper system. The compliance behaviour of this reference model can be tuned by the selection of K_r, K_i and G_f.

Model reference compliance control simulation using ISMC

In this section a compliance control scheme realised via ISMC is simulated on the BERT II arm. The is a similar Cartesian space (x, y and z) scheme to the one presented and

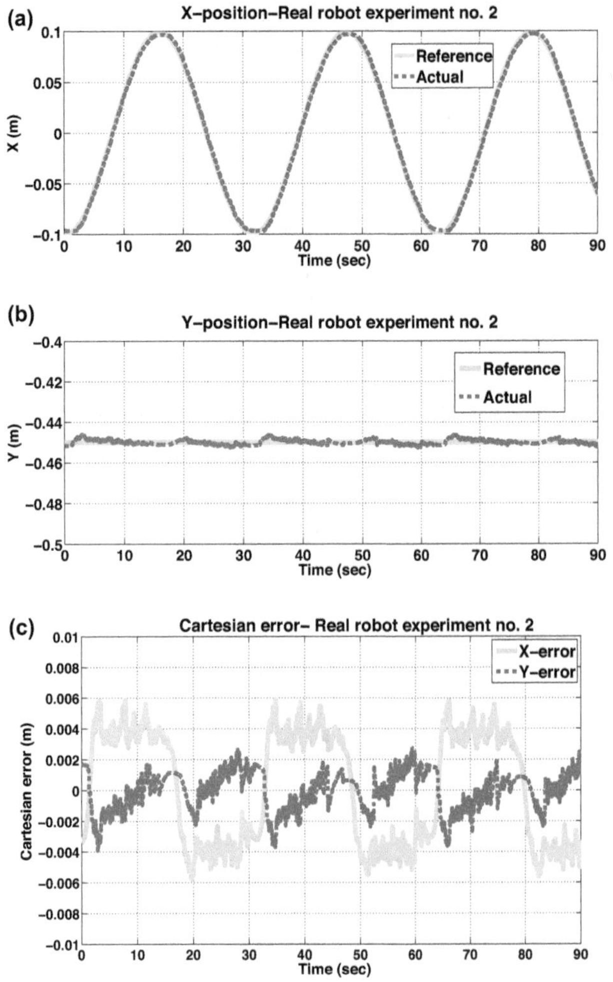

Figure 6. Position x **(a)** and y **(b)** and Position errors **(c)** experiment 2).

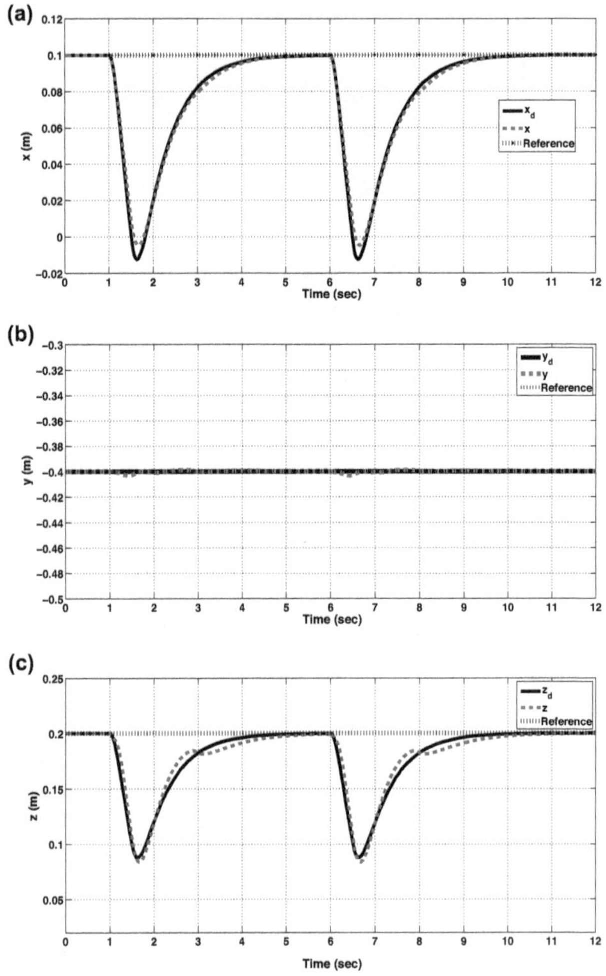

Figure 7. Cartesian position x **(a)** and y **(b)** and z **(c)** for $K_r = 8\,\mathrm{N\,s\,m^{-1}}$ and $K_i = 10\,\mathrm{N\,m^{-1}}$ ($Fx = Fz = 5\,\mathrm{N}$ and $Fy = 0$).

experimental result were presented in Sect. 2. However, here is focus on compliant behaviour. Figure 7 shows the compliance simulation results. External forces of 5 N are applied on x and z direction while $K_r = 8\,\mathrm{N\,s\,m^{-1}}$ and $K_i = 10\,\mathrm{N\,m^{-1}}$. In the absence of external forces, the controller will track the reference position demands (in this case a constant demand for x, y and z). In the presence of external forces robot will follow a modified trajectory (x_d, y_d and z_d) to compensate for the forces and behave compliantly. The compliance level is created by choosing the values K_r and K_i. Similarly, in Fig. 8, another set of compliance results are shown. The scheme can be made more compliant by selecting lower value for K_r. It should be noted that joint torques sensors/end effector force sensor are essential for real implementation of this controller (Herrmann et al., 2014). It is worth pointing out that decoupling external torques from robot body-own torques may be challenging. However, we have addressed this issue to some extent in our previous work (Khan et al., 2014b, 2010a). It is to be noted that the geometric splitting (i.e. decoupling torques) of the control task of the end-effector and the control of the redundant degrees of freedom has been introduced by (De Sapio et al., 2005). The idea is that multiple configurations were avoided by introducing and minimizing cost which resolves the redundancy issues. Other applications by using decoupling technique can be found in Herrmann et al. (2014) and Spiers et al. (2009a, b).

4 Conclusions

Safe physical human robot interaction lies in the core of social robotics. Safety needs multi-dimensional approach and compliance alone cannot solve safety problem. However, compliance can play an important role in providing immediate layer of safety. In this paper, compliance control has been realised through a model reference integral sliding mode control. An ISMC tracking control scheme is turned into a model reference compliance controller, by introducing external force feedback into the sliding mode element. The de-

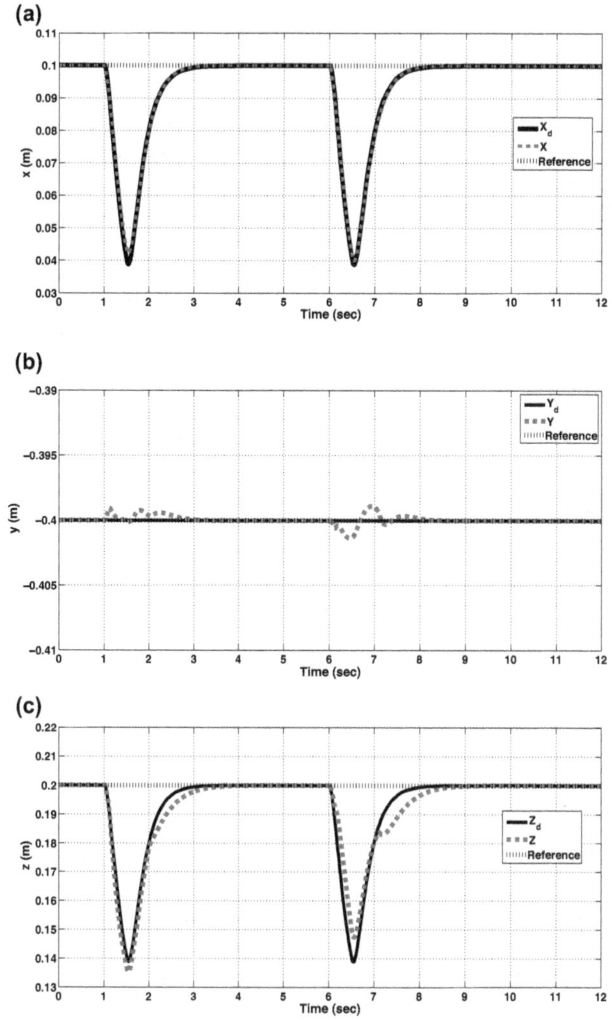

Figure 8. Cartesian position x **(a)** and y **(b)** and z **(c)** for $K_r = 12\,\mathrm{N\,s\,m^{-1}}$ and $K_i = 30\,\mathrm{N\,m^{-1}}$ ($Fx = Fz = 5$ N and $Fy = 0$).

sired compliance level can be achieved simply by changing the derivative and integral gain in the sliding mode variable r. ISMC is simple and very easy to implement. The tuning of the control scheme for the desired performance is also very straightforward as there are not many parameters to be tuned. Real-time Cartesian position tracking results and compliance control simulation included here highlights the flexibility of ISMC based control and its effectiveness to be used as model reference compliance control scheme. The ultimate goal of this work is to use the ISMC based control scheme for safe human-robot interaction.

Acknowledgements. Both the authors would like to thank Bristol Robotics Laboratory (BRL) UK for the great support they got during their PhD studies. The experimental results presented here were produced in BRL, UK.

References

Ajwad, S. A., Ullah, M. I., Baizid, K., and Iqbal, J.: A comprehensive state-of-the-art on control of industrial articulated robots, J. Balkan Tribolog. Assoc., 20, 1310–4772, 2014.

Ajwad, S. A., Iqbal, J., Ullah, M. I., and Mehmood, A.: A systematic review of current and emergent manipulator control approaches, in: Frontiers of Mechanical Engineering, Springer, Berlin, Heidelberg, Germany, 198–210, 2015.

Braun, D., Petit, F., Huber, F., Haddadin, S., van der Smagt, P., Albu-Schäffer, A., and Vijayakumar, S.: Robots Driven by Compliant Actuators, IEEE T. Opt. Control Actuat. Constr. Robot., 29, 1085–1101, 2013.

Castaños, F. and Fridman, L.: Analysis and Design of Integral Sliding Manifolds for Systems With Unmatched Perturbations, IEEE T. Autom. Contr., 51, 853–858, 2006.

Colbaugh, R., Glass, K., and Wedeward, K.: Adaptive Compliance Control of Electrically-Driven Manipulators, in: Proceedings of the 35th Conference on Decision and Control, Kobe, Japan, 394–399, 1996.

De Sapio, V., Khatib, O., and Delp, S.: Simulating the task level control of human motion: a methodology and framework for implementation, Visual Comput., 21, 289–302, 2005.

Di Natali, C. and Valdastri, P.: Surgical robotics and instrumentation, Surgery, 245, 379–384, 2012.

Ding, X. and Fang, C.: A Novel Method of Motion Planning for an Anthropomorphic Arm Based on Movement Primitives, IEEE/ASME T. Mechatron., 18, 624–636, doi:10.1109/TMECH.2012.2197405, 2013.

Eker, I. and Akinal, S.: Sliding mode control with integral augmented sliding surface: design and experimental application to an electromechanical system, Elect. Eng., 90, 189–197, 2008.

Ficuciello, F. and Villani, L.: Compliant hand-arm control with soft fingers and force sensing for human-robot interaction, in: Biomedical Robotics and Biomechatronics (BioRob), 2012 4th IEEE RAS & EMBS International Conference Rome, Italy, 1961–1966, 2012.

Herrmann, G., Jalani, J., Mahyuddin, M. N., Khan, S. G., and Melhuish, C.: Robotic hand posture and compliant grasping control using operational space and integral sliding mode control, Robotica, 1–23, doi:10.1017/S0263574714002811, 2014.

Iqbal, U., Samad, A., Nissa, Z., and Iqbal, J.: Embedded control system for AUTAREP – A novel AUTonomous Articulated Robotic Educational Platform, Tehnicki Vjesnik-Technical Gazette, 21, 1255–1261, 2014.

Islam, R. U., Iqbal, J., and Khan, Q.: Design and comparison of two control strategies for multi-DOF articulated robotic arm manipulator, in: vol. 02, Control Engineering and Applied Informatics (CEAI), Bucharest, Romania, 1454–8658, 2014.

Jalani, J., Herrmann, G., and Melhuish, C.: Robust Trajectory Following for Underactuated Robot Fingers, in: UKACC International Conference on CONTROL, Conventry, UK, 2010a.

Jalani, J., Herrmann, G., and Melhuish, C.: Concept for Robust Compliance Control of Robot Fingers, in: 11th Conference Towards Autonomous Robotic Systems, Plymouth, UK, 97–102, 2010b.

Jalani, J., Mahyuddin, N., Herrmann, G., and Melhuish, C.: Active robot hand compliance using operational space and Integral Sliding Mode Control, in: vol. 01, IEEE/ASME International Conference on Advanced Intelligent Mechatronics (AIM), Wollongong, NSW Australia, 1749–1754, 2013.

Khan, S.: Adaptive and reinforcement learning control methods for active compliance control of a humanoid robot arm, PhD Thesis, University of the West of England, Bristol, UK, 2012.

Khan, S. and Herrmann, G.: NRCGHRIComplianceControl, https://www.youtube.com/watch?v=IKE8Rrtr-Ow (last access: 11 January 2016), 2014.

Khan, S., Herrmann, G., Pipe, T., and Melhuish, C.: Adaptive multi-dimensional compliance control of a humanoid robotic arm with anti-windup compensation, in: IEEE/RSJ International Conference on Intelligent Robots and Systems (IROS), 2218–2223, 2010a.

Khan, S., Herrmann, G., Pipe, T., Melhuish, C., and Spiers, A.: Safe Adaptive Compliance Control of a Humanoid Robotic Arm with Anti-Windup Compensation and Posture Control, Int. J. Social Robot., 2, 305–319, 2010b.

Khan, S. G., Herrmann, G., Al Grafi, M., Pipe, T., and Melhuish, C.: Compliance Control and Human Robot Interaction Part I Survey, Int. J. Human. Robot., 11, 1430001-1–1430001-28, 2014a.

Khan, S. G., Herrmann, G., Lenz, A., Al Grafi, M., Pipe, T., and Melhuish, C.: Compliance Control and Human Robot Interaction Part II Experimental Examples, Int. J. Human. Robot., 11, 1430002-1–430002-21, 2014b.

Khatib, O.: A Unified Approach for Motion and Force Control of Robot Manipulators: The Operational Space Formulation, IEEE J. Robot. Autom., RA3, 43–53, 1987.

Kim, H., Kim, I., Cho, C., and Song, J.: Safe joint module for safe robot arm based on passive and active compliance method, Mechatronics, 22, 1023–1030, 2012.

Lewis, F., Dawson, D., and Abdallah, C.: Robot Manipulator Control: Theory and Practice, Marcel Dekker Inc., New York, USA, 2003.

Makoto, Y., Gyu-Nam, K., and Masahiko, T.: Integral Sliding Mode Control with Anti-windup Compensation and Its Application to a Power Assist System, J. Vibrat. Control, 16, 503–512, doi:10.1177/1077546309106143, 2010.

Maplesoft: MapleSim, http://www.maplesoft.com/products/maplesim/modelgallery/detail.aspx?id=108 (last access: 11 January 2016), 2015.

Nemec, B. and Zlajpah, L.: Null space velocity control with dynamically consistent pseudo-inverse, Robotica, 18, 513–518, 2000.

Potkonjak, V., Jovanovic, K., Svetozarevic, B., Holland, O., and Mikicic, D.: Modeling and Control of a Compliantly Engineered Anthropomimetic Robot in Contact Tasks, in: 35th mechanisms and robotics conference, ASME, Portland, Oregon, USA, 2011.

Rezoug, A. B. T. and Hamerlain, M.: Experimental Study of Non-singular Terminal Sliding Mode Controller for Robot Arm Actuated by Pneumatic Artificial Muscles, 19th World Congress The International Federation of Automatic Control (IFAC), 24–29 August 2014, Cape Town, South Africa, 10113–10118, 2014.

Shi, J., Liu, H., and Bajcinca, N.: Robust control of robotic manipulators based on integral sliding mode, Int. J. Control, 81, 1537–1548, 2008.

Spiers, A., Herrmann, G., and Melhuish, C.: Implementing ?Discomfort? in Operational Space: Practical Application of a Human Motion Inspired Robot Controller, TAROS conference: Towards Autonomous Robotic Systems, Derry, Northern Ireland, UK, 2009a.

Spiers, A., Herrmann, G., Melhuish, C., Pipe, T., and Lenz, A.: Robotic Implementation of Realistic Reaching Motion using a Sliding Mode/Operational Space Controller, Lecture Notes in Computer Science (ICSR '09), Springer-Verlag, Berlin, Heidelberg, Germany, 230–238, 2009b.

Utkin, V. and Shi, J.: Integral sliding mode in systems operating under uncertainty conditions, IEEE Proceedings of the 35th Decision and Control, Kobe, Japan, 4591–4596, 1996.

10

Kinematic analysis of a novel 3-CRU translational parallel mechanism

B. Li[1,2], Y. M. Li[1,2], X. H. Zhao[2], and W. M. Ge[2]

[1]Faculty of Science and Technology, University of Macau, Taipa, Macau, China

[2]Tianjin Key Laboratory for Advanced Mechatronic System Design and Intelligent Control, Tianjin University of Technology, Tianjin, China

Correspondence to: Y. M. Li (ymli@umac.mo)

Abstract. In this paper, a modified 3-DOF (degrees of freedom) translational parallel mechanism (TPM) three-CRU (C, R, and U represent the cylindrical, revolute, and universal joints, respectively) structure is proposed. The architecture of the TPM is comprised of a moving platform attached to a base through three CRU jointed serial linkages. The prismatic motions of the cylindrical joints are considered to be actively actuated. Kinematics and performance of the TPM are studied systematically. Firstly, the structural characteristics of the mechanism are described, and then some comparisons are made with the existing 3-CRU parallel mechanisms. Although these two 3-CRU parallel mechanisms are both composed of the same CRU limbs, the types of freedoms are completely different due to the different arrangements of limbs. The DOFs of this TPM are analyzed by means of screw theory. Secondly, both the inverse and forward displacements are derived in closed form, and then these two problems are calculated directly in explicit form. Thereafter, the Jacobian matrix of the mechanism is derived, the performances of the mechanism are evaluated based on the conditioning index, and the performance of a 3-CRU TPM changing with the actuator layout angle is investigated. Thirdly, the workspace of the mechanism is obtained based on the forward position analysis, and the reachable workspace volume is derived when the actuator layout angle is changed. Finally, some conclusions are given and the potential applications of the mechanism are pointed out.

1 Introduction

In recent years, lower-mobility translational parallel mechanisms (TPMs) have been extensively studied. Compared with 6-DOF (degrees of freedom) parallel manipulators, lower-mobility mechanisms possess many merits in terms of simpler mechanical design, larger workspace, and lower manufacturing cost, in addition to the inherent advantages of the general parallel manipulators, such as high accuracy, high stiffness, high velocity, high dynamic performance, large load-to-weight ratio, low moving inertia, and little accumulation of positional errors. A lower-mobility TPM with 3 DOFs is a focus of the current trend in the research community, and various forms of 3-DOF TPMs have been designed. A 3-DOF TPM is a 3-DOF parallel mechanism whose moving platform can achieve three independent orthogonal translational motions with respect to its fixed base.

In past related research literature and industrial applications, the well-known 3-DOF TPM Delta robot (Clavel, 1988), one of the most successful parallel manipulators in market, was commonly used in pick-and-place applications. A pure translational 3-UPU PM (P represents the prismatic joint) was proposed in Tsai and Joshi (2000); this mechanism has been the subject of much study and was previously widely used in practice, and several variants of this mechanism were designed, such as 3-PUU TPM and the FlexPLP tripod. A translational 3-URC mechanism was proposed in Gregorio (2004); the position and velocity of this PM were written in explicit form. A kind of 3-DOF translational parallel cube manipulator was presented in the study of Li et al. (2003), in which the kinematics and workspace of the manipulator were investigated. Compared with other 3-DOF TPMs, the parallel cube manipulator possessed some obvious merits

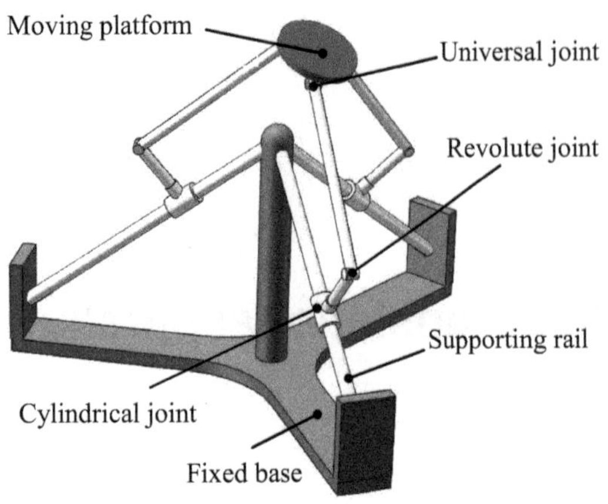

Figure 1. The existing 3-CRU spherical wrist.

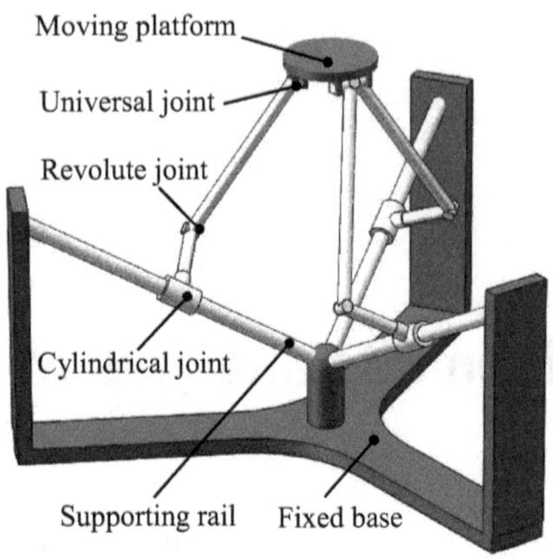

Figure 2. The modified 3-CRU TPM.

in terms of higher compactness and stiffness and no singularities in the workspace. Lou and Li (2006) proposed a novel 3-DOF purely TPM, named Orthotripod. The mechanism possessed a nearly ball-shaped workspace, and had a much better conditioning index than that of the tripod-based mechanism. The CICABOT was presented in Ruiz et al. (2012), which was a novel 3-DOF TPM based on two five-bar mechanisms. The workspace of the CICABOT was large and its workspace volume was limited by the size of the links, and both the inverse and direct kinematics were very simple to determine. The dynamics modeling of a translational 3-CPU was investigated via screw theory (Carbonari et al., 2013). The operation performance of a LARM leg mechanism with 3-UPU parallel architecture was studied in Wang et al. (2015). Besides these new 3-DOF TPMs mentioned above, some other 3-DOF architectures can be found in the literature (Li and Xu, 2006; Simoni et al., 2013; Gregorio and Parenti-Castelli, 2002; Ji and Wu, 2003; Chung and Hervé, 2006; Kim and Chung, 2003; Merlet, 2006; Li and Xu, 2005a, b; Tsai and Joshi, 2002; Chablat and Wenger, 2001).

In this paper, a modified 3-DOF TPM (3-CRU, where the letters C, R, and U represent the cylindrical, revolute, and universal joints, respectively) is proposed. The structure of the paper is arranged as follows. In Sect. 2, the proposed mechanism is compared with other existing mechanisms. The mobility of the mechanism is analyzed by means of screw theory in Sect. 3. In Sect. 4, an analytical model for the kinematics of the mechanism is established, and the exact analytical solutions are found both for the inverse and forward kinematics problems. In Sect. 5, the Jacobian matrix of the mechanism is derived. The reachable workspace of the mechanism is obtained based on the forward position analysis in Sect. 6. Conclusions and areas of future research are given in Sect. 7.

2 A modified 3-CRU TPM and its structural characteristics

With regard to previous works, a 3-CRU rotational parallel manipulator, as shown in Fig. 1, was first introduced in Fang and Tsai (2004); subsequent to this, the kinematics, dynamics and kineto-elasto-static synthesis of a 3-CRU spherical wrist were studied by Callegari et al. (2007a, b).

In this paper, a modified structure of 3-CRU TPM, shown in Fig. 2, is proposed. The orientational mechanism (Fig. 1) and the positional mechanism (Fig. 2) differ from each other in the axes of the revolute joints and universal joints. The different joint arrangements of CRU limbs in these two types of 3-CRU parallel mechanisms are demonstrated in Fig. 3. The 3-CRU TPM consists of a base platform, a moving platform, three supporting rails, and three limbs with identical kinematic structures. Each limb connects the fixed base to the moving platform via a cylindrical joint, a revolute joint, and a universal joint in sequence. The cylindrical joint, actuated by a linear actuator, can move along the supporting rail and rotate on the rail simultaneously (Fig. 4), and the rails are symmetrically arranged 120° apart. Thus, the moving platform is attached to the base by three identical CRU linkages.

For the sake of analysis, a fixed Cartesian reference coordinate frame O-xyz is attached at the centered point O of the intersection point of three supporting rails as shown in Fig. 5. A_i is the center of the cylindrical joint, B_i is the center of the revolute joint, C_i is the center of the universal joint, and point P is the center of the moving platform. Angle α_i is measured from the base platform to rails OA_i and is defined as the actuator layout angle. In order to ensure the isotropic property of the mechanism, we assume $\alpha_1 = \alpha_2 = \alpha_3 = \alpha$.

3 Mobility analysis via screw theory

Considering that the general Grübler–Kutzbach criterion can only obtain the number of DOFs for some mechanisms but cannot indicate the properties of the DOF (i.e., whether they

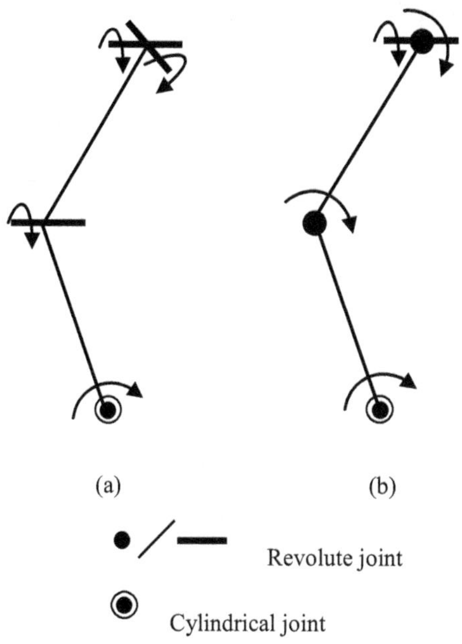

Figure 3. Different joint arrangement of a CRU limb. **(a)** Joint arrangement of the CRU limb of an existing 3-CRU spherical wrist. **(b)** Joint arrangement of a CRU limb of a modified 3-CRU TPM.

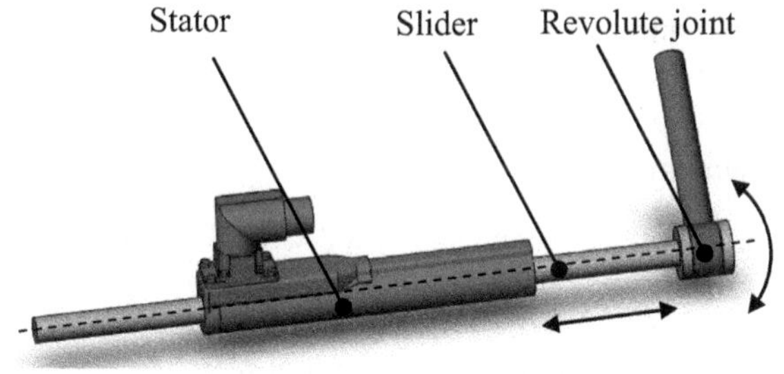

Figure 4. A CAD model of a cylindrical joint.

are translational or rotational DOF), screw theory is employed to analyze the mobility of a 3-CRU parallel manipulator, which is a convenient tool for studying instantaneous motion systems that include both rotational and translational motions in three-dimensional space.

The mobility of the 3-CRU TPM is determined by the combined effect of the three limb constraint forces/couples. Here, the reciprocal screw theory is used to analyze constraint forces exerted on the moving platform in order to give a complete description of how the mobility of TPMs is computed (Dai et al., 2006). Without losing generality, a local coordinate system $A_i - x_i y_i z_i$, $(i = 1\text{–}3)$ is established for each limb and twist system of the ith CRU limb as shown in

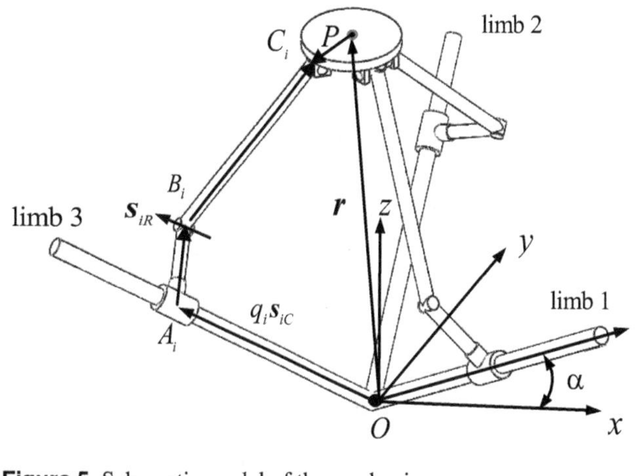

Figure 5. Schematic model of the mechanism.

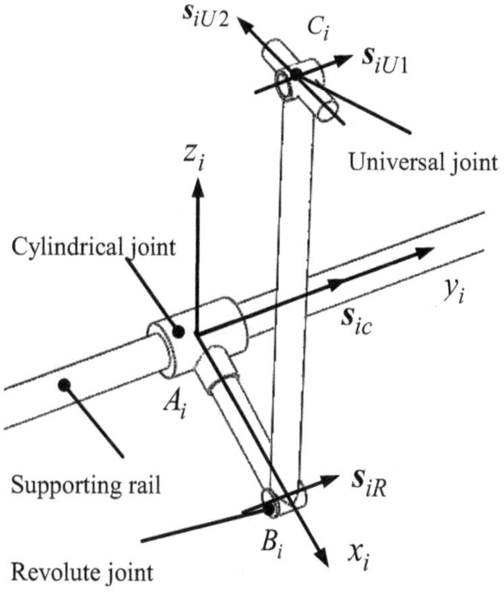

Figure 6. Twist systems of a CRU limb.

Fig. 6:

$$\begin{aligned}
\$_1 &= (\mathbf{0} \quad \boldsymbol{s}_{i\mathrm{C}}) \\
\$_2 &= (\boldsymbol{s}_{i\mathrm{C}} \quad \mathbf{0}) \\
\$_3 &= (\boldsymbol{s}_{i\mathrm{R}} \quad \boldsymbol{r}_{i\mathrm{R}} \times \boldsymbol{s}_{i\mathrm{R}}) \\
\$_4 &= (\boldsymbol{s}_{i\mathrm{U}1} \quad \boldsymbol{r}_{i\mathrm{U}} \times \boldsymbol{s}_{i\mathrm{U}1}) \\
\$_5 &= (\boldsymbol{s}_{i\mathrm{U}2} \quad \boldsymbol{r}_{i\mathrm{U}} \times \boldsymbol{s}_{i\mathrm{U}2}),
\end{aligned} \tag{1}$$

where $\boldsymbol{s}_{i\mathrm{C}}$ and $\boldsymbol{s}_{i\mathrm{R}}$ stand for the unit direction vector of cylindrical and revolute joints, respectively, and $\boldsymbol{s}_{i\mathrm{C}} = \boldsymbol{s}_{i\mathrm{R}}$. $\boldsymbol{s}_{i\mathrm{U}1}$ and $\boldsymbol{s}_{i\mathrm{U}2}$ are the unit direction vectors of universal joints.

Using the reciprocity between twist and wrench, the CRU limb constraint system can be calculated by

$$\$_1^r = (\mathbf{0} \quad \boldsymbol{n}_i), \tag{2}$$

where $\boldsymbol{n}_i = \boldsymbol{s}_{i\mathrm{U}1} \times \boldsymbol{s}_{i\mathrm{U}2}$. $\$_1^r$ denotes a constraint couple whose direction is perpendicular to the axes of joint screws $\$_4$ and $\$_5$.

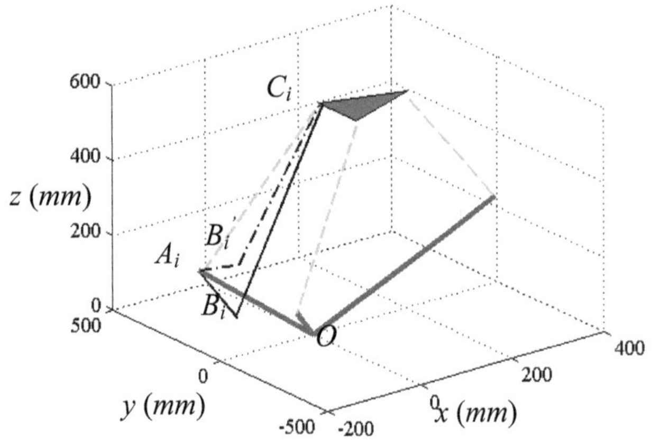

Figure 7. Different configurations for inverse position.

In Eq. (2), three CRU limbs provide three constraint couples restricting rotational DOFs around the axis of the constraint couples. Therefore the moving platform of the 3-CRU TPM behaves as if it has only three translational DOFs.

4 Inverse and forward position analyses

4.1 Inverse position analysis

Inverse position analysis of the 3-CRU TPM involves the determination of the position of A_i given the position of the moving platform.

In the O-xyz frame, the position vector $\boldsymbol{r}=(xyz)^T$ of position P can be expressed as

$$\boldsymbol{r}=q_i\boldsymbol{s}_{i\mathrm{C}}+\overline{\boldsymbol{A}_i\boldsymbol{B}_i}+\overline{\boldsymbol{B}_i\boldsymbol{C}_i}-\overline{\boldsymbol{PC}_i},\quad (i=1,2,3), \tag{3}$$

where $q_i=\|OA_i\|$ stands for the position of A_i, $\boldsymbol{s}_{i\mathrm{C}}$ is the unit vector of $\overline{OA_i}$, $\overline{\boldsymbol{A}_i\boldsymbol{B}_i}$ is the vector from point A_i to point B_i, $\overline{\boldsymbol{B}_i\boldsymbol{C}_i}$ is the vector from point B_i to point C_i, and $\overline{\boldsymbol{PC}_i}$ is the vector from point P to point C_i.

Note that, for the CRU limb, the constraint imposed by the revolute joint restricts both $\overline{\boldsymbol{A}_i\boldsymbol{B}_i}$ and $\overline{\boldsymbol{B}_i\boldsymbol{C}_i}$ so that they are normal to the unit vector $\boldsymbol{s}_{i\mathrm{R}}$ of the revolute joint axis. Thus, taking the dot product with $\boldsymbol{s}_{i\mathrm{R}}$ on both sides of Eq. (3) leads to

$$q_i=\left(\boldsymbol{r}+\overline{PC_i}\right)^T\boldsymbol{s}_{i\mathrm{R}},\quad (i=1,2,3). \tag{4}$$

Thus, for a given position vector $\boldsymbol{r}=(x\ \ y\ \ z)^T$ of the moving platform, the position of A_i can be obtained directly using Eq. (4).

4.2 Forward position analysis

Forward position analysis of the mechanism is concerned with the determination of the moving platform position given the position of A_i.

Expanding Eq. (4) yields

$$q_1=\cos(\alpha)x+\sin(\alpha)z+\cos(\alpha)a, \tag{5}$$

$$q_2=-\frac{1}{2}\cos(\alpha)x+\frac{\sqrt{3}}{2}\cos(\alpha)y+\sin(\alpha)z+\cos(\alpha)a, \tag{6}$$

$$q_3=-\frac{1}{2}\cos(\alpha)x-\frac{\sqrt{3}}{2}\cos(\alpha)y+\sin(\alpha)z+\cos(\alpha)a. \tag{7}$$

Subtracting Eq. (7) from Eq. (6) yields

$$y=\frac{q_2-q_3}{\sqrt{3}\cos(\alpha)}. \tag{8}$$

Adding Eqs. (5), (6), and (7) together yields

$$z=\frac{q_1+q_2+q_3-3\cos(\alpha)a}{3\sin(\alpha)}. \tag{9}$$

Then, substituting Eq. (9) into Eq. (5) gives

$$x=\frac{2q_1-q_2-q_3}{3\cos(\alpha)}. \tag{10}$$

Lastly, given a set of $(q_1\ \ q_2\ \ q_3)$, x, y, and z can be solved by using Eqs. (8), (9), and (10).

It should be pointed out that the actuator layout angle should be set in the range of 0–90° to ensure that the robot has real solutions for the forward position analysis.

For the proposed 3-CRU TPM in this paper, both the inverse and the forward position analyses of the mechanism can be calculated directly in explicit form as shown in Sects. 4.1 and 4.2, which is extremely significant for the possible practical applications of the mechanism.

4.3 Numerical examples

The architectural parameters of a 3-CRU TPM are selected as $a=100$ mm, $l_1=300$ mm, $l_2=500$ mm, and $\alpha=30°$; here, $a=\|\boldsymbol{PC}_i\|$, $l_1=\|\boldsymbol{A}_i\boldsymbol{B}_i\|$, and $l_2=\|\boldsymbol{B}_i\boldsymbol{C}_i\|$. For the inverse position analysis, given a set of inputs $(x\ \ y\ \ z)$, output parameters can be calculated as shown in Table 1, and Fig. 7 depicts the configurations associated with these solutions. For the forward position analysis, given a set of inputs $(q_1\ \ q_2\ \ q_3)$, output parameters can be calculated as shown in Table 2, and the configurations associated with these solutions are described in Fig. 8.

For a given position vector $r=(x\ \ y\ \ z)^T$ of the moving platform, only one solution of the A_i positions can be obtained using Eq. (4) directly, as shown in Fig. 7. However, the locations of point B_i have two possibilities for a CRU limb, as shown in Fig. 9. Therefore, there are in total eight configurations for the inverse position analysis. Accordingly, the forward position analysis also has eight solutions in a similar manner. But for the 3-CRU orientation parallel mechanism, each limb has four feasible solutions, leading to a total of 64 possibilities for the inverse and forward position analyses (Callegari et al., 2007b). Obviously, the kinematics of the modified positional parallel mechanism 3-CRU are much simpler than those of the orientational mechanism.

Table 1. Inverse position analysis of the mechanism (unit: mm).

	Inputs			Outputs		
	x	y	z	q_1	q_2	q_3
Case (a)	80	−50	600	455.8846	314.4615	389.4615

Table 2. Forward position analysis of the mechanism (unit: mm).

	Inputs			Outputs		
	q_1	q_2	q_3	x	y	z
Case (b)	300	500	500	−153.9601	0	693.4616

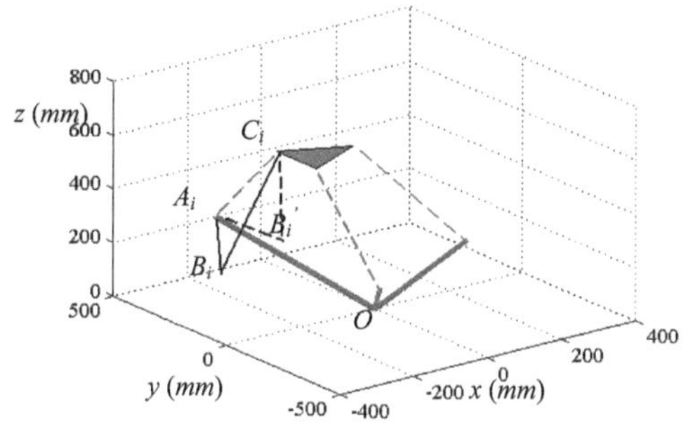

Figure 8. Different configurations for forward position analysis.

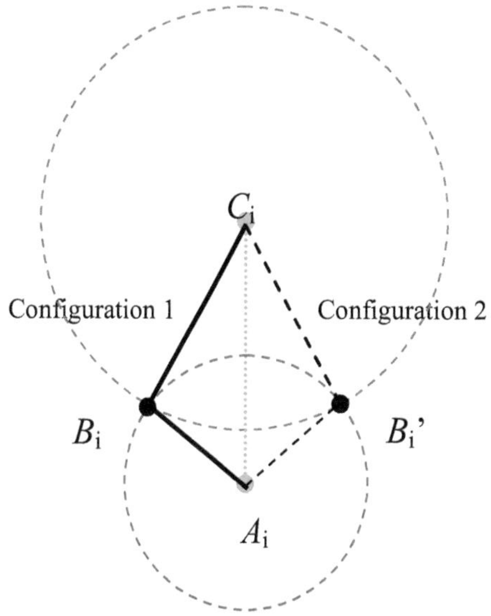

Figure 9. Two possible configurations of a CRU limb.

5 Jacobian matrix of the mechanism and performance analysis

5.1 Jacobian matrix of the mechanism

Differentiating Eqs. (5), (6), and (7) with respect to time respectively yields

$$\dot{q}_1 = \cos(\alpha)\dot{x} + \sin(\alpha)\dot{z}, \tag{11}$$

$$\dot{q}_2 = -\frac{1}{2}\cos(\alpha)\dot{x} + \frac{\sqrt{3}}{2}\cos(\alpha)\dot{y} + \sin(\alpha)\dot{z}, \tag{12}$$

$$\dot{q}_3 = -\frac{1}{2}\cos(\alpha)\dot{x} - \frac{\sqrt{3}}{2}\cos(\alpha)\dot{y} + \sin(\alpha)\dot{z}. \tag{13}$$

Equations (11), (12), and (13) can be written in matrix form. Only when the manipulator is away from singularities is the matrix invertible.

$$\dot{\boldsymbol{q}} = \mathbf{J}\dot{\boldsymbol{X}}, \tag{14}$$

where $\dot{\boldsymbol{q}} = \begin{bmatrix}\dot{q}_1 & \dot{q}_2 & \dot{q}_3\end{bmatrix}^T$, $\dot{q}_i$ is the velocity of the ith linear actuator, and $\dot{\boldsymbol{X}} = \begin{bmatrix}\dot{x} & \dot{y} & \dot{z}\end{bmatrix}^T$ represents the three-dimensional linear velocity of the moving platform.

$$\mathbf{J} = \begin{bmatrix} \cos(\alpha) & 0 & \sin(\alpha) \\ -\frac{1}{2}\cos(\alpha) & \frac{\sqrt{3}}{2}\cos(\alpha) & \sin(\alpha) \\ -\frac{1}{2}\cos(\alpha) & -\frac{\sqrt{3}}{2}\cos(\alpha) & \sin(\alpha) \end{bmatrix}$$

is the Jacobian matrix of the mechanism.

When $\mathbf{J}$ is invertible, Eq. (14) can be written as

$$\dot{\boldsymbol{X}} = \mathbf{J}^{-1}\dot{\boldsymbol{q}}. \tag{15}$$

Equation (15) represents the forward velocity solution for a 3-CRU TPM.

5.2 Performance analysis

With respect to performance evaluation and optimization, the most used parameter is the Jacobian matrix, which is the matrix map of the velocity of the end effector onto the vector of actuated joint velocities. The conditional number of the Jacobian matrix, called the local conditioning index (LCI) (Li et al., 2005), was applied for performance evaluation of parallel manipulators. The conditioning index can be defined as the ratio of the smallest $\lambda_{\min}$ to the largest $\lambda_{\max}$ singular values of $\mathbf{J}$, i.e.,

$$\kappa = \frac{\lambda_{\min}}{\lambda_{\max}}. \tag{16}$$

For the proposed 3-CRU TPM, according to Eq. (14), the conditioning index of the mechanism is only related to the actuator layout angle α. Naturally, how the output characteristics of a 3-CRU TPM vary with differences in actuator layout angle is studied. The mechanical parameters of the

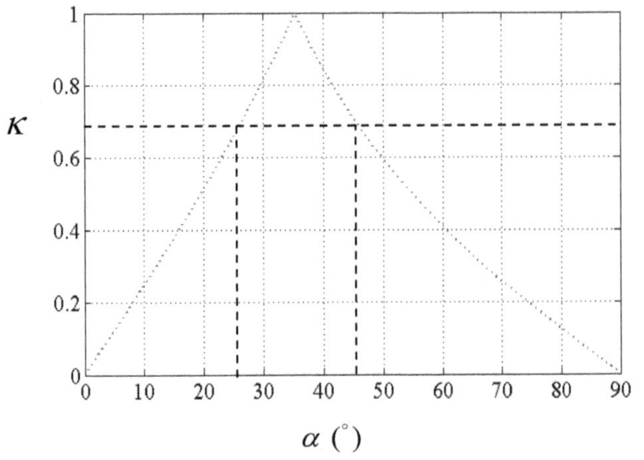

Figure 10. Relationship between the performance and the actuator layout angle.

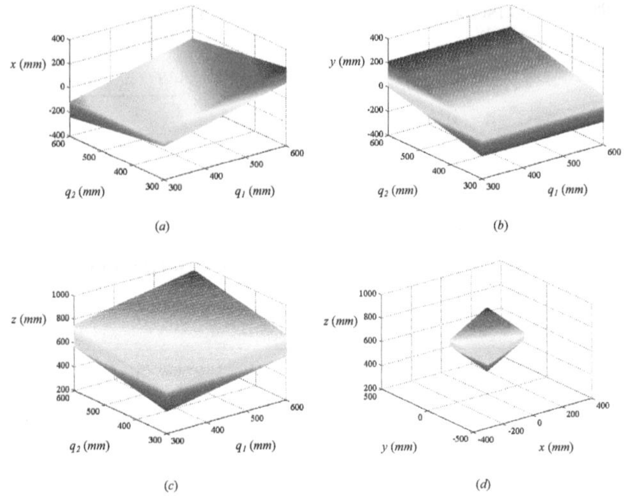

Figure 11. The workspace of the manipulator.

mechanism are set up in the same way as those in Sect. 4.3. Then, the relationship between the performance of the mechanism and the actuator layout angle α is obtained as shown in Fig. 10, and the actuator layout angle should be given in the range 25–45° to ensure good kinematic performance of the manipulator.

6 Workspace analysis

In this section, the reachable workspace of the 3-CRU TPM is obtained based on the forward position analysis. Given a set of limb lengths (q_1 q_2 q_3), the position of the moving platform can be calculated directly by corresponding equations as shown in Section 4. Thus when the restrictions to the limb lengths are set up, the reachable workspace of the mechanism can be obtained. It should be mentioned that few TPMs can obtain the workspace through forward position analysis; this is also one novel contribution in this paper.

Compared with serial ones, parallel manipulators have a relatively small workspace. The characteristics change with the variation in the actuator layout angle in the reachable workspace of a 3-CRU TPM..

The restrictions to the limb lengths are defined as $300\,\text{mm} \leq q_1 \leq 600\,\text{mm}$, $300\,\text{mm} \leq q_2 \leq 600\,\text{mm}$, and $300\,\text{mm} \leq q_3 \leq 600\,\text{mm}$. The workspace of the manipulator can be generated by a MATLAB program, the results of which are shown in Fig. 11.

In order to investigate the reachable workspace volume of a 3-CRU TPM with the changing of the actuator layout angle, the workspace volume is illustrated in Fig. 12, from which it can be observed that the maximum workspace volume occurs when the actuator layout angle α is around 35°. The x–y section of the workspace for $\alpha = 35°$ is shown in Fig. 13.

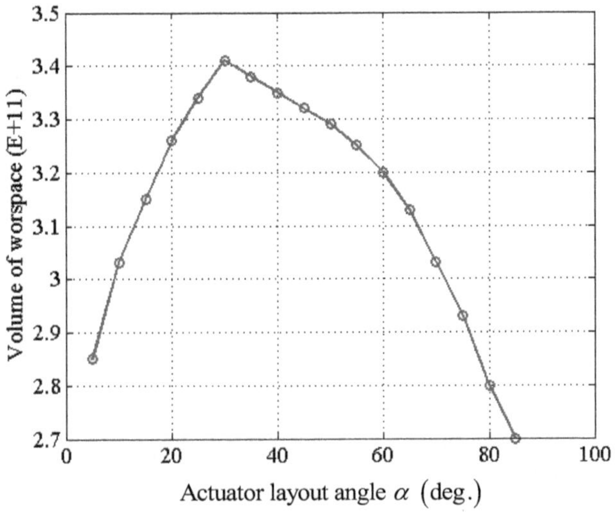

Figure 12. Reachable workspace volume versus actuator layout angle.

7 Conclusions

This paper proposes a modified 3-DOF TPM 3-CRU. The mobility of the mechanism is analyzed based on screw theory. Each CRU limb exerts one constraint couple on the platform. Both inverse and forward position analyses are performed, and the analytical solutions are obtained with respect to these two problems. Unlike most parallel robots, the proposed TPM has explicit solutions for inverse and forward kinematics issues. Therefore, both the path planning and control problems of the mechanism are very simple. Additionally, the Jacobian matrix of the mechanism is obtained, the performance is evaluated through a conditioning index, and the performance of a 3-CRU TPM along with the various actuator layout angles is investigated. Furthermore, the

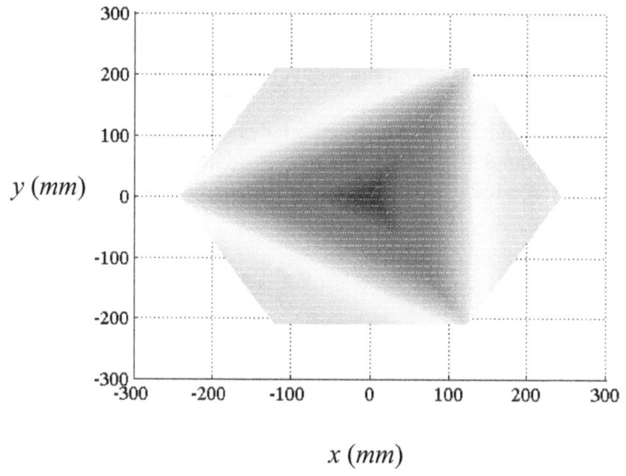

Figure 13. x–y section of the workspace at $\alpha = 35°$.

reachable workspace is obtained based on the forward position analysis, and the reachable workspace volume is obtained when the actuator layout angle is varied. On the basis of the kinematics analysis of the mechanism, analyses of inverse/forward dynamics and stiffness performance as well as kinematic optimization of the mechanism will be investigated in future work.

Acknowledgements. This work was supported by the National Natural Science Foundation of China (51205289, 51275353), the Natural Science Foundation of Tianjin (14JCZDJC39100), the Macao Science and Technology Development Fund (108/2012/A3, 110/2013/A3), and the Research Committee of University of Macau (MYRG183(Y1-L3)FST11-LYM, MYRG203(Y1-L4)-FST11-LYM).

References

Callegari, M., Cammarata, A., Gabrielli, A., and Sinatra, R.: Kinematics and dynamics of a 3-CRU spherical parallel robot, in: Proceedings of the ASME International Design Engineering Technical Conferences & Computers and Information in Engineering Conference, Las Vegas, Nevada, USA, 91–101, 2007.

Callegari, M., Cammarata, A., and Ruggiu, M.: Kinetostatic synthesis of a 3-CRU spherical wrist for miniaturized assembly tasks, Proc. 18th AIMeTA Congress of Theoretical and Applied Mechanics, Brescia, Italy, 10–26, 2007.

Carbonari, L., Battistelli, M., Callegari, M., and Palpacelli, M.-C.: Dynamic modelling of a 3-CPU parallel robot via screw theory, Mech. Sci., 4, 185–197, doi:10.5194/ms-4-185-2013, 2013.

Chablat, D. and Wenger, P.: Architecture optimization of a 3-DOF translational parallel mechanism for machining applications, the Orthoglide, IEEE Trans. Robot. Auto., 255, 1–8, 2001.

Chung, C. L. and Hervé, J. M.: Translational parallel manipulators with doubly planar limbs, Mech. Mach. Theory, 41, 433–455, 2006.

Clavel, R.: Delta, A fast robot with parallel geometry, in: Proceedings of 18th International Symposium on Industrial Robots, Lausanne, France, 91–100, 1988.

Dai, J. S., Huang, Z., and Lipkin, H.: Mobility of overconstrained parallel mechanisms, ASME J. Mech. Des., 128, 220–222, 2006.

Di Gregorio, R.: Kinematics of the translational 3-URC mechanism, ASME J. Mech. Des., 126, 1113–1117, 2004.

Di Gregorio, R. and Parenti-Castelli, V.: Mobility analysis of the 3-UPU parallel mechanism assembled for a pure translational motion, ASME J. Mech. Des., 124, 259–264, 2002.

Fang, Y. F. and Tsai, L. W.: Analytical identification of limb structures for translational parallel manipulators, J. Robot. Syst., 21, 209–218, 2004.

Ji, P. and Wu, H. T.: Kinematics analysis of an offset 3-UPU translational parallel robotic manipulator, Robot. Auton. Syst., 42, 117–123, 2003.

Kim, D. and Chung, W. K.: Kinematic condition analysis of three-DOF pure translational parallel manipulators, ASME J. Mech. Des., 125, 323–331, 2003.

Li, W., Gao, F., and Zhang, J.: R–CUBE, A decoupled parallel manipulator only with revolute joints, Mechan. Mach. Theory, 41, 433–455, 2006.

Li, Y. M. and Xu, Q. S.: Kinematic analysis and design of a new 3-DOF translational parallel manipulator, ASME J. Mech. Des., 128, 729–737, 2006.

Li, Y. M. and Xu, Q. S.: Design and Analysis of a New 3-DOF Compliant Parallel Positioning Platform for Nanomanipulation, in: Proceedings of 5th IEEE Conference on Nanotechnology, 126–129, 2005.

Li, Y. M. and Xu, Q. S., Kinematics and dexterity analysis for a novel 3-DOF translational parallel manipulator, in: Proceedings of IEEE International Conference on Robotics and Automation, 2955–2960, 2005.

Liu, X. J., Jeong, J., and Kim, J.: A three translational DOFs parallel cube-manipulator, Robotica, 21, 645–653, 2003.

Lou, Y. J. and Li, Z. X.: A novel 3-DOF purely translational parallel mechanism, in: Proceedings of the IEEE/RSJ, International Conference on Intelligent Robots and Systems, Beijing, China, 2144–2149, 2006.

Martins, D. and Carboni, A. P., Variety and connectivity in kinematic chains, Mech. Mach. Theory, 43, 1236–1252, 2008.

Merlet, J.-P.: Jacobian, manipulability, condition number, and accuracy of parallel robots, ASME J. Mech. Des., 128, 199–206, 2006.

Ruiz-Torres, M. F., Castillo-Castaneda, E., and Briones-Leon, J. A.: Design and analysis of CICABOT: a novel translational parallel manipulator based on two 5-bar mechanisms, Robotica, 30, 449–456, 2012.

Simoni, R., Doria, C. M., and Martins, D.:, Symmetry and invariants of kinematic chains and parallel manipulators, Robotica, 31, 61–70, 2013.

Tsai, L. W. and Joshi, S.: Kinematics and optimization of a spatial 3-UPU parallel manipulator, ASME J. Mech. Des., 122, 439–446, 2000.

Tsai, L. W. and Joshi, S.: Kinematics analysis of 3-DOF position mechanisms for use in hybrid kinematic machines, ASME J. Mech. Des., 124, 245–253, 2002.

Wang, M. F., Ceccarelli, M., and Carbone, G.: Experimental tests on operation performance of a LARM leg mechanism with 3-DOF parallel architecture, Mech. Sci., 6, 1–8, doi:10.5194/ms-6-1-2015, 2015.

Xiao, S. L., Li, Y. M., and Meng, Q. L.: Mobility analysis of a 3-PUU flexure-based manipulator based on screw theory and compliance matrix method, Int. J. Precision Eng. Manuf., 14, 1345–1353, 2013.

Xu, Q. S. and Li, Y. M.: A novel design of a 3-PRC translational compliant parallel micromanipulator for nanomanipulation, Robotica, 24, 527–528, 2006.

11

Experimental comparison of five friction models on the same test-bed of the micro stick-slip motion system

Y. F. Liu[1], J. Li[1], Z. M. Zhang[2], X. H. Hu[2], and W. J. Zhang[1,2]

[1]Complex and Intelligent System Laboratory, School of Mechanical and Power Engineering, East China University of Science and Technology, Shanghai, China

[2]Department of Mechanical Engineering, University of Saskatchewan, Saskatoon, Canada

Correspondence to: W. J. Zhang (chris.zhang@usask.ca)

Abstract. The micro stick-slip motion systems, such as piezoelectric stick-slip actuators (PE-SSAs), can provide high resolution motions yet with a long motion range. In these systems, friction force plays an active role. Although numerous friction models have been developed for the control of micro motion systems, behaviors of these models in micro stick-slip motion systems are not well understood. This study (1) gives a survey of the basic friction models and (2) tests and compares 5 friction models in the literature, including Coulomb friction model, Stribeck friction model, Dahl model, LuGre model, and the elastoplastic friction model on the same test-bed (i.e. the PE-SSA system). The experiments and simulations were done and the reasons for the difference in the performance of these models were investigated. The study concluded that for the micro stick-slip motion system, (1) Stribeck model, Dahl model and LuGre model all work, but LuGre model has the best accuracy and (2) Coulomb friction model and the elastoplastic model does not work. The study provides contributions to motion control systems with friction, especially for micro stick-slip or step motion systems as well as general micro-motion systems.

1 Introduction

Micro stick-slip motion systems can provide high resolution motions yet with a long motion range. The piezoelectric stick-slip actuator (PE-SSA), which is a hybridization of the piezoelectric actuator (PEA) and the stick slip actuator (SSA), is a typical example in these systems. By hybridization, it is meant that PE and SS are complementary to each other, according to the hybridization design principle proposed by Zhang et al. (2010).

The working process of the PE-SSA is demonstrated in Fig. 1. At position (1), a voltage is applied to the PEA and leads to an (relatively slow) expansion of the PEA, which pushes the stage moving to the right. The friction between the stage and the end effectorbrings the end effector to the position (2) (stick motion). When the applied voltage is shut down quickly, the PEA contracts quickly and the slip between the end effector and stage takes place due to the inertia of the end effector, which overcomes the friction resistance. The relative displacement S, with respect to its initial position (1), is thus generated at position (3) (slip motion). If the aforementioned process is repeated periodically, the end effector will keep moving to the right as long as the physical system allows. The back forward motion of the end effector can also be obtained by reversing the actuation potential signal applied on the PEA.

It can be found from the aforementioned working process of the PE-SSA that the friction force between the end effector and stage plays both an active role in the forward stroke motion and a passive role in the backward stroke in the stick-slip motion system. The modeling of the friction for prediction and control of such a system is extremely important as well as difficult (Makkar et al., 2007; Li et al., 2008a). The difficulty lies in the possibility that the motion direction on the two contact surfaces may change the frictional effect (Zhang et al., 2011). Although numerous friction models have been studied in the context of macro, micro motion, such as Coulomb friction model, viscous friction model, Stribeck friction model, Dahl model, LuGre model,

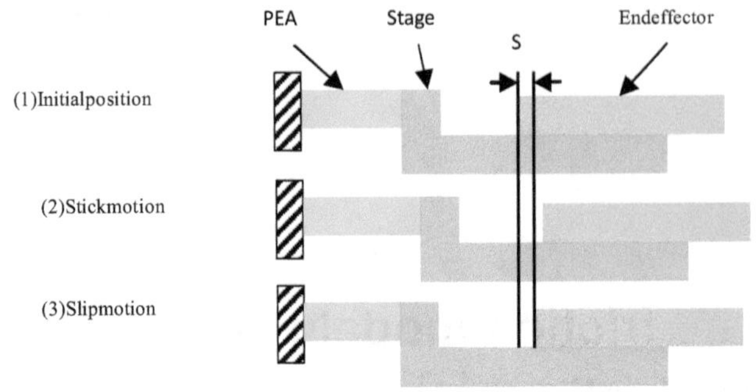

Figure 1. The working process of the PE-SSA.

Table 1. Friction characteristics captured by different models.

	Viscous	Stribeckeffect	Pre-sliding	Hysteresis
Coulomb	No	No	No	No
Viscous	Yes	No	No	No
Stribeck	Yes	Yes	No	No
Dahl	No	No	Yes	Yes
LuGre	Yes	Yes	Yes	Yes
Elastoplastic	Yes	Yes	Yes	–
Leuven	Yes	Yes	Yes	Yes
GMS	Yes	Yes	Yes	Yes

and elastoplastic friction model, the performances of these models in micro stick-slip motion systems are not well understood. This paper aims to investigate the performance of different friction models inmicrostick-slip motion systems. A survey of basic friction models is presented first and five of the mare selected for comparison. A common micro stick-slip motion system- the PE-SSA system which was introduced in our previous study (Li et al., 2008a) is used as a test-bed. Parameters in each friction model will be determined using the system identification technique. Performances of these models are then compared. Leuven and GMS models are out of the scope of the study because they are not commonly used in the step motion system such as the stick-slip motion system.

The remainder of the paper is organized as follows. Section 2 gives an overview and history of the five friction models. The mathematical descriptions of these friction models are presented in Sect. 3. The experimental setup is described in Sect. 4. Section 5 presents the analysis methodology. Section 6 presents the experimental results along with discussions, followed by conclusions in Sect. 7.

2 Overview of the friction models

The first friction model is Coulomb friction model (or called Amontons–Coulomb friction model), referring to the work done by Guillaume Amontons and Charles-Augustin de Coulomb in 1699 and 1785, respectively. In the Coulomb friction model, the friction force is the function of load and direction of the velocity. Morin (1833) found that the static friction (i.e. friction at zero sliding speed) is larger than the Coulomb friction. With respect to the static friction, the Coulomb friction is also called dynamic friction. Viscous friction was later introduced in relation to lubricants by Reynolds and it is often combined with Coulomb friction model. Stribeck (1902) experimentally observed that friction force decreases with the increase of the sliding speed from the static friction to Coulomb friction. The phenomenon is thus called Stribeck effect. The integration of the Coulomb friction, viscous friction, and Stribeck effect is often an idea to obtain a more accurate friction model, which is called Stribeck model in literature.

Dahl (1968) first modeled friction as a function of the relative displacement of two contact surfaces, and it is thus called Dahl model. The model is based on the fact that friction force is dependent on the "micro motion" in ball bearings. The "micro motion", later called pre-sliding behavior, is that when the external force is not large enough to overcome the static friction, the asperities on two contact surfaces will experience deformation that results in the pre-sliding motion. The asperities form a kind of spring-damping system. When the external force is sufficiently large, the spring is broken, leading to a relative sliding between two contact surfaces. The Dahlmodel successfully describes the so-called break-away phenomenon.

Canudas et al. (1995) developed a friction model called LuGre model, named after the two universities, namely Lund and Grenoble. LuGre model incorporates the viscous friction and Stribeck effect into Dahl model. The problem of incorporating the pre-sliding behavior in the friction model is that both Dahl and LuGre models experience "drift" when there is an arbitrarily small bias force or vibration. The reason for this drift is that both Dahl and LuGre models only include a "plastic" component in their model when they describe the pre-sliding phenomenon.

To overcome this drift, Dupont et al. (2000) proposed a friction model based on LuGre model, in which the pre-sliding was defined as the elastoplastic deformation of asperities; i.e. the relative displacement is elastic (reversible) first and then it transits to the plastic (irreversible) stage. The model of Dupont et al. (2000) is thus called elastoplastic friction model.

Leuven friction model and generalized Maxwell-slip (GMS) model were proposed by Swevers et al. (2000) and Lampaert et al. (2002), respectively. The two models were developed based on the experimental findings that the friction force in the pre-sliding regime has a hysteresis characteristic with respect to the position. It is reported that the two models can improve the hysteresis behavior of the friction predicted with LuGre model.

The friction characteristics that are captured by the aforementioned models are listed in Table 1. To control a dynamic system for high accuracy, a common sense seems to go with the friction model that can capture more friction characteristics. However, such may not always be the case. Generally speaking, for a complex dynamic system such as frictional systems, a (complete) model may be viewed as an integration of a couple of sub-models, each of which captures one or more characteristics. While being integrated, each of them may produce "side effects" to the modeling of other characteristics (Li et al., 2008b, 2009), because these characteristics are coupled, changing with time and perhaps, the physical structure of the frictional system changes as well.

3 The mathematical description of friction models

3.1 Coulomb model

Coulomb friction model is represented using the following equation:

$$F = \begin{cases} F_c \cdot Sgn(\dot{x}) & \text{if } \dot{x} \neq 0 \\ F_{app} & \text{if } \dot{x} = 0 \text{ and } F_{app} < F_c \end{cases} \quad (1)$$

where F is friction force, $\dot{x}$ is sliding speed, F_{app} is applied force, F_C is the Coulomb friction force, which is defined as

$$F_c = \mu F_N \quad (2)$$

where μ is the Coulomb friction coefficient (or called the dynamic friction coefficient), and F_N is a normal load between two contact surfaces.

The Coulomb model is illustrated in Fig. 2a. When $F_{app} < F_c$, there is no sliding (i.e. $\dot{x} = 0$ in the "macro" sense) between two contact surfaces, and the Coulomb friction can take any value from zero up to F_c. If $\dot{x} \neq 0$, Coulomb friction only takes F_c or $-F_c$, depending on the direction of the sliding.

Coulomb friction model is commonly used in the applications such as the prediction of temperature distribution in bearing design and calculation of cutting force in machine tools due to its simplicity. However, it is often troublesome to use the Coulomb friction model in micro motion systems because of the "undefined" friction force at $\dot{x} = 0$.

3.2 Viscous friction model

Viscous friction model is given by

$$F = k_v \dot{x} \quad (3)$$

where F is the friction force, k_v the viscous coefficient, and $\dot{x}$ the sliding speed.

The viscous friction model is illustrated in Fig. 2b. In the viscous model, the friction force is a linear equation of the sliding speed. The application of the viscous model is limited because it has a poor representation in regions where there is no lubricant (Andersson et al., 2007).

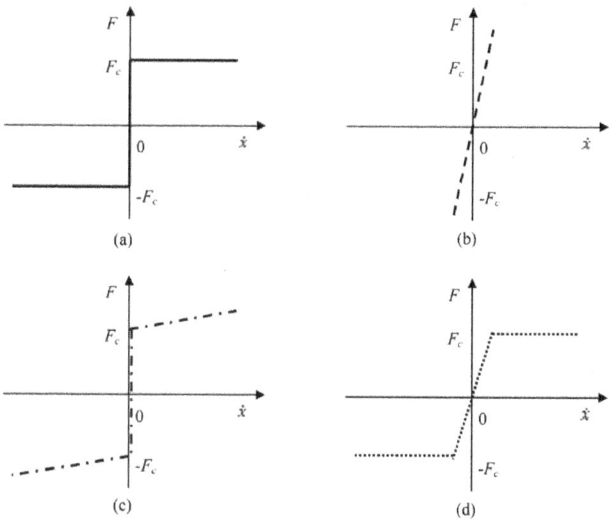

Figure 2. Friction force vs. sliding speed.

3.3 Integrated Coulomb and viscous model

There are two ways to combine viscous friction model and Coulomb model, leading to two different integrated Coulomb and viscous models. One is described by

$$F = \begin{cases} F_c \cdot Sgn(\dot{x}) + k_v\dot{x} & \text{if } \dot{x} \neq 0 \\ F_{app} & \text{if } \dot{x} = 0 \text{ and } F_{app} < F_c \end{cases}. \quad (4)$$

This model (Eq. 4) is illustrated in Fig. 2c. The problem with this model is that the friction force at is still "undefined". To overcome this problem, the idea is to integrate the Coulomb model and the viscous model near $\dot{x} = 0$. This comes with the second model given by (Andersson et al., 2007)

$$F = \begin{cases} \min\ (F_c, k_v\dot{x}) & \text{if } \dot{x} \geq 0 \\ \max\ (-F_c, k_v\dot{x}) & \text{if } \dot{x} < 0 \end{cases}. \quad (5)$$

This model (Eq. 5) is illustrated in Fig. 2d. The viscous coefficient determines the speed of the friction force transition from $-$ to $+$.

3.4 Stribeck friction model

Stribeck friction model is described by

$$F = \left(F_c + (F_s - F_c)e^{-\left(\left|\frac{\dot{x}}{v_s}\right|\right)^i}\right) Sgn(\dot{x}) + k_v\dot{x} \quad (6)$$

where F is friction force, $\dot{x}$ Sliding speed, F_c the Coulomb friction force, F_s the static friction force, v_s the Stribeck velocity, k_v the viscous friction coefficient, and i an exponent. It is clear from Eq. (6) that the Stribeck friction force takes F_s as the upper limit and F_c as the lower limit. The relation of friction versus sliding speed in Stribeck model is illustrated in Fig. 3.

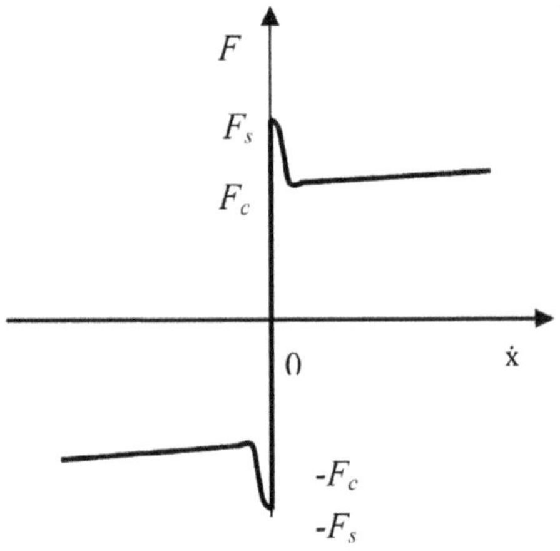

Figure 3. Stribeck friction model (Armstrong-Helouvry, 1993).

3.5 Dahl model

In Stribeck model, friction force is a function of the sliding speed. However, according to this model, friction force is "undefined" when it is less than F_s. Dahl developed a model to describe friction at this pre-sliding stage, which is given by

$$\frac{dF(x)}{dt} = \frac{dF(x)}{dx}\frac{dx}{dt} \tag{7}$$

with

$$\frac{dF}{dx} = \sigma_0 \left| - \frac{F}{F_c} sgn(\dot{x}) \right|^i sgn\left(1 - \frac{F}{F_c} sgn(\dot{x})\right) \tag{8}$$

where F is the friction force, σ_0 the stiffness coefficient, and i the exponent which determines the shape of the hysteresis. In literature, Dahl model is often simplified with the exponent $i = 1$ and given by

$$\frac{dF}{dx} = \sigma_0 \left(1 - \frac{F}{F_c} sgn(\dot{x})\right). \tag{9}$$

Dahl model does not describe the Stribeck effect (Canudas et al., 1995).

3.6 LuGre model

LuGre model has the following form (Canudas et al., 1995)

$$\begin{cases} F = \sigma_0 z + \sigma_1 \dot{z} + \sigma_2 \dot{x} \\ \dot{z} = \dot{x} - \sigma_0 \frac{\dot{x}}{g(\dot{x})} z \\ g(\dot{x}) = F_c + (F_s - F_c) e^{-\left(\left|\frac{\dot{x}}{v_s}\right|\right)^j} \end{cases} \tag{10}$$

where F is the friction force, σ_0 the contact stiffness, z the average deflection of the contacting asperities, σ_1 the damping coefficient of the bristle, σ_2 the viscous friction coefficient, x the relative displacement, F_c the Coulomb friction force, F_s the static friction force, $\dot{x}$ the sliding velocity, $g(\dot{x})$ the Stribeck effect, v_s the Stribeck velocity, j the Stribeck shape factor ($j = 2$ is often used in the literature). LuGre model integrates pre-sliding friction ($\sigma_0 z$), viscous friction ($\sigma_2 \dot{x}$), and Stribeck effect ($g(\dot{x})$) into one single model.

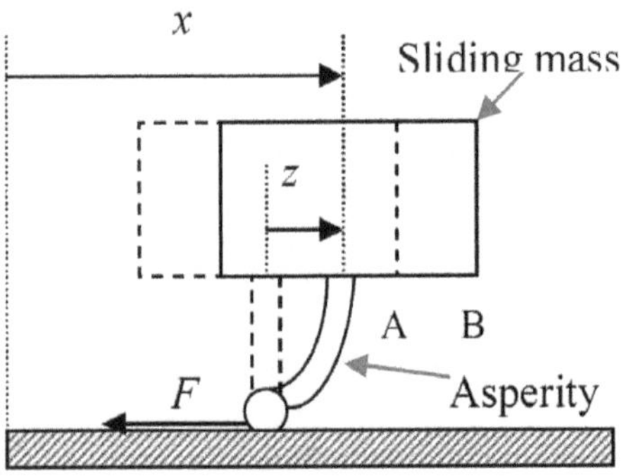

Figure 4. Illustration of deformation of asperity on the frictional surface.

3.7 Elastoplastic friction model

The elastoplastic friction model is described by

$$\begin{cases} F = \sigma_0 z + \sigma_1 \dot{z} + \sigma_2 \dot{x} \\ \dot{z} = \dot{x}\left(1 - \sigma(z, \dot{x}) \frac{z}{z_{ss}(\dot{x})}\right) \end{cases} \tag{11}$$

with

$$z_{ss}(\dot{x}) = \frac{g(\dot{x})}{\sigma_0} \tag{12}$$

$$g(\dot{x}) = F_c + (F_s - F_c) e^{-\left(\left|\frac{\dot{x}}{v_s}\right|\right)^j}. \tag{13}$$

In Eq. (11), $\sigma(z, \dot{x})$ is used to define the zones of the elastic and plastic deformation of asperities and given by

$$\begin{cases} \alpha(z, \dot{x}) = \begin{cases} 0, & |z| \le z_{ba} \\ \alpha_m(*) & z_{ba} < |z| < z_{ss} \\ 1, & |z| \ge z_{ss} \end{cases} & Sgn(\dot{x}) \ne Sgn(z) \\ \alpha(z, x) = 0, & Sgn(\dot{x}) = Sgn(z) \end{cases} \tag{14}$$

where $Sgn(\dot{x}) = Sgn(z)$ represents that the sliding mass moves from position A to position B, as shown in Fig. 4, and $Sgn(\dot{x}) \ne Sgn(z)$ represents that the sliding mass moves from position B to A.

According to Equation (14), when $Sgn(\dot{x}) \ne Sgn(z)$, $\alpha(z, \dot{x}) = 0$. This represents that no slip occurs; the "sliding" mass is in an elastic deformation region, or two contact objects are in a stick phase. When $Sgn(\dot{x}) \ne Sgn(z)$, if $|z| \le z_{ba}$, $\alpha(z, \dot{x}) = 0$. This represents that no slip occurs; the "sliding" mass is in an elastic deformation region, or two contact objects are in a stick phase. When $Sgn(\dot{x}) \ne Sgn(z)$, if $z_{ba} < |z| < z_{ss}$, $\alpha(z, \dot{x}) = \alpha_m(*)$. This represents that elastic deformation of the asperities starts to transit to the plastic deformation, i.e. transition phase. When $Sgn(\dot{x}) \ne Sgn(z)$, if $|z| \ge z_{ss}$, $\alpha(z, \dot{x}) = 1$. This represents that slip occurs; the

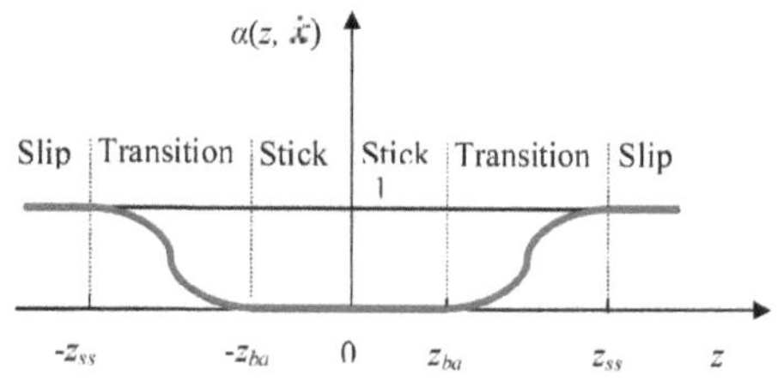

Figure 5. Dependency of $\alpha(z, \dot{x})$ on deformation of the asperity.

"sliding" mass is in a plastic deformation, or two contact objects are in a slip phase. z_{ba} and z_{ss} are given by

$$0 < z_{\mathrm{ba}} \leq z_{\mathrm{ss}} \text{ and } z_{\mathrm{ss}} = \mathrm{Max}\left(z_{\mathrm{ss}}(\dot{x})\right) = \frac{F_{\mathrm{s}}}{\sigma_0}, \ \dot{x} \in R. \tag{15}$$

$\alpha_m(*)$ is typically in the following form:

$$\alpha_m(*) = \frac{1}{2}\sin\left(\pi \frac{z - \left(\frac{z_{\mathrm{ss}}+z_{\mathrm{ba}}}{2}\right)}{z_{\mathrm{ss}} - z_{\mathrm{ba}}}\right) + \frac{1}{2}. \tag{16}$$

The case when $Sgn(\dot{x}) = Sgn(z)$ in Eq. (14) are illustrated in Fig. 5.

From Fig. 5 it can be seen that in the elastoplastic friction model, $|z| \leq z_{\mathrm{ba}}$ is defined as the stiction zone. The stiction zone is to overcome the drift problem of Dahl and LuGre models. The transitions between stick and slip is given by $\alpha_m(*)$, and the slip zone is represented by $|z| \geq z_{\mathrm{ss}}$ when it returns to LuGre model.

4 Experimental setup

The PE-SSA prototype is shown in Fig. 6. This system is composed of 1: frames, 2: friction plates, 3: temperature sensor, 4: weights, 5: end effector, 6: displacement sensor, 7: stage, 8: wheel, 9: PEA, and 10: vibration-isolated test bed. The PEA (Model: AE0505D16) purchased from NEC/TOKIN Corp. Is connected to the frame at one end, and its other end is connected to the stage. Friction plates were placed between the end effector and stage, and they were connected with the stage by screws. The weights were used to adjust the pressure between the end effector and stage. The wheel was used to support the stage. The temperature sensors are installed inside of the stage to measure temperature change in the system, which is not used in this work. The control system for the PE-SSA was designed as an open-loop system and implemented with dSPACE and Matlab/Simulink. The system was placed on the vibration isolated test bed in order to reduce disturbance to the system. More details about this prototype can be found in our previous work (Li et al., 2008a). During the experiments, the applied voltage to the PEA was a repeating saw tooth wave with amplitude of 30 V and frequency of 5 Hz. The displacements of the end effector were then measured with a KAMAN instrument (SMU 9000, Kaman) based on the eddy-current inductive principle and with a resolution of 0.01 μm.

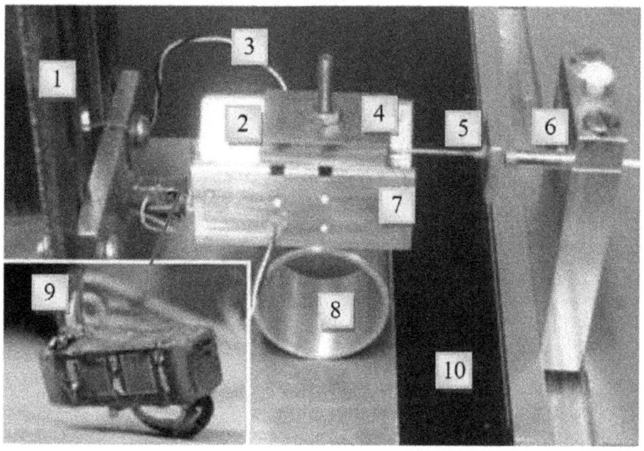

Figure 6. Experimental system of the PE-SSA (Li et al., 2008a).

5 The method for analysis

The method used to compare the performance of different friction models consists of the following steps.

- Step 1: get experimental data, i.e. the displacement of the PE-SSA with respect to time under a certain driven voltage and frequency.
- Step 2: model the PEA and stage and determine the parameters in the model (see Sect. 5.1).
- Step 3: model the PE-SSA by integrating the friction of the stick-slip motion into the model of the PEA and stage (see Sect. 5.2).
- Step 4: identify friction model parameters (see Sect. 5.3), including (a) use different friction models to represent friction in the PE-SSA model and (b) determine the parameters for each friction model.
- Step 5: compare the performance of the friction models in terms of their prediction of displacement, friction force, and sliding speed (i.e. relative speed between end effector and stage). The details are discussed in Sect. 6.

5.1 Modeling of the PEA and stage

Adriaens et al. (2000) showed that the PEA and stage can be modeled as a spring-mass-damper system which is shown in Fig. 7.

The governing equations of this spring-mass-damper system are given in Eqs. (17)–(19) as follows:

$$m\ddot{x}_{\mathrm{p}} + c\dot{x}_{\mathrm{p}} + kx_{\mathrm{p}} = F_{\mathrm{p}} \tag{17}$$

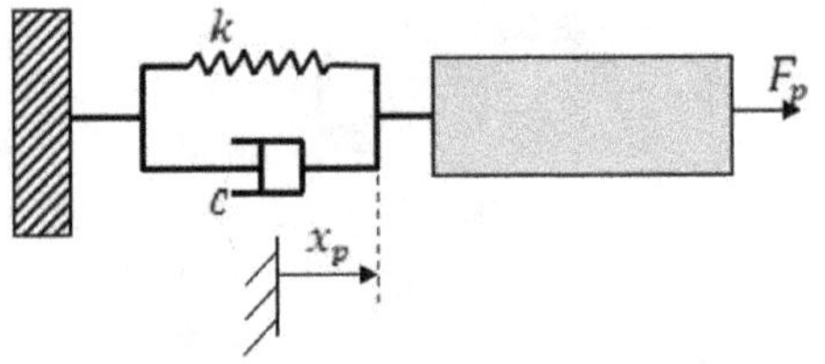

Figure 7. Physical model of the PE-SSA system (without friction) (Adriaens et al., 2000).

$$\begin{cases} m = \frac{4m_p}{\pi^2} + m_s \\ c = c_p + c_s \\ k = k_p + k_s \end{cases} \quad (18)$$

$$F_p = T_{em} u_p \quad (19)$$

where x_p is the displacement of the PEA, m_p the mass of PEA, c_p the damping coefficient of PEA, k_p the stiffness of PEA, m_s the mass of stage, c_s the damping coefficient of the stage, k_s the stiffness of stage, F_p the transducer force from the electrical side, T_{em} the electromechanical transducer ratio, and u_p the applied voltage on the PEA. In this study, Adriaens' model is taken due to its simplicity, which was validated by our previous study (Li et al., 2008b; Kang, 2007).

Equation(17) can be further written as

$$\ddot{x}_p + 2\xi\omega_n\dot{x}_p + \omega_n^2 x_p = K\omega_n^2 u_p \quad (20)$$

with

$$\begin{cases} \frac{c}{m} = 2\xi\omega_n \\ \frac{k}{m} = \omega_n^2 \\ \frac{T_{em}}{m} = K\omega_n^2 \end{cases} \quad (21)$$

where ξ is the damping ratio, ω_n the natural frequency, K and the amplified coefficient. The transfer function of the spring-mass-damper system can be written as,

$$G(s) = \frac{x_p(s)}{U(s)} = \frac{K\omega_n^2}{s^2 + 2\xi\omega_n s + \omega_n^2}. \quad (22)$$

In this study, the parameters ξ and are ω_n calculated from the following equations,

$$\begin{cases} \xi = \frac{\ln(\text{os} \times 100)}{\sqrt{\pi^2 + \ln^2(\text{os} \times 100)}} \\ \omega_n = \frac{\pi}{T_p\sqrt{1-\xi^2}} \end{cases} \quad (23)$$

where os is the overshoot of a step response and T_p is the peak time of a step response. The os and T_p are determined from the step response of the system. K is the ratio of output and input in the steady state of the step response. In our PEA system, we have $\xi = 0.2488$, $\omega_n = 6685.5\,\text{rad s}^{-1}$, and $K = 0.096 \times 10^{-6}\,\text{m v}^{-1}$, and they were obtained from the measured step responses of the system.

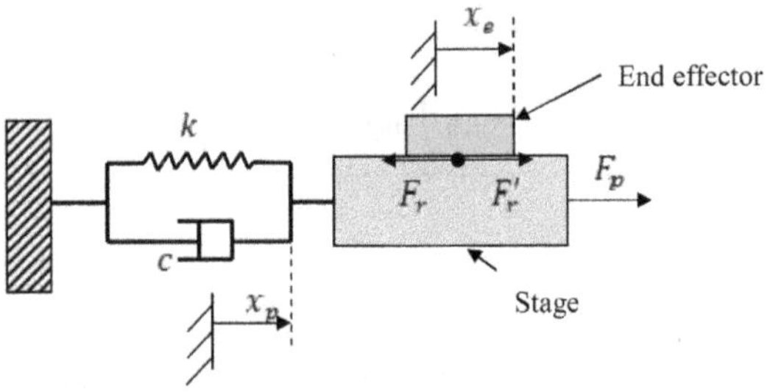

Figure 8. Physical model of the PE-SSA system (with friction) (Adriaens et al., 2000).

5.2 Modeling of the PE-SSA

In the PE-SSA, the friction force F_r between the end effector and stage is applied on the stage, as shown in Fig. 8.

Taking into account F_r, the following equation can be obtained from Eq. (17),

$$m\ddot{x}_p + c\dot{x}_p + kx_p + F_r = F_p. \quad (24)$$

On the end effector, friction force, denoted by that pushes the end effector move to the right, is given by

$$\begin{cases} F_r' = m_e\ddot{x}_e \\ F_r' = -F_r \\ x_e = x_p + x_{pe} \end{cases} \quad (25)$$

where x_e is the displacement of the end effector, and x_{pe} is the relative displacement between the end effector and stage. The stick-slip induced friction force F_r can be represented by any one of the aforementioned friction models.

5.3 Identification of parameters

The applied voltage to the PEA was a repeating sawtooth wave with an amplitude of 30 V and frequency of 5 Hz, and the displacements of the end effector were then measured by the displacement sensor. To identify parameters of in a particular friction model, the stick-slip induced friction force F_r in Eq. (25) is substituted by this friction model described in Sect. 3, and based on the experimental data the parameters in this friction model was identified using *Simulink* and *nonlinear fitting* (functions in Matlab software). The details for parameter identification technique were discussed in our previous study (Li et al., 2008b). The parameters determined for each friction model are listed in Table 2.

6 Results with discussion

The friction force and sliding speed predicted by Coulomb friction model are shown in Figs. 9 and 10, respectively. From Fig. 9 it can be seen that friction is either F_c or $-F_c$. From Figs. 9 and 10 it can be seen that the sign of the friction force is determined by the sign of the sliding speed. The

Table 2. Parameters determined for each friction model.

Parameter	Coulomb	Stribeck	Dahl	LuGre	Elastoplastic
μ	0.0587	N/A	N/A	N/A	N/A
k_v (N/mm)	N/A	40 101	N/A	N/A	N/A
F_c (N)	N/A	4.477	3.387	2.5	2.5
F_s (N)	N/A	5	N/A	3	3
v_s (m s^{-1})	N/A	1.6×10^{-8}	N/A	1.6×10^{-8}	1.6×10^{-8}
α_0 (N/mm)	N/A	N/A	1674	1670	1670
α_1 (Ns mm^{-1})	N/A	N/A	N/A	6	6
α_2 (Ns mm^{-1})	N/A	N/A	N/A	26	26
Z_{ba} (mm)	N/A	N/A	N/A	N/A	1×10^{-7}

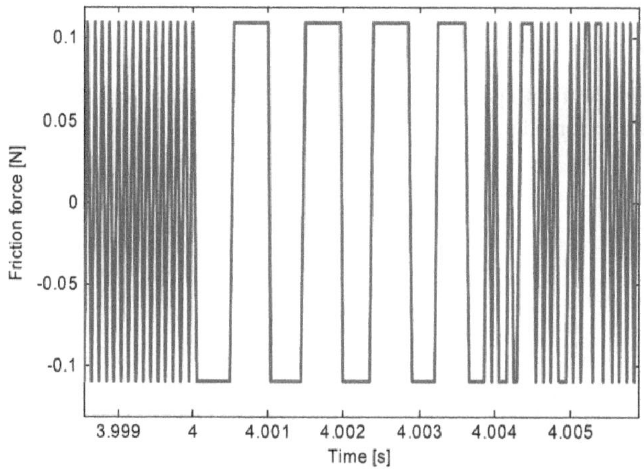

Figure 9. Friction predicted using Coulomb friction model (at around 4th second).

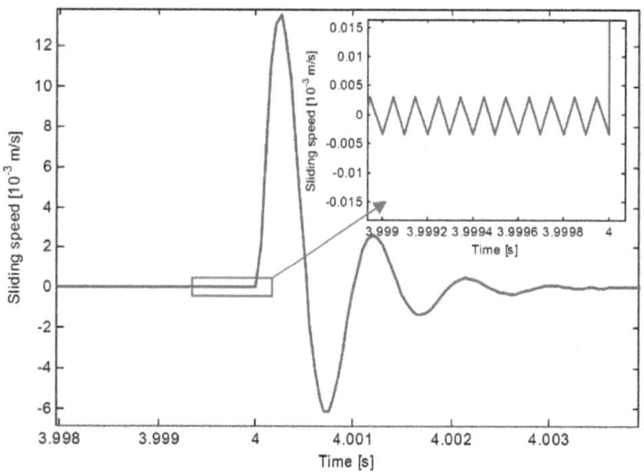

Figure 10. Sliding speed predicted using Coulomb friction model (at around the 4th second).

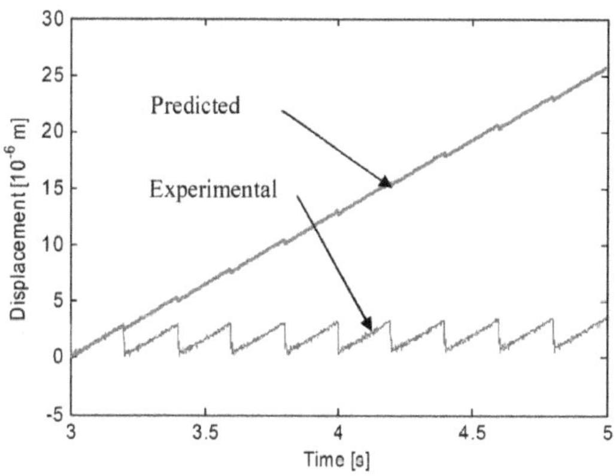

Figure 11. Displacement predicted by Coulomb model ($\mu = 0.0587$).

magnitude of F_c or $-F_c$ is dependent on the Coulomb friction coefficient.

It was found that the displacement predicted by the model fits either stick motion or slip motion but fails to fit both of them no matter what friction coefficient is chosen. Figure 11 shows a typical example of the displacement predicted by the model when $\mu = 0.0587$. It can be seen from this figure that the displacement in the stick motion period is well predicted, but the displacement predicted for the slip motion period cannot fit the experimental data and leads to huge errors after only a few cycles. The reason for this is analyzed as follows.

As previously discussed, the displacement is related to the friction force using the Newton's second law (see Eq. 25). The friction force in Coulomb friction model is a function of (1) the sign of sliding speed and (2) friction coefficient. The sliding speed (shown in Fig. 10) consists of two parts, i.e. sliding speed in stick motion and sliding speed in slip motion. Notice that the friction coefficient is the only effective variable in the Coulomb friction model. Once the friction coefficient required in stick motion is not the one required in slip motion, the Coulomb friction model will not work for the stick-slip motion system. This problem could be solved by considering an ideal situation, as reported in reported in Chang and Li (1999), where in the stick motion period, there is no sliding; correspondingly, in the stick motion period, the displacement of the end effector is determined by the displacement of the stage. In the slip motion period, the displacement of the end effector is determined by the Coulomb

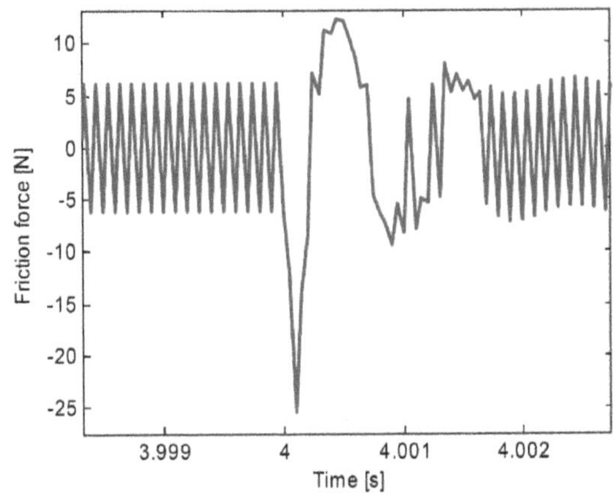

Figure 12. Friction force predicted using Stribeck model (at around the 4th second).

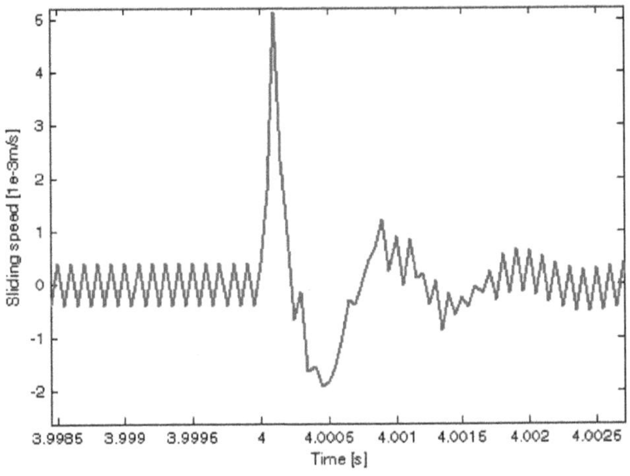

Figure 13. Sliding speed predicted using Stribeck model (at around the 4th second).

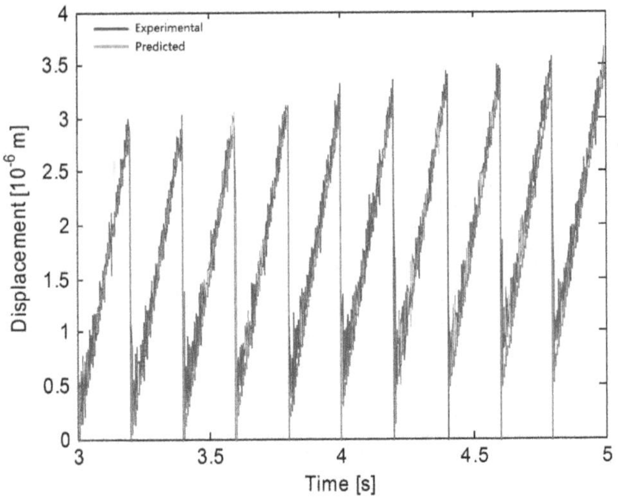

Figure 14. Experimental data and displacements predicted using Stribeck model.

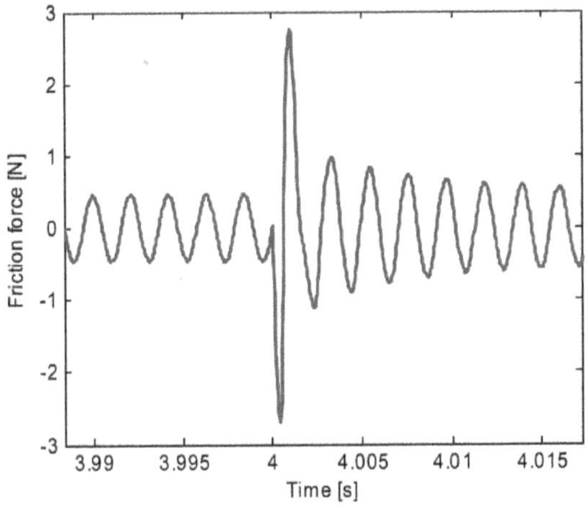

Figure 15. Friction predicted using Dahl model.

friction model. In other words, Coulomb friction model only models friction force in the slip motion period, instead of the entire stick-slip motion cycle. Another problem with the Coulomb friction model is the fluctuation in the sliding speed in stick motion shown in Fig. 10., which is supposed to be zero in theory. The fluctuation is caused by the use of *Sgn* function in the model (see Eq. 1).

6.1 Results of Stribeck friction model

The friction force and sliding speed predicted using the Stribeck friction model are shown in Figs. 12 and 13, respectively. The displacement predicted using the Stribeck friction model is shown in Fig. 14 from which it can be seen that Stribeck friction model can well predict friction both in stick and slip motion. Compared to the Coulomb friction model, Stribeck friction model can be used to model the friction force in the stick-slip motion. The reason for this might be that in the Stribeck model friction force is not only dependent on the sign of sliding, but also on the sliding speed (see Figs. 12 and 13). From Fig. 12 it can be seen that the friction force predicted by the model fluctuates around zero in the stick motion, which further causes the sliding speed fluctuation (see Fig. 13) and displacement fluctuation (see Fig. 14). Such fluctuation results from the *Sgn* function in the Stribeck model.

6.2 Results of Dahl model

The friction force and sliding speed predicted using the Dahl model are shown in Figs. 15 and 16, respectively. From the two figures it can be seen that the friction force is not only dependent on the sign of sliding but also on the sliding speed. The displacement predicted using the Dahl model is shown in Fig. 17 from which it can be seen that Dahl model can well predict friction both in stick and slip motion. To observe the details, a scaled up picture from Fig. 17 (at around

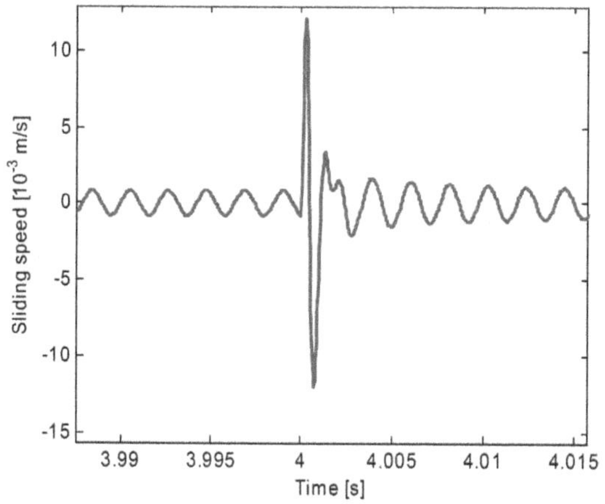

Figure 16. Sliding speed predicted using Dahl model.

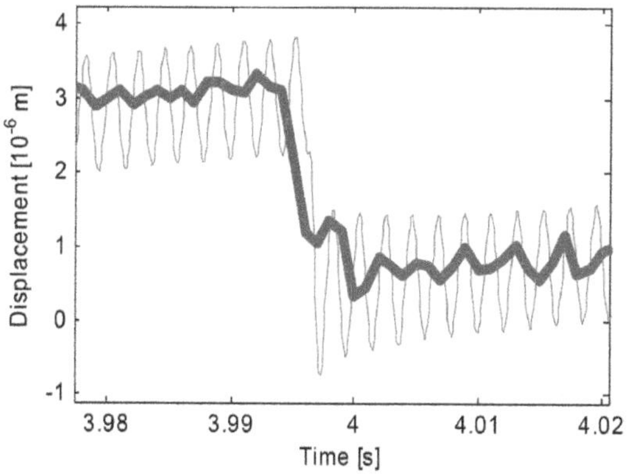

Figure 18. Scaled up picture from Fig. 16 (at around the 4th second).

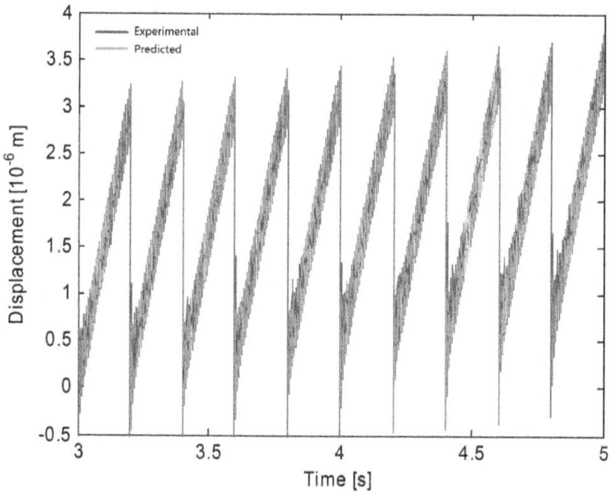

Figure 17. Experimental displacement and displacement predicted using Dahl model.

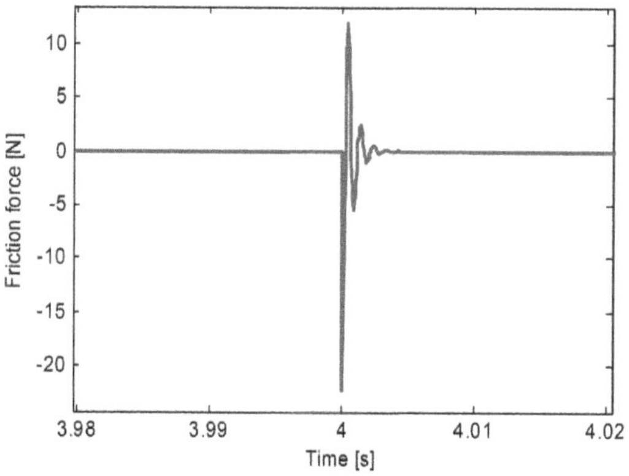

Figure 19. Friction predicted using LuGre model.

the 4th second) is shown in Fig. 18. From Fig. 18 it can be seen that the displacement predicted by the Dahl model is fluctuated around the experimental data. This is due to the sliding speed fluctuation, as shown in Fig. 16. The fluctuation (both in friction and sliding speed) results from the function of *Sgn* in the model. Compared to Stribeck model, the fluctuation (both in friction and sliding speed) in the Dahl model seems to be smoother than that in Stribeck model. The reason for this is that the friction force in the Dahl model is the integral of the *Sgn* function; see Eq. (9).

6.3 Results of LuGre model

The friction force and sliding speed predicted using the LuGre model are shown in Figs. 19 and 20, respectively. The displacement predicted using LuGre model is shown in Fig. 21 and its scaled up picture is shown in Fig. 22. From Figs. 19 and 20, it can be seen that the friction force is not only dependent on the sign of sliding but also on the sliding speed, which is constant with the mathematical model. From Figs. 21 and 22 it can be seen that LuGre friction model can well predict friction both in stick and slip motion. Compared to Coulomb model, Stribeck model, and Dahl model (see Figs. 9, 10, 12, 13, 15 and 16), the friction force and sliding speed predicted by the LuGre model for stick motion are zero (see Figs. 19 and 20) instead of fluctuation around zero. From Figs. 21 and 22 it can be seen that the displacement predicted by the model during slip motion does not have any fluctuation, compared to Stribeck and Dahl models which are shown in Figs. 14 and 18, respectively. As we discussed before, the unwanted fluctuations in prediction of the Coulomb model, Stribeck model, and Dahl model are due to the *Sgn* function in the mathematical model. While this problem seems to be completely solved by the LuGre friction model, in which there is no *Sgn* function.

6.4 Results of the elastoplastic friction model

The displacement predicted by the elastoplastic model cannot fit the slip motion period no matter what initial parameters are chosen in order to determine the parameters in the

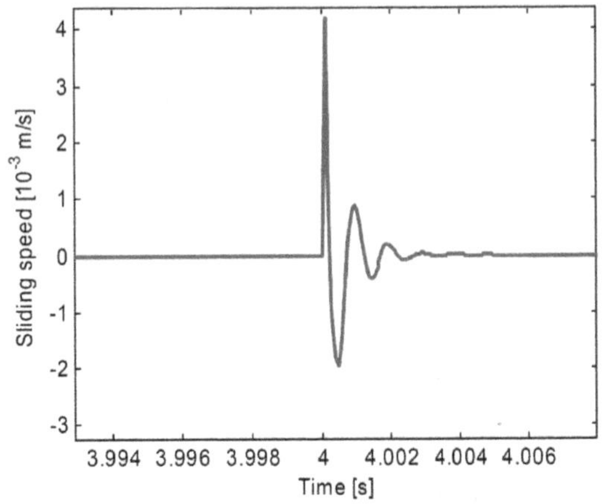

Figure 20. Sliding speed predicted using LuGre model.

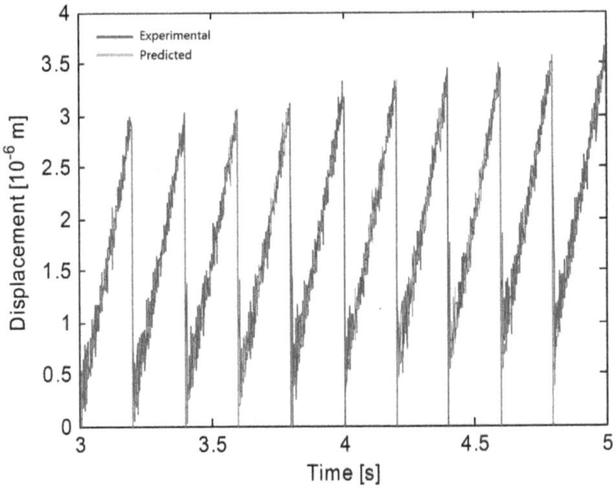

Figure 21. Displacement predicted using LuGre model.

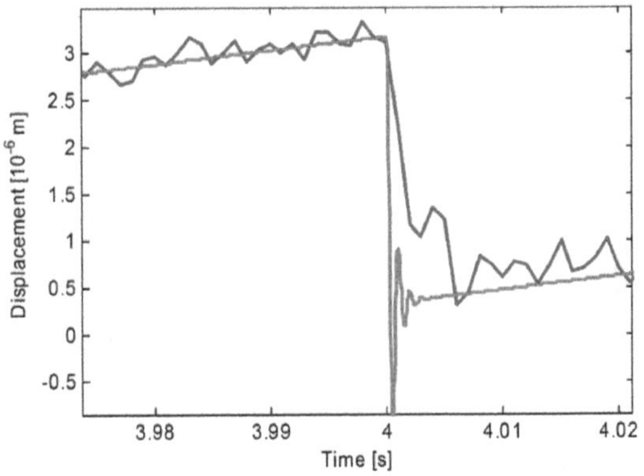

Figure 22. Scaled up picture from Fig. 21 (at around the 4th second).

model using the system identification technique. The details are as follows.

- Measure (1): the parameters reported in Dupont et al. (2000) for the elastoplastic model were used as initial values in identification. However, the identification results (not shown in this paper) show that the displacement predicted by the model cannot fit the experimental data.

- Measure (2): the elastoplastic model was developed based on LuGre model, and it only has one more parameter Z_{ba} than LuGre model to be identified. Z_{ba} was initially set as 33 % of Z_{ss}, and the remaining parameters were initially set as the same as those in LuGre model. The identification results (not shown in this paper) show that the displacement predicted by the model still cannot fit the experimental data.

- Measure (3): similar with Step (2), Z_{ba} was initially set as 33 % of Z_{ss} and determined by identification, but the remaining parameters were assumed (not determined by identification) as the same as those (which were already determined in Sect. 6.4) in LuGre model in order to find out the role of Z_{ba}. The identification results (not shown in this paper) show that the displacement predicted by the model still cannot fit the experimental data regardless of what Z_{ba} was chosen.

- Measure (4): comparing the elastoplastic model with LuGre model, it can be seen that if Z_{ss} and Z_{ba} approach to zero, the eleastoplastic model should turn back to LuGre model. In order to test the elastoplastic model, the parameters Z_{ss} and Z_{ba} were set as $Z_{ss} = 0.4$ nm and $Z_{ba} = 0.1$ nm, and the remaining parameters were determined as the same as those in LuGre model (which were already determined in Sect. 6.4 and listed in Table 2).

The friction force, sliding speed, and displacement predicted by the model identified in the Measure (4) are shown in Figs. 23–25, respectively. From Figs. 23 and 24 it can be seen that the friction force and sliding speed predicted by the elastoplastic model have the same characteristic with those predicted by LuGre model (see Figs. 19 and 20), i.e. no fluctuations in stick motion. However, the displacement predicted by the elastoplastic model cannot fit the slip motion period, as shown in Fig. 25. According to identification Measure (3) and Measure (4), it seems that Z_{ba} makes the elastoplastic model can not turn back to LuGre model, which is thus the cause that the elastoplastic model fails to model friction force in the micro stick-slip motion system. The plausible reason for this is discussed as follows.

The elastoplastic model expects that during slip motion, $sgn(\dot{x})$ should always be equal to $sgn(z)$ according to Eq. (14); in other words, when $sgn(\dot{x}) \neq sgn(z)$, the model predicts that stick motion occurs. The relation between $sgn(\dot{x})$ and $sgn(z)$ are shown in Figs. 26 and 27 (which is the amplified figure of Fig. 26) in which 0 represents $sgn(\dot{x}) = sgn(z)$ and 2 represents $sgn(\dot{x}) \neq sgn(z)$. It can be seen that during the stick motion period, $sgn(\dot{x})$ is

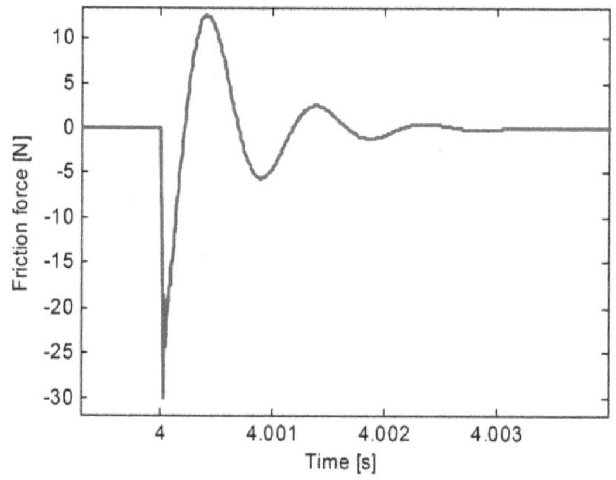

Figure 23. Friction predicted using the elastoplastic model.

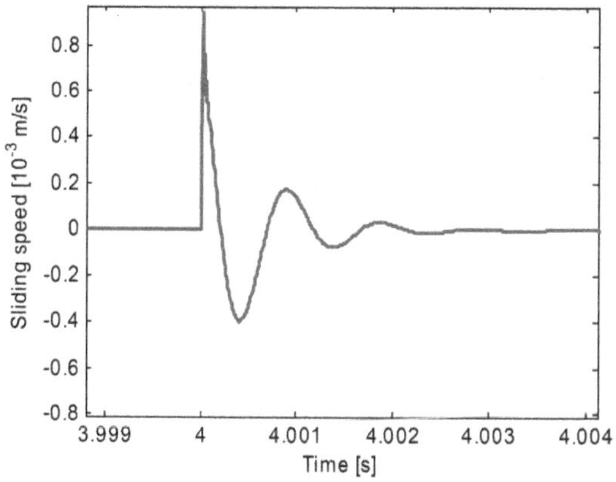

Figure 24. Sliding speed predicted using the elastoplastic model.

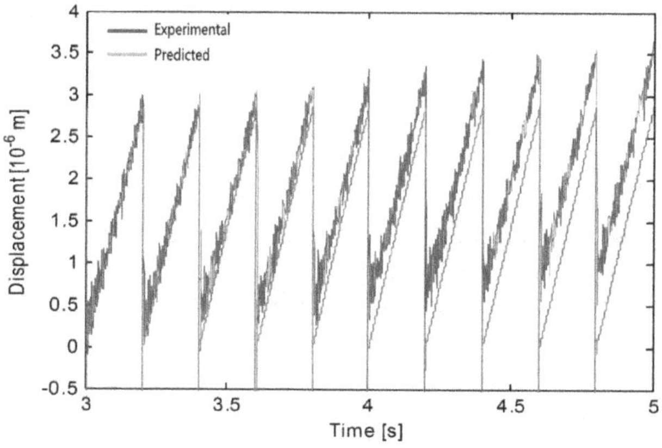

Figure 25. Experimental data and displacement predicted using the elastoplastic model.

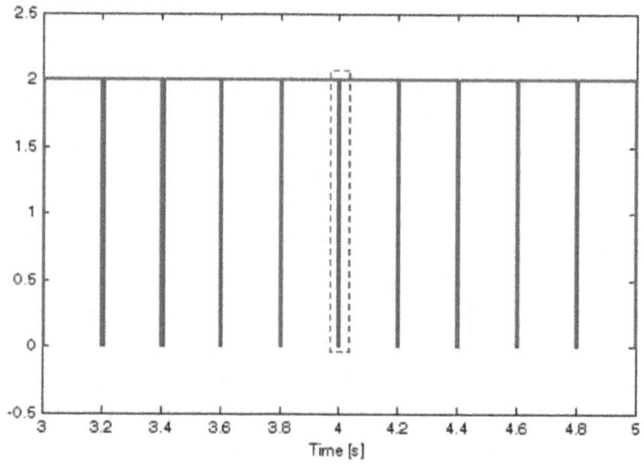

Figure 26. Relation of $sgn(\dot{x})$ and $sgn(z)$.

not equal to $sgn(z)$ over the whole period; as such, the model predicts that the stage and end-effector stick "together". During the slip motion period, however, it can be seen that $sgn(\dot{x})$ is not always equal to $sgn(z)$; in other words, the model predicts more stick motion than it really has. Therefore, the elastoplastic friction model leads to a larger static friction force and results in no relative motion in the slip motion period, as shown in Fig. 25. The reason that $sgn(\dot{x})$ is never always equal to $sgn(z)$ during the slip motion period is discussed as follows.

In stick-slip motion systems such as PE-SSA systems, the stage periodically extends and contract which makes $sgn(\dot{x})$ reverse when sliding occurs. However, according to the elastoplastic friction model, when $sgn(\dot{x})$ reverses, i.e. the end effector begins to move from B to A (Fig. 4), the deformation of contact as perities starts to decrease but still remain its initial direction until it reaches the original position, which leads to $sgn(\dot{x}) \neq sgn(z)$; correspondingly, the elastaplastic friction model predicts that no sliding occurs according to Eq. (14). This is the reason that the displacement predicted using the elastoplastic model (see Fig. 25) seems not to move at all in the slip motion period, which is not consistent with the observed actual stick-slip motion phenomena.

6.5 Comparison of Stribeck, Dahl, and LuGre models

To compare Stribeck, Dahl, and LuGre models in terms of accuracy, an error index representing the deviation between the displacement predicted and experimental data is given by

$$e = \frac{1}{n}\sum_{k=1}^{n}|e_k| \tag{26}$$

where $|e_k|$ is the absolute value of the error between experimental data and the displacement predicted, n the number of displacement data, and $k = 1, 2, \ldots, n$.

The error index for each friction model is listed in Table 3.

It is shown that LuGre model has the least error index, and Stribeck model and Dahl model have larger error indexes. The reason is that the displacements predicted by Stribeck model and Dahl model fluctuate around the experimental data (see Figs. 14 and 17 or Fig. 18), which is caused

Table 3. Error index for friction models.

Models	Stribeck	Dahl	LuGre
Error index	0.176 μm	0.232 μm	0.135 μm

by the *Sgn* function in the two models. LuGre model integrates, pre-sliding friction ($\sigma_0 z$) that is captured by Dahl model and viscous friction ($\upsilon_2 \dot{x}$) and Stribeck effect ($g(\dot{x})$) that are captured by Stribeck friction model, into one single model. Moreover, LuGre model does not have *Sgn* function and it has no fluctuation in displacement prediction (see Fig. 22). Thus LuGre model results in a better accuracy of friction force prediction in the micro stick-slip motion system.

7 Conclusions with further discussion

This paper first reviewed the friction models, i.e. Coulomb, viscous, combined Coulomb and viscous model, Stribeck, Dahl, LuGre, and the elastoplastic friction models. Five of them were applied to model friction force in the micro stick-slip motion. Parameters involved in each model were determined using the system identification technique. The performances of these models were compared. The plausible reasons for the difference in performance among these models applied in micro stick-slip motion were discussed.

This study concludes that Coulomb friction model is not adequate for describing the friction in the micro stick-slip motion. Stribeck model, Dahl model and LuGre model can all predict friction force in the micro stick-slip motion, and LuGre model has the best accuracy among the three. The elastoplastic friction model fails to model friction in the micro stick-slip motion because it introduces a larger static friction during the slip motion period that makes two contact surfaces seem to stick together. The failure of the elastoplastic model for the micro stick-slip motion system demonstrates our observation of the so-called “side effect” for complex dynamic systems. In the case of the elastoplastic model, it has more parameters to capture the friction phenomena, especially capturing the so-called “drift” phenomenon when LuGre model is used for the application of positioning system, but the model which resolves the drift problem also brings the side effect in the sense that the same modeling elements, i.e. elastic and plastic deformation zones, causes trouble to model friction force in the micro stick-slip motion.

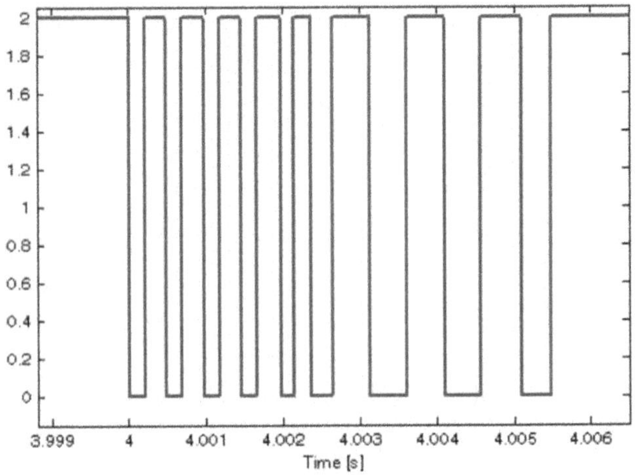

Figure 27. Scaled up picture from Fig. 26.

This study may also demonstrate an idea of using displacement measurement of micro motion system and system identification technique to investigate the dynamic friction in micro motion level. Further validation of this idea calls for some future study.

Appendix A

Table A1. Nomenclature.

c_p	the damping coefficient of PEA, $\mathrm{Ns\,m^{-1}}$
c_s	the damping coefficient of the stage, $\mathrm{Ns\,m^{-1}}$
F	the friction force, N
F_{app}	the applied force, N
F_c	the Coulomb sliding friction force, N
F_N	the normal load, N
F_r	friction force on the stage, N
F_r'	friction force on the end effector, N
F_s	the maximum static friction force, N
F_p	the transduced force from the electrical side, N
$g(\dot{x})$	the Stribeck curve
$G(s)$	the transfer function
i	an exponent
j	the Stribeck shape factor
K	the amplified coefficient, $\mathrm{m\,v^{-1}}$
k_p	the stiffness of PEA, $\mathrm{N\,m^{-1}}$
k_s	the stiffness of stage, $\mathrm{N\,m^{-1}}$
k_v	the viscous coefficient, $\mathrm{Ns\,m^{-1}}$
m_p	the mass of PEA, kg
m_s	the mass of stage, kg
os	the overshoot of a step response
T_{em}	the electromechanical transducer ratio, $\mathrm{N\,v^{-1}}$
T_p	the peak time of a step response, s
u_p	the applied voltage on the PEA, V
v_s	the Stribeck velocity, $\mathrm{m\,s^{-1}}$
x	the relative displacement, m
$\dot{x}$	the sliding speed, $\mathrm{m\,s^{-1}}$
x_e	the displacement of the end effector, m
x_p	the displacement of the PEA, m
x_{pe}	the relative displacement between the end effector and stage, m
z	the average bristle deflection, m
z_{ba}	the stiction zone, m
z_{ss}	the parameter for transition from sticking to slipping, m
$\alpha(z, \dot{x})$	function of elastic and plastic deformation
$\alpha_m(*)$	the function of transition from sticking to slipping
μ	Coulomb friction coefficient
ξ	the damping ratio
ω_n	the natural frequency, $\mathrm{rad\,s^{-1}}$
σ_0	the stiffness of the bristles, $\mathrm{N\,mm^{-1}}$
σ_1	the damping coefficient of the bristles, $\mathrm{Ns\,mm^{-1}}$
σ_2	the viscous friction coefficient, $\mathrm{Ns\,mm^{-1}}$

Acknowledgements. The authors want to thank J. W. Li, Q. S. Zhang for their performing the experiments. This work has been supported by The Natural Sciences and Engineering Research Council of Canada (NSERC), National Natural Science Foundation of China (NSFC) (grant number: 51375166) and China Scholarship Council (CSC), and these organizations are thanked for the support.

References

Adriaens, H. J. M. T. A., De Koning, W. L., and Banning, R.: Modeling piezoelectric actuators, IEEE/ASME T. Mechatron., 5, 331–341, 2000.

Andersson, S., Söderberg, A., and Björklund, S.: Friction models for sliding dry, boundary and mixed lubricated contacts, Tribol. Int., 40, 580–587, 2007.

Armstrong-Helouvry, B.: Stick slip and control in low-speed motion. IEEE T. Automat. Contr., 38, 1483–1496, 1993.

Chang, S. H. and Li, S. S.: A high resolution long travel friction-driven micropositioner with programmable step size, Rev. Scient. Instrum., 70, 2776–2782, 1999.

Canudas de Wit, C., Olsson, H., Åström, K. J., and Lischinsky, P.: A New Model for Control of Systems with Friction, IEEE T. Automat. Contr., 40, 419–425, 1995.

Dahl, P. R.: A solid friction model, Technical Report Tor-0158(3107-18)-1, The Aerospace Corporation, El Segundo, CA, 1968.

Dupont, P., Armstrong, B., and Hayward, V.: Elasto-plastic friction model: contact compliance and stiction, Proc. Am. Control Conf., Chiacago, IL, 1072–1077, 2000.

Kang, D.: Modeling of the Piezoelectric-Driven Stick-Slip Actuators, Thesis of Master of Science, University of Saskatchewan, Saskatchewan, Canada, 2007.

Lampaert, V., Swevers, J., and Al-Bender, F.: Modification of the Leuven integrated friction model structure, IEEE T. Automat. Contr., 47, 683–687, 2002.

Li, J. W., Yang, G. S., Zhang, W. J., Tu, S. D., and Chen, X. B.: Thermal effect on piezoelectric stick-slip actuator systems, Rev. Scient. Instrum., 79, 046108, doi:10.1063/1.2908162, 2008a.

Li, J. W., Chen, X. B., An, Q., Tu, S. D., and Zhang, W. J.: Friction models incorporating thermal effects in highly precision actuators, Rev. Scient. Instrum., 80, 045104, doi:10.1063/1.3115208, 2008b.

Li, J. W., Zhang, W. J., Yang, G. S., Tu, S. D., and Chen, X. B.: Thermal-error modeling for complex physical systems: the-state-of-arts review, Int. J. Adv. Manufact. Technol., 42, 168–179, 2009.

Makkar, C., Hu, G., Sawyer, W. G., and Dixon, W. E.: Lyapunov-Based Tracking Control in the Presence of Uncertain Nonlinear Parameterizable Friction, IEEE T. Automat. Contr., 52, 1988–1994, 2007.

Morin, A.: New friction experiments carried out at Metz in 1831–1833, Proc. French Roy. Acad. Sci., 4, 1–128, 1833.

Stribeck, R.: Die wesentlichen Eigenschaften der Gleit- und Rollenlager, Zeitschrift des Vereins Deutscher Ingenieure, Duesseldorf, Germany, 36 Band 46, 1341–1348, 1432–1438, 1463–1470, 1902.

Swevers, J., Al-Bender, F., Ganseman, C. G., and Prajogo, T.: An integrated friction model structure with improved presliding behaviour for accurate friction compensation, IEEE T. Automat. Contr., 45, 675–686, 2000.

Zhang, W. J., Ouyang, P. R., and Sun, Z.: A novel hybridization design principle for intelligent mechatronics systems, Proceedings of International Conference on Advanced Mechatronics (ICAM2010), Osaka University Convention Toyonaka, Japan, 4–6, 2010.

Zhang, Z. M., An, Q., Zhang, W. J., Yang, Q., Tang, Y. J., and Chen, X. B.: Modeling of directional friction on a fully lubricated surface with regular anisotropic asperities, Meccanica Springer Netherlands, 2011.

12

Solving the dynamic equations of a 3-PRS Parallel Manipulator for efficient model-based designs

M. Díaz-Rodríguez[1], J. A. Carretero[2], and R. Bautista-Quintero[2,3]

[1]Departamento de Tecnología y Diseño, Facultad de Ingeniería, Universidad de los Andes, Mérida, 5101, Venezuela

[2]Department of Mechanical Engineering, University of New Brunswick, Fredericton, NB, E3A 5A3, Canada

[3]Departamento de Ingeniería Mecánica, Instituto Tecnológico de Culiacán, Sinaloa, 80220, Mexico

Correspondence to: M. Díaz-Rodríguez (dmiguel@ula.ve)

Abstract. Introduction of parallel manipulator systems for different applications areas has influenced many researchers to develop techniques for obtaining accurate and computational efficient inverse dynamic models. Some subject areas make use of these models, such as, optimal design, parameter identification, model based control and even actuation redundancy approaches. In this context, by revisiting some of the current computationally-efficient solutions for obtaining the inverse dynamic model of parallel manipulators, this paper compares three different methods for inverse dynamic modelling of a general, lower mobility, 3-PRS parallel manipulator. The first method obtains the inverse dynamic model by describing the manipulator as three open kinematic chains. Then, vector-loop closure constraints are introduced for obtaining the relationship between the dynamics of the open kinematic chains (such as a serial robot) and the closed chains (such as a parallel robot). The second method exploits certain characteristics of parallel manipulators such that the platform and the links are considered as independent subsystems. The proposed third method is similar to the second method but it uses a different Jacobian matrix formulation in order to reduce computational complexity. Analysis of these numerical formulations will provide fundamental software support for efficient model-based designs. In addition, computational cost reduction presented in this paper can also be an effective guideline for optimal design of this type of manipulator and for real-time embedded control.

1 Introduction

Seminal research in Parallel Manipulators (PMs) described architectures of 6 degrees of freedom (DOF) which are mainly used to perform industrial tasks. Nevertheless, not all applications (e.g., commercial, space exploration, entertainment or even industrial) require full 6-DOF capabilities, thus, cost-effective PMs with less than 6 DOF (i.e., lower-mobility) have been developed. One such architecture is the 3-PRS manipulator which has a platform and a fixed base connected through three identical sets of links and joints (i.e., legs). Each leg has a slider attached to the base by an actuated prismatic joint (P), a coupler connected to the slider by a passive rotational joint (R) and to the platform by a passive spherical joint (S). The 3-PRS manipulator was first described in Carretero et al. (2000b) for a telescope application, and then it was proposed as a machining centre in Fan et al. (2003) and for a medical application in Merlet (2001). The kinematics and workspace analyses of the manipulator have been extensively studied in Carretero et al. (2000a), Carretero et al. (2000b), Tsai et al. (2002), Li and Xu (2007), to name a few. On the other hand, despite the fact that inverse dynamic modelling is essential for optimal design, parameter identification (Mata et al., 2008), model-based control (Díaz-Rodríguez et al., 2013), and internal redundancy (Parsa et al., 2013) amongst others, few papers have focused on revisiting and comparing computationally-inexpensive methods in order to obtain the dynamic models that are cost-effective for real-time applications.

Lagrangian formulations allowed to develop the inverse dynamics model of the 3-PRS manipulator (Li and Xu, 2004). The formulation uses the Lagrange multiplier to include the constraints forces that lead to a modelling approach not only intricate but also computationally complex. Li and Xu (2004) applied the Principle of Virtual Work (PVW), but they simplify the dynamics of the coupler link by dividing its mass into two portions located at its extremes. Tsai and Yuan (2010) solved the inverse dynamic model along with the reaction forces through a special decomposition of the reaction forces at the joints that connect the leg with the platform. A similar approach was used in Yuan and Tsai (2014) for solving direct dynamics including friction effect. However, the later approach considers the calculation of reaction forces which are may be needed for structural design of a manipulator but its computation increases computational complexity which is unnecessary for parameter identification or model-based control. Staicu (2012) analyses and compares the power consumption of the 3-PRS vs. the 3-PRS configuration using the PVW with recursive modelling. The method obtains the Jacobian by differentiating the vector loop equation. Carbonari et al. (2013) solved the inverse dynamics of a 3-DOF parallel manipulator via screw theory and the PVW. On the other hand, the 3-RPS manipulator presents similar characteristics to the 3-PRS manipulator, in this respect, Mata et al. (2008) implement recursive velocity equations used in serial manipulator analysis to find the Jacobian of the manipulator for the inverse dynamic modelling, and Ibrahim and Khalil (2007) exploit architectural characteristics of the 3-RPS to give a closed form solution for the inverse and direct dynamics modelling.

The inherent complexity of the dynamic models lies on the way the system is modelled and how the Jacobian matrix is put forward. In this context, this paper compares the computational number of operation of three formulations for inverse dynamic modelling of a 3-PRS. The first formulation applies the general solution of PMs dynamic modelling proposed in Khalil and Ibrahim (2007). The second method considers the manipulator as a set of open kinematic chains and finds the Jacobian in joint space coordinates by taking into account the vector loop constraints at the split joints (Mata et al., 2008). The third method relies on the modelling approach originally presented in Li and Xu (2004).

The ultimate goal of contrasting these numerical formulations is focused on supporting the implementation of emerging model-based designs which not only depends of the discrete inverse dynamic model but also the numerical finite realization in a given computational platform (Williamson, 1991). In fact, software architecture, for model based control (Díaz-Rodríguez et al., 2013), relies on minimizing computational task timing commonly constrained by fast sampling periods (Goodwin et al., 1992). Similarly, in other applications, such parameter identification (Mata et al., 2008), and internal redundancy (Parsa et al., 2013) few papers have focused on revisiting and comparing computationally-inexpensive methods in order to obtain the dynamic model of the 3-PRS configuration for cost-effective real-time applications.

To this end, this paper is organized as follows: Sect. 2 presents in general terms how the dynamic model for a parallel manipulator is developed for the three approaches investigated in this paper. Section 3 presents the implementation of the approaches for solving the dynamics problem of a 3-PRS spatial parallel manipulator. Section 4 summarizes and discusses the complexity and the computational load of these three formulations. Finally, the conclusions are drawn.

2 Development of the dynamic models

The dynamic model of a closed chain mechanical system such as a parallel manipulator can be obtained by virtually cutting or splitting the manipulator at one or more of its joints until the complete dynamic model of a tree-like system with several open chains is obtained. Newton–Euler formulations are then used for solving the dynamics of each serial chain. Finally, constraint equations obtained by means of the Lagrange Multipliers are incorporated to include the necessary forces at the splitting points as to ensure the kinematic chains remain closed. On the other hand, the Lagrangian approach can be applied by using the Lagrange equation with respect to a minimum set of generalized coordinates. Yiu et al. (2001) showed that either application of Lagrange Equation or tree-like system analysis are equivalent to one another and lead to the same set of equations when applied to a parallel manipulator. Moreover, similar results were obtained by Murray and Lovell (1989) using the D'Alembert's principle and the principle of virtual work.

Regardless, of the dynamics equation (Newton–Euler, Lagrange or Principle of Virtual Work) used for developing the model, the dynamics of closed chain system can be written as:

$$\boldsymbol{\tau} = \mathbf{G}^T \boldsymbol{h}. \tag{1}$$

This equation establishes that the generalized forces corresponding to an open chain system ($\boldsymbol{h}$) are related to the actuated forces of a closed chain system ($\boldsymbol{\tau}$) by a linear transformation $\left(\mathbf{G}^T\right)$. This linear transformation is based on the Jacobian matrix. Thus, the dynamics equation of a parallel manipulator essentially relies on finding the Jacobian matrix that relate the passive generalized coordinates to the actuated ones.

The following subsections revisit three approaches for solving the dynamics equations of a 3-PRS parallel manipulator. The differences in the approaches are based on two facts: (1) how the parallel manipulator is split into open chains and (2) the type of generalized coordinates used for modelling the manipulator. Each approach leads to establishing different formulations of the Jacobian matrix. Since each Jacobian matrix holds different formulation there are differences in the number of algebraic operations to compute each. The objective of this work is mostly revisiting these approaches to find the cost-effectiveness of each when solving the inverse dynamics through an example. This is of particular interest as inexpensive computer models are essential when adaptive control algorithms are used to control manipulator at fast update rates. For instance, in the fields of parameter identification and actuation redundancy, a fast estimation of the inverse model implies cost-effectiveness for real-time applications.

2.1 Dynamics considering the legs and platform as subsystems

This approach is based on the general formulation for modelling parallel manipulators presented by Khalil and Ibrahim (2007), which is based on the following aspects: (1) the manipulator is split open at the spherical joints so that the moving platform is separated from the legs and (2) the local joint coordinates systems $\boldsymbol{q}$ can be used to develop the dynamic equations of each leg $\boldsymbol{h}_i$ while the Cartesian coordinates $\boldsymbol{x}$ are used to obtain the dynamic equations of the platform $\boldsymbol{h}_{\mathrm{p}}$. Then, the dynamic equations are combined and projected onto the active joint space as follows:

$$\boldsymbol{\tau} = \mathbf{J}_{\mathrm{p}}^T \boldsymbol{h}_{\mathrm{p}} + \sum_{i=1}^{m} \left(\frac{\delta \dot{q}_i}{\delta \dot{q}_{\mathrm{a}}} \right)^T \boldsymbol{h}_i, \tag{2}$$

where $\mathbf{J}_{\mathrm{p}}$ is the Jacobian projecting the task space coordinates (6 in the general case) to the n active joint coordinates, while m is the number of joints for each leg. Likewise,

$$\boldsymbol{h}_{\mathrm{p}} = \begin{bmatrix} m_{\mathrm{p}}\,\boldsymbol{g} - m_{\mathrm{p}}\,\boldsymbol{a}_{\mathrm{p}} \\ -\mathbf{I}_{\mathrm{p}}\dot{\omega}_{\mathrm{p}} - \omega_{\mathrm{p}} \times (\mathbf{I}_{\mathrm{p}}\omega_{\mathrm{p}}) \end{bmatrix}, \tag{3}$$

where m_{p} is the mass of the platform, $\mathbf{I}_{\mathrm{p}}$ denotes the inertial matrix of the moving platform about its centre of mass, $\boldsymbol{a}_{\mathrm{p}}$ stands for the acceleration of the end effector, and $\boldsymbol{\omega}_{\mathrm{p}}, \dot{\boldsymbol{\omega}}_{\mathrm{p}}$ respectively denote the angular velocity and angular acceleration of the platform.

2.1.1 Method I

In order to develop the model in actuated joint space one has to project the passive joint variables to the active ones. That is:

$$\boldsymbol{\tau} = \boldsymbol{h}_i^{\mathrm{a}} + \mathbf{J}_{\mathrm{p}}^T \boldsymbol{h}_{\mathrm{p}} + \mathbf{G}_{\mathrm{I}}^T \boldsymbol{h}_i^{\mathrm{p}}, \tag{4}$$

where indices a and p stand for the active and passive joints, respectively, while $\mathbf{G}_{\mathrm{I}}$ is a $l \times n$ matrix projecting the dynamics from the passive to the active joints. Here, l represents the number of passive joints while n is the number of active joints on each leg.

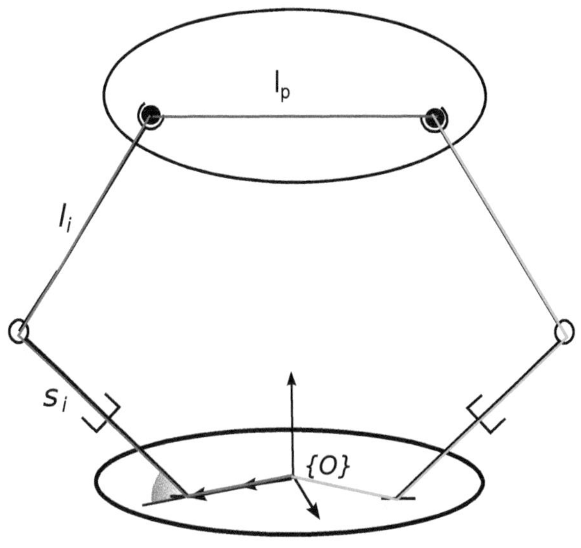

Figure 1. Closed chain equation for finding $\mathbf{G}_{\mathrm{I}}$.

Equation (4) can be written in the form of Eq. (1). That is:

$$\boldsymbol{\tau} = \begin{bmatrix} \mathbf{I} & \mathbf{J}_{\mathrm{p}}^T & \mathbf{G}_{\mathrm{I}}^T \end{bmatrix} \begin{bmatrix} \boldsymbol{h}_i^{\mathrm{a}} \\ \boldsymbol{h}_{\mathrm{p}} \\ \boldsymbol{h}_i^{\mathrm{p}} \end{bmatrix} = \mathbf{G}^T \boldsymbol{h}, \tag{5}$$

where $\mathbf{I}$ is the identity matrix with dimension equal to the degree of freedom of the manipulator.

In Eq. (4), matrix $\mathbf{G}_{\mathrm{I}}$ can be obtained by considering the fact that the distance among spherical joints at the platform is constant due to the rigid body assumption. This distance is calculated based on the norm of the vector obtained by subtracting the position vector identifying the location of the spherical joints. The partial derivatives of each equation with respect to the joints coordinates yields matrix $\mathbf{G}_{\mathrm{I}}$. Figure 1 shows how to establish the closed chain equations.

Note that in Eq. (4), $\boldsymbol{h}_{\mathrm{p}}$ is a 6×1 vector and, in the particular case of a 3-PRS parallel manipulator, $\boldsymbol{\tau}$ is a 3×1 vector. Therefore, matrix $\mathbf{J}_{\mathrm{p}}$ is not square and cannot be obtained using previous methods for solving the inverse kinematics of this kind of manipulator. For instance, when developing the inverse dynamics model using the Principle of Virtual Work, Li and Xu (2004) found a 3×3 square matrix $\mathbf{J}_{\mathrm{p}}$. They did so by only considering three of the components of the platform inertial forces; they considered those associated with the desired degrees of freedom of the end effector. In Sect. 3 a method obtaining $\mathbf{J}_{\mathrm{p}}$ considering the 6 components of $\boldsymbol{h}_{\mathrm{p}}$ for the particular case of the 3-PRS manipulator is shown.

2.1.2 Method II

Another approach to formulate the dynamic model is to find the Jacobian matrices according to the approach presented by Khalil and Ibrahim (2007). The method is based on projecting the dynamics equation of the passive joint onto the task space, and then, project them back to the active joint space so that:

$$\boldsymbol{\tau}=\boldsymbol{h}_i^{\mathrm{a}}+\mathbf{J}_{\mathrm{p}}^{T}\left[\boldsymbol{h}_{\mathrm{p}}+\mathbf{G}_{\mathrm{II}}^{T}\boldsymbol{h}_i^{\mathrm{p}}\right], \tag{6}$$

where $\mathbf{G}_{\mathrm{II}}$ is a $l \times 6$ matrix that holds new definition that can be written as follow:

$$\mathbf{G}_{\mathrm{II}}^{T}=\mathbf{J}_{v_i}^{T}\mathbf{J}_{q_i}^{-T}, \tag{7}$$

where $\mathbf{J}_{v_i}$ and $\mathbf{J}_{q_i}^{-T}$ can be obtained respectively from the direct Jacobian $\mathbf{J}_x$ and the inverse Jacobian $\mathbf{J}_q$ of the manipulator.

Equation (6) can be written in the form of Eq. (1). That is:

$$\boldsymbol{\tau}=\begin{bmatrix}\mathbf{I} & \mathbf{J}_{\mathrm{p}}^{T} & (\mathbf{G}_{\mathrm{II}}\mathbf{J}_{\mathrm{p}})^{T}\end{bmatrix}\begin{bmatrix}\boldsymbol{h}_i^{\mathrm{a}}\\ \boldsymbol{h}_{\mathrm{p}}\\ \boldsymbol{h}_i^{\mathrm{p}}\end{bmatrix}=\mathbf{G}^{T}\boldsymbol{h}. \tag{8}$$

2.2 Dynamics considering the manipulator as open kinematic chains

A parallel manipulator can be split open into $m-1$ joints yielding m open chain systems. In this approach the platform is attached to one of the legs, see Fig. 2. Algorithms for obtaining the dynamics model of serial manipulators may now be applied to obtain the dynamics of each leg.

The cut joints introduce constraint forces, which can be included into the model by means of the Lagrange multipliers:

$$\boldsymbol{\tau}=\boldsymbol{h}_i+\mathbf{A}\lambda_i, \tag{9}$$

where $\mathbf{A}$ is the Jacobian that can be found by analysing the constraint equations. Mata et al. (2008) presented a method for the Jacobian matrix by considering the linear velocity at the split joints. The velocities can be computed through the Jacobian analysis of each leg. Then, the velocity obtained at the split joint following each leg are the same. In this approach, recursive modelling of the velocity analysis from conventional serial manipulator methods can be applied for each leg.

Once the Jacobian matrix is found, the Lagrange multipliers in Eq. (9) are eliminated by multiplying matrix $\mathbf{C}$, so that, $\mathbf{CA}=0$. One way to find $\mathbf{C}$ is by obtaining the natural orthogonal complement of $\mathbf{C}$. On the other hand, the matrix can be found by separating matrix $\mathbf{A}=0$ into the passive and the active joints. That is:

$$\boldsymbol{\tau}=\boldsymbol{h}^{\mathrm{a}}+\mathbf{G}_{\mathrm{III}}^{T}\boldsymbol{h}^{\mathrm{p}}, \tag{10}$$

where $\mathbf{G}_{\mathrm{III}}$ has dimensions $l \times n$ and can be computed as follows:

$$\mathbf{G}_{\mathrm{III}}=\mathbf{A}_{\mathrm{p}}^{-1}\mathbf{A}_{\mathrm{a}}, \tag{11}$$

where subscripts p and a respectively refer to the passive and active variable terms in Jacobian $\mathbf{A}$. Equation (10) can be written in the form of Eq. (1). That is:

$$\boldsymbol{\tau}=\begin{bmatrix}\mathbf{I} & \mathbf{G}_{\mathrm{III}}\end{bmatrix}\begin{bmatrix}\boldsymbol{h}_i^{\mathrm{a}}\\ \boldsymbol{h}_i^{\mathrm{p}}\end{bmatrix}=\mathbf{G}^{T}\boldsymbol{h}. \tag{12}$$

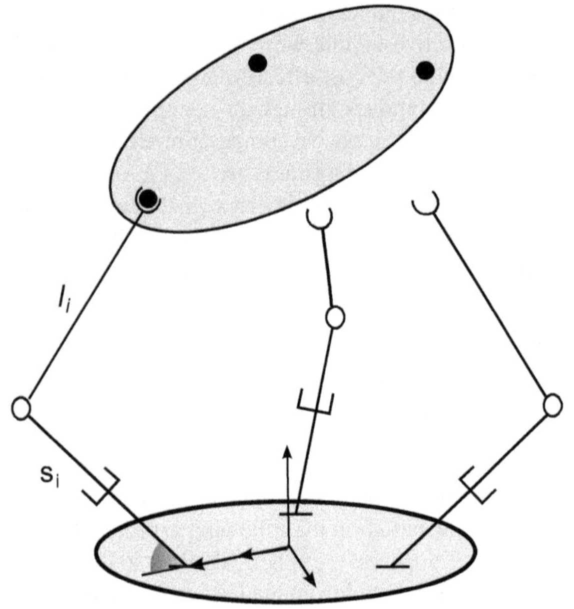

Figure 2. Several open chains obtained after cutting open the parallel manipulator.

As Eqs. (5), (8) and (12) show, the three methods establish a linear relation between the cut-open model ($\boldsymbol{h}$) to the original system. The different lies in how matrix $\mathbf{G}^T$ and vector $\boldsymbol{h}$ are solved in each model. This is illustrated in the next section for the particular case of a general 3-P̲RS parallel manipulator.

3 Inverse dynamics of the 3-P̲RS PM

A schematic representation of the 3-P̲RS PM is shown in Fig. 3 where the 7 moving rigid bodies are shown. There, links 0, 1, and 2 can be seen as 2-DOF serial manipulator with PR joints, also it applies to links 0, 4, and 5 as well as links 0, 6, and 7. The platform is indicated by the number 3. The manipulator is a lower mobility (i.e., less that 6-DOF) spacial PM with 3-DOF. This manipulator holds the characteristic of zero torsion at its platform because the three spherical joints move in vertical planes intersecting at a common line (Liu and Bonev, 2008). In addition, the topology of its legs provides 2-DOF of angular rotation (2R) in two axes (A/B axis, rolling and pitching) and 1-DOF translation (1T) motion (heave) at the end effector.

As Fig. 3 shows, reference frame $\{O\}$ defines the global coordinate system (x, y, z) while $\{p\}$ defines the local coordinate system attached to the moving platform. In the typical symmetrical configuration, the centre of the spherical joints

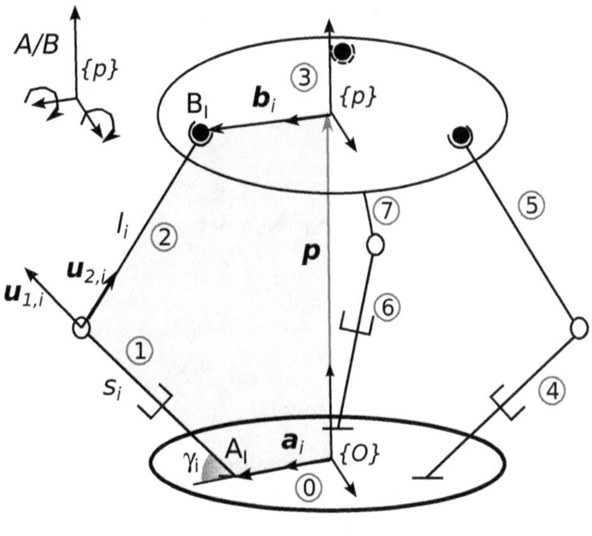

Figure 3. Sketch of a general 3-PRS manipulator.

at the platform form an equilateral triangle circumscribed in a circle with centre p and radius r_p. The line of action of the prismatic joints intersects the base Oxy plane at A_i also forming an equilateral triangle. The distance from O to A_i is the base platform radius r_b.

In order to take advantage of the dynamic equation algorithms developed for serial manipulators, when modelling the manipulator with leg and platform as subsystem, the joint coordinates $\boldsymbol{q}$ can be used to develop the dynamics equations of each leg (Mata et al., 2002). To this end, a local coordinate system $\{O_{i,0}\}$ is defined at the bottom of leg i where the leg meets the based plane. Table 1 lists the Denavit–Hartenberg (D–H) parameters, according to Craig's notation (Craig, 2005), of each leg i from $\{O_{i,0}\}$ up to the location of the axis $\{O_{i,2}\}$ at the revolute joint.

The rotation matrix ${}^{O_{i,0}}\mathbf{R}_o$ can be found as:

$$ {}^{O_{i,0}}\mathbf{R}_O = \begin{bmatrix} -\sin(\xi_i) & 0 & \cos(\xi_i) \\ \cos(\xi_i) & 0 & \sin(\xi_i) \\ 0 & 1 & 0 \end{bmatrix}, \quad (13) $$

where $\xi = \begin{bmatrix} 0 & 2/3\pi & 4/3\pi \end{bmatrix}^T$.

In addition, the roll-pitch-yaw (α, β and ϕ) Euler angles represent the orientation of the moving frame $\{p\}$ with respect to the global coordinate system $\{O\}$. The rotation matrix ${}^O\mathbf{R}_\text{p}$ is defined as:

$$ {}^O\mathbf{R}_\text{p} = \begin{bmatrix} c_\alpha c_\beta & c_\alpha s_\beta s_\phi - s_\alpha c_\phi & c_\alpha s_\beta c_\phi + s_\alpha s_\phi \\ s_\alpha c_\beta & s_\alpha s_\beta s_\phi + c_\alpha c_\phi & s_\alpha s_\beta c_\phi - c_\alpha s_\phi \\ -s_\beta & c_\beta s_\phi & c_\beta c_\phi \end{bmatrix}, \quad (14) $$

where $c_* = \cos(*)$ and $s_* = \sin(*)$.

Table 1. D–H parameters for a 3-PRS PM when modelling the manipulator with leg and platform as subsystem.

i	$\theta_{i,1}$	$d_{i,1}$	$a_{i,1}$	$\alpha_{i,1}$	$\theta_{i,2}$	$d_{i,2}$	$a_{i,2}$	$\alpha_{i,2}$
1	$\pi/2$	s_1	0	γ_1	θ_1	0	0	$-\pi/2$
2	$\pi/2$	s_2	0	γ_2	θ_2	0	0	$-\pi/2$
3	$\pi/2$	s_3	0	γ_3	θ_3	0	0	$-\pi/2$

The task space ($\boldsymbol{x}$) and joint space ($\boldsymbol{q}_i$) coordinates are given by:

$$ \boldsymbol{x} = \begin{bmatrix} x_\text{p} & y_\text{p} & z_\text{p} & \phi & \beta & \alpha \end{bmatrix}^T, \quad (15) $$

$$ \boldsymbol{q} = \begin{bmatrix} \boldsymbol{q}_1^T & \boldsymbol{q}_2^T & \boldsymbol{q}_3^T \end{bmatrix}, \quad (16) $$

$$ \boldsymbol{q}_1 = \begin{bmatrix} s_1 \\ \theta_1 \end{bmatrix}, \quad \boldsymbol{q}_2 = \begin{bmatrix} s_2 \\ \theta_2 \end{bmatrix}, \quad \text{and} \quad \boldsymbol{q}_3 = \begin{bmatrix} s_3 \\ \theta_3 \end{bmatrix}, \quad (17) $$

where s_i represents the displacement along the axis of the prismatic joint i, and θ_i the angle of the link 2 and the axis of the prismatic joint i in the plane of movement of the leg i.

The components of $\boldsymbol{b}_i$ with respect to the local coordinate system $\{p\}$ are given by:

$$ {}^p\boldsymbol{b}_1 = \begin{bmatrix} r_\text{p} \\ 0 \\ 0 \end{bmatrix}, \quad {}^p\boldsymbol{b}_2 = \begin{bmatrix} -\frac{1}{2} r_\text{p} \\ \frac{\sqrt{3}}{2} r_\text{p} \\ 0 \end{bmatrix}, \text{and } {}^p\boldsymbol{b}_3 = \begin{bmatrix} -\frac{1}{2} r_\text{p} \\ -\frac{\sqrt{3}}{2} r_\text{p} \\ 0 \end{bmatrix} \quad (18) $$

while the components of $\boldsymbol{a}_i$ with respect to the global coordinate system $\{O\}$ are given by:

$$ \boldsymbol{a}_1 = \begin{bmatrix} r_\text{b} \\ 0 \\ 0 \end{bmatrix}, \quad \boldsymbol{a}_2 = \begin{bmatrix} -\frac{1}{2} r_\text{b} \\ \frac{\sqrt{3}}{2} r_\text{b} \\ 0 \end{bmatrix}, \text{and } \boldsymbol{a}_3 = \begin{bmatrix} -\frac{1}{2} r_\text{b} \\ -\frac{\sqrt{3}}{2} r_\text{b} \\ 0 \end{bmatrix}. \quad (19) $$

The position problem for the considered PM is not included in this paper since it can be found in Carretero et al. (2000b), Tsai et al. (2002), Li and Xu (2004), Mata et al. (2008). The following subsections focus on the computation of the Jacobian matrix for the aforementioned approaches.

3.1 Jacobian matrix for Model I

Matrices $\mathbf{J}_\text{p}$ and $\mathbf{G}$ for computing Eq. (4) are found following the approach presented in Li and Xu (2004). The vector loop equation of the ith leg can be written as:

$$ \boldsymbol{p} + \boldsymbol{b}_i = \boldsymbol{a}_i + s_i \boldsymbol{u}_{1_i} + l_i \boldsymbol{u}_{2_i}, \quad (20) $$

where $\boldsymbol{u}_{1_i}$ and $\boldsymbol{u}_{2_i}$ are unit vectors, l_i is the distance between the rotational joint and the spherical joint, $\boldsymbol{b}_i$ is the position vector from the origin O to the centre of the spherical joint. Vector $\boldsymbol{a}_i$ is the position vector between O and local frame of each leg. Differentiating Eq. (20) with respect to time and after some algebraic manipulation, the following equation can be obtained:

$$\mathbf{J}_q \dot{\boldsymbol{q}} = \mathbf{J}_x \dot{\boldsymbol{x}} \tag{21}$$

where:

$$\mathbf{J}_q = \begin{bmatrix} \boldsymbol{u}_{21}^T \boldsymbol{u}_{11} & 0 & 0 \\ 0 & \boldsymbol{u}_{22}^T \boldsymbol{u}_{12} & 0 \\ 0 & 0 & \boldsymbol{u}_{23}^T \boldsymbol{u}_{13} \end{bmatrix} \tag{22}$$

and

$$\mathbf{J}_x = \begin{bmatrix} \boldsymbol{u}_{21}^T & (\boldsymbol{b}_1 \times \boldsymbol{u}_{21})^T \\ \boldsymbol{u}_{22}^T & (\boldsymbol{b}_2 \times \boldsymbol{u}_{22})^T \\ \boldsymbol{u}_{23}^T & (\boldsymbol{b}_3 \times \boldsymbol{u}_{23})^T \end{bmatrix}. \tag{23}$$

Due to the constraints imposed by the fact that each legs moves on a plane, the following set of equations holds:

$$\begin{aligned} x_\mathrm{p} &= -r_\mathrm{p} s_\alpha c_\beta, \\ y_\mathrm{p} &= -\frac{1}{2} r_\mathrm{p} \left(c_\alpha c_\beta - s_\alpha s_\beta s_\phi - c_\alpha c_\phi \right), \\ \tan(\alpha) &= \left(s_\beta s_\phi \right) / \left(c_\phi + c_\beta \right). \end{aligned} \tag{24}$$

From these equations, a 6×3 Jacobian matrix mapping the dependent task space coordinates to the independent ones can be found such that:

$$\begin{bmatrix} \dot{x}_\mathrm{p} & \dot{y}_\mathrm{p} & \dot{z}_\mathrm{p} & \dot{\phi} & \dot{\beta} & \dot{\alpha} \end{bmatrix}^T = \mathbf{J}_r^* \begin{bmatrix} \dot{z}_\mathrm{p} & \dot{\phi} & \dot{\beta} \end{bmatrix}^T \tag{25}$$

It is important to note that $\dot{\boldsymbol{x}} = \begin{bmatrix} \dot{x}_\mathrm{p} & \dot{y}_\mathrm{p} & \dot{z}_\mathrm{p} & \omega_{p_x} & \omega_{y_\mathrm{p}} & \omega_{z_\mathrm{p}} \end{bmatrix}^T$. In order to apply Eq. (25), one has to find the angular velocity of the platform through the rate of change of the generalised coordinates $\begin{bmatrix} \dot{\alpha} & \dot{\beta} & \dot{\phi} \end{bmatrix}^T$. That is:

$$\begin{bmatrix} \omega_{p_x} \\ \omega_{y_\mathrm{p}} \\ \omega_{z_\mathrm{p}} \end{bmatrix} = \begin{bmatrix} c_\alpha c_\beta & -s_\alpha & 0 \\ s_\alpha c_\beta & c_\alpha & 0 \\ -s_\beta & 0 & 1 \end{bmatrix} \begin{bmatrix} \dot{\phi} \\ \dot{\beta} \\ \dot{\alpha} \end{bmatrix}. \tag{26}$$

After some algebraic manipulation the following equation can be obtained:

$$\mathbf{J}_\mathrm{p} = \mathbf{J}_r \mathbf{J}_c^{-1} \tag{27}$$

where $\mathbf{J}_c = \left[\mathbf{J}_q^{-1} \mathbf{J}_x \mathbf{J}_r \right]$.

The 3×3 matrix $\mathbf{G}_\mathrm{I}$ is found by considering the fact that the distance among spherical joints at the platform is constant due to the rigid body assumption. That is:

$$\begin{aligned} &l_\mathrm{p}^2 - \| \boldsymbol{a}_i + s_i \boldsymbol{u}_{1i} + l_i \boldsymbol{u}_{2i} - \boldsymbol{a}_{i+1} - s_{i+1} \boldsymbol{u}_{1i+1} \\ &- l_{i+1} \boldsymbol{u}_{2i+1} \| = 0 \end{aligned} \tag{28}$$

with $i = 1, 2, 3$ and when $i = 3, i+1 = 1$.

Thus, by obtaining the partial derivatives of the above set of equation matrix $\mathbf{G}_\mathrm{I}$ can be written as:

$$\begin{bmatrix} \frac{\delta \dot{\theta}_1}{\delta \dot{s}_1} & \frac{\delta \dot{\theta}_2}{\delta \dot{s}_1} & \frac{\delta \dot{\theta}_3}{\delta \dot{s}_1} \\ \frac{\delta \dot{\theta}_1}{\delta \dot{s}_2} & \frac{\delta \dot{\theta}_2}{\delta \dot{s}_2} & \frac{\delta \dot{\theta}_2}{\delta \dot{s}_2} \\ \frac{\delta \dot{\theta}_1}{\delta \dot{s}_3} & \frac{\delta \dot{\theta}_2}{\delta \dot{s}_3} & \frac{\delta \dot{\theta}_3}{\delta \dot{s}_3} \end{bmatrix} = \mathbf{X}_\mathrm{p}^{-1} \mathbf{X}_a = \mathbf{G}_\mathrm{I}. \tag{29}$$

3.2 Jacobian matrix for Model II

Another approach to compute matrices $\mathbf{J}_\mathrm{p}$ and $\mathbf{G}$ is to consider that the spherical joints in each leg is constrained to move on a plane normal to the revolute joint. The motion of each leg at point B_i can be found in terms of the joint coordinates. Moreover, it can also be expressed with respect to $O_{i,2}$. Due to the constraints provided by the P–R pair, the third row of the Jacobian matrix consist of zero entries. That is:

$${}^{i,2}\boldsymbol{v}_{B_i} = {}^i\mathbf{J}_q \, \boldsymbol{q}_i, \tag{30}$$

where

$${}^i\mathbf{J}_q = \begin{bmatrix} -\sin(\theta_i) & 0 \\ -\cos(\theta_i) & l_i \\ 0 & 0 \end{bmatrix}. \tag{31}$$

The linear velocity of the end effector can be related to the linear velocity of points B_i as follows:

$${}^{i,2}\boldsymbol{v}_{B_i} = {}^i\mathbf{J}_v \boldsymbol{v} = \begin{bmatrix} {}^{i,2}\mathbf{R}_O & -{}^{i,2}\mathbf{R}_O \, {}^O\mathbf{R}_\mathrm{p} \, {}^p\widetilde{b_i} \end{bmatrix} \boldsymbol{v}, \tag{32}$$

where $\boldsymbol{v} = \begin{bmatrix} \boldsymbol{v}_\mathrm{p}^T & \boldsymbol{\omega}_\mathrm{p}^T \end{bmatrix}^T = \begin{bmatrix} v_{p_x} & v_{p_y} & v_{p_z} & \omega_x & \omega_y & \omega_z \end{bmatrix}^T$, $\widetilde{b}$ stands for the skew symmetric matrix substituting the cross product $\boldsymbol{b}_i \times$, and

$${}^{i,2}\mathbf{R}_\mathrm{p} = \begin{bmatrix} -c_{\sigma_i} s_{\gamma_i + \theta_i} & -c_{\sigma_i} c_{\gamma_i + \theta_i} & s_{\sigma_i} \\ -s_{\sigma_i} s_{\gamma_1 + \theta_i} & -s_{\sigma_i} c_{\gamma_1 + \theta_i} & -c_{\sigma_i} \\ c_{\gamma + \theta_i} & -s_{\gamma_i + \theta_i} & 0 \end{bmatrix}. \tag{33}$$

The first two rows of Eqs. (31) and (32) relate the task space to the joint spaces coordinates. That is:

$$\mathbf{J}_q \dot{\boldsymbol{q}} = \mathbf{J}_x \boldsymbol{v}, \tag{34}$$

where

$$\mathbf{J}_q = \begin{bmatrix} {}^1\mathbf{J}_q & \mathbf{0} & \mathbf{0} \\ \mathbf{0} & {}^2\mathbf{J}_q & \mathbf{0} \\ \mathbf{0} & \mathbf{0} & {}^3\mathbf{J}_q \end{bmatrix} \tag{35}$$

Table 2. D–H Parameters for the first leg of the 3-PRS PM when modelling as a three open chains.

j	$\theta_{1,j}$	$d_{1,1}$	$a_{1,j}$	$\alpha_{1,j}$
1	$\pi/2$	s_1	0	$-\gamma_1$
2	θ_1	0	0	$-\pi/2$
3	θ_4	l_1	0	0
4	θ_5	0	0	$\pi/2$
5	θ_6	0	0	$\pi/2$

and,

$$\mathbf{J}_x = \begin{bmatrix} \boldsymbol{x}^{T\,1,2}\mathbf{R}_O & \boldsymbol{x}^T\left(-^{1,2}\mathbf{R}_O^O\mathbf{R}_\mathrm{p}^p\widetilde{b}_1\right) \\ \boldsymbol{y}^{T\,1,2}\mathbf{R}_O & \boldsymbol{y}^T\left(-^{1,2}\mathbf{R}_O^O\mathbf{R}_\mathrm{p}^p\widetilde{b}_1\right) \\ \boldsymbol{x}^{T\,2,2}\mathbf{R}_O & \boldsymbol{x}^T\left(-^{2,2}\mathbf{R}_O^O\mathbf{R}_\mathrm{p}^p\widetilde{b}_1\right) \\ \boldsymbol{y}^{T\,2,2}\mathbf{R}_O & \boldsymbol{y}^T\left(-^{2,2}\mathbf{R}_O^O\mathbf{R}_\mathrm{p}^p\widetilde{b}_1\right) \\ \boldsymbol{x}^{T\,3,2}\mathbf{R}_O & \boldsymbol{x}^T\left(-^{3,2}\mathbf{R}_O^O\mathbf{R}_\mathrm{p}^p\widetilde{b}_1\right) \\ \boldsymbol{y}^{T\,3,2}\mathbf{R}_O & \boldsymbol{y}^T\left(-^{3,2}\mathbf{R}_O^O\mathbf{R}_\mathrm{p}^p\widetilde{b}_1\right) \end{bmatrix}. \tag{36}$$

In the above equations $\boldsymbol{x} = \begin{bmatrix} 1 & 0 & 0 \end{bmatrix}^T$, $\boldsymbol{y} = \begin{bmatrix} 0 & 1 & 0 \end{bmatrix}^T$, and $\mathbf{0}$ is a 2×2 zero matrix.

From the above equation, matrix $\mathbf{G}_{\mathrm{II}}$ can be obtained as follows:

$$\mathbf{J}_q^{-1}\mathbf{J}_x = \begin{bmatrix} \mathbf{J}_{\mathrm{q1}}^{-1}\mathbf{J}_{\mathrm{v1}} \\ \mathbf{J}_{\mathrm{q2}}^{-1}\,\mathbf{J}_{\mathrm{v2}} \\ \mathbf{J}_{\mathrm{q3}}^{-1}\mathbf{J}_{\mathrm{v2}} \end{bmatrix} \tag{37}$$

$$\mathbf{G}_{\mathrm{II}}^T = \begin{bmatrix} \mathbf{J}_{\mathrm{v1}}^T\mathbf{J}_{\mathrm{q1}}^{-T}(:,2) & \mathbf{J}_{\mathrm{v2}}^T\mathbf{J}_{\mathrm{q1}}^{-T}(:,2) & \mathbf{J}_{\mathrm{v3}}^T\mathbf{J}_{\mathrm{q3}}^{-T}(:,2) \end{bmatrix} \tag{38}$$

where $\mathbf{A}(:,2)$ denotes the 2nd column of matrix $\mathbf{A}$.

The Jacobian matrix, relating the task space coordinates $\mathbf{J}_r$, is found by considering the third column of velocity equation following each leg and the platform. The Jacobian matrix is obtained by following method 2 which is graphically represented in Fig. 4.

3.3 Jacobian matrix for Model III

The inverse dynamic is computed as a function of three open chains which are obtained after disassembling two of the three spherical joints. The platform is attached to one of the legs, and the spherical joint is modelled as three intersecting revolute joints with the three axes mutually perpendicular to one another. Therefore, the chain with the end effector platform is modelled by using the set of D–H parameters presented in Table 2. The remaining legs have only sets of two variables which are the same as those presented in Table 1.

One of the advantages of considering the manipulator as tree-like serial chains is that the Jacobian for each leg can be computed by using well-known recursive modelling from serial manipulator. In this respect, the velocities at the cut joints can be computed through recursive formulation (Angeles, 2002). That is:

$$\mathbf{A}_i\dot{\boldsymbol{q}}_i = V_{B_{i,i+1}} = \mathbf{A}_{i+1}\dot{\boldsymbol{q}}_{i+1}, \tag{39}$$

where $i = 1,2,3$ and when $i = 3, i+1 = 1$, $\mathbf{A}_i$ is the Jacobian matrix for the ith leg, and $V_{B_{i,i+1}}$ is the velocity at the cut joint connecting leg i and $i+1$.

From Eq. (39) the following equation can be obtained:

$$\mathbf{A}_i\,\dot{\boldsymbol{q}}_i - \mathbf{A}_{i+1}\,\dot{\boldsymbol{q}}_{i+1} = 0. \tag{40}$$

This equation provides a set of three linear systems relating the joint coordinates following each loop. If the set of linear equations is appended together the following equation holds:

$$\mathbf{X} = \begin{bmatrix} \mathbf{A}_1 & -\mathbf{A}_2 & \mathbf{0}_{3\times3} \\ \mathbf{0}_{3\times3} & \mathbf{A}_2 & -\mathbf{A}_3 \\ \mathbf{A}_1 & \mathbf{0}_{3\times3} & -\mathbf{A}_3 \end{bmatrix}\dot{\boldsymbol{q}} = 0. \tag{41}$$

The relationship between the active and passive coordinates can be obtained from Eq. (41) as follows:

$$\mathbf{G}_{\mathrm{III}} = \mathbf{X}_\mathrm{p}^{-1}\mathbf{X}_\mathrm{a}. \tag{42}$$

4 Results and discussion

In order to solve the inverse dynamics of the 3-PRS parallel manipulator, each term of Eqs. (4), (6), and (10) are found in closed form by using a Computer Algebra Symbolic (CAS) program, such as Maple. One of the advantages of developing the model in a CAS program relies on the fact that the mathematical operations can be performed symbolically and simplified. In the present case, built-in functions such as `simplify` and `combine` (with `option=trig`) of Maple programming environment were used to reduce the number of operations for solving each model.

A second advantage of obtaining the model in closed form is that the code can be written automatically for `Matlab` by using the code generation capabilities of the software. The `Matlab` procedure with `optimize=tryhard` option of the package `CodeGeneration` were used in this case to develop Matlab code. The number of algebraic operations (i.e., additions/subtractions and divisions/multiplications) necessary for solving the dynamic problem was obtained through the `cost` function of the `codegen` package. Without any loss of generality, the number of operations for matrix inversion (i.e., when obtaining $\mathbf{J}_\mathrm{p}$) were computed by considering the number of operations conventional LU decomposition takes to solve a linear system: about $+, -n^3/3 - n^2/2 + 5n/6$ and $\times$ $n^3/3 + n^2 - n/3$ (Chapra and Canale, 2006). The objective of this paper is to evaluate the complexity and the computational load of these three formulations which is presented in Table 3.

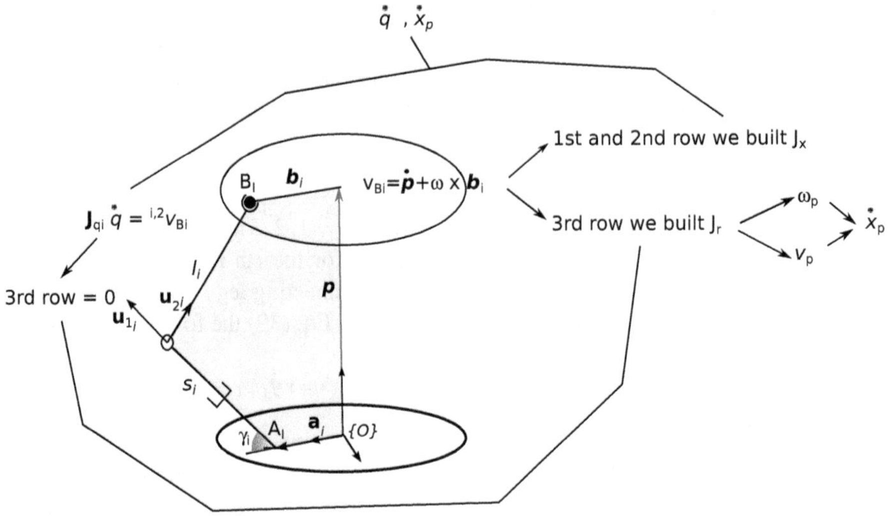

Figure 4. Formulation of the Jacobian matrix using Method II.

Table 3. Computational effort for solving the inverse dynamic problem.

Model Eq. (6)			Model Eq. (4)			Model Eq. (10)		
Term	×/÷	+/−	Term	×/÷	+/−	Term	×/÷	+/−
$\boldsymbol{h}^{\mathrm{a}}$	21	15	$\boldsymbol{h}^{\mathrm{a}}$	21	15	$\boldsymbol{h}^{\mathrm{a}}$	97	67
$\boldsymbol{h}^{\mathrm{p}}$	27	18	$\boldsymbol{h}^{\mathrm{p}}$	27	18	$\boldsymbol{h}^{\mathrm{p}}$	600	380
$\boldsymbol{h}_{\mathrm{p}}$	404	253	$\boldsymbol{h}_{\mathrm{p}}$	404	253			
$\mathbf{X}_{\mathrm{p}}$	–	–	$\mathbf{X}_{\mathrm{p}}$	6	12	$\mathbf{X}_{\mathrm{p}}$	150	56
$\mathbf{X}_{\mathrm{a}}$	–	–	$\mathbf{X}_{\mathrm{a}}$	31	27	$\mathbf{X}_{\mathrm{a}}$	0	0
$\mathbf{G}_{\mathrm{II}}$	76	17	$\mathbf{G}_{\mathrm{I}}$	109	72	$\mathbf{G}_{\mathrm{III}}$	478	308
$\mathbf{J}_q^{-1}\mathbf{J}_x$	90	50	$\mathbf{J}_q^{-1}\mathbf{J}_x$	70	28			
$\mathbf{J}_r$	52	26	$\mathbf{J}_r$	52	26			
$\mathbf{J}_c$	196	121	$\mathbf{J}_c$	196	113			
$\mathbf{J}_{\mathrm{p}}$	265	226	$\mathbf{J}_{\mathrm{p}}$	245	204			
$\mathbf{G}^T\boldsymbol{h}$	832	568	$\mathbf{G}^T\boldsymbol{h}$	833	574	$\mathbf{G}^T\boldsymbol{h}$	1193	773

As Table 3 shows, the models in Eqs. (4) and (6), which are based on splitting the platform from the legs, hold fewer number of operation than those of the model considering the platform attached to one of the legs. This fact is due to the inversion of a 6×6 matrix when finding the matrix $\mathbf{G}_{\mathrm{III}}$. On the other hand, by having the platform attached to one leg, the projection of the platform generalised forces onto the actuated joints is cumbersome. The result shows that either using Eqs. (4) or (6) a reduction of about 30 % in the number of multiplication and about 25 % in additions is obtained when comparing to the model given in Eq. 10). These results indicate that considering the platform and legs as subsystem can improve speed in the calculation of dynamics for applications such as model-based control, see for example Díaz-Rodríguez et al. (2013).

5 Conclusions

Three strategies for dynamic modelling of parallel manipulators were revisited and applied for developing the inverse dynamic model of a general 3-PRS parallel manipulator. The first method exploited parallel manipulators particularities such that the platform and the links are considered as subsystems, namely: the legs and the platform, thus, joint coordinates were used for modelling the legs dynamics and task space coordinates for the platform. The core concept of the second method is based on solving the dynamic model through this approach reduces to how the Jacobian matrix relating the task and the joint space coordinates is computed and it was explained in this paper. A third method was presented which is similar to the second method but uses a different formulation of the Jacobian matrix. In the third approach the dynamic model was obtained by describing the manipulator as three open kinematic chains. The vector-loop closure constraints introduced the relationship between the dynamics of the open kinematic chains and the original closed chains. A Computer Algebraic Software allowed to find each term of the dynamic in symbolic form and to compute the computational burden of each model.

The results showed that the approaches based on splitting the manipulator in two sub-systems (platform and legs) require about 30 % in the number of multiplication and about 25 % in additions are obtained when comparing to the model given by Eq. (10). These results indicate that in problems when the model is needed to be computed on-line at high rates of speed, method 1 and 2 can be useful. This work has provided numerical guidelines for implementing computationally efficient models for use in numerically intensive optimal mechanical synthesis problems or in resource-constraint embedded computers, particularly for control and model identification. The software support presented for solving the inverse dynamic problem efficiently also provides some insight on some of the advantages and/or disadvantages on these revisited methods.

Acknowledgements. The authors acknowledge the financial support from the Natural Science and Engineering Research council of Canada (NSERC), the New Brunswick Innovation Foundation (NBIF), Fondo Nacional de Ciencia, Tecnología e Innovación (FONACIT-Venezuela), CONACYT scholarship 326912/381134 and also the SNI-México.

Disclaimer. Conflict of interests – the authors declare that the research was conducted in the absence of any commercial, financial, or personal relationships that could be construed as a potential conflict of interests.

References

Angeles, J.: Fundamentals of Robotic Mechanical Systems, Chapter 4, 2nd Edn., Springer, 2002.

Carbonari, L., Battistelli, M., Callegari, M., and Palpacelli, M.-C.: Dynamic modelling of a 3-CPU parallel robot via screw theory, Mech. Sci., 4, 185–197, doi:10.5194/ms-4-185-2013, 2013.

Carretero, J. A., Nahon, M. A., and Podhorodeski, R. P.: Workspace analysis and optimization of a novel 3-DOF parallel manipulator, Int. J. Robot. Autom., 14, 178–188, 2000a.

Carretero, J. A., Podhodeski, R. P., Nahon, M. A., and Gosselin, C. M.: Kinematic analysis and optimization of a new three degree-of-freedom spatial parallel manipulator, J. Mech. Design, 122, 17–24, 2000b.

Chapra, S. C. and Canale, R.: Numerical Methods for Engineers, 5th Edn., McGraw-Hill, Inc., New York, NY, USA, 2006.

Craig, J. J.: Introduction to Robotics: Mechanics and Control, 3rd Edn., Pearson Education, Upper Saddle River, NJ, USA, 2005.

Díaz-Rodríguez, M., Valera, A., Mata, V., and Valles, M.: Model-Based Control of a 3-DOF Parallel Robot Based on Identified Relevant Parameters, IEEE-ASME T. Mech., 18, 1737–1744, 2013.

Fan, K. C., Wang, H., and Chang, T. H.: Sensitivity analysis of the 3-PRS parallel kinematic spindle platform of a serial-parallel machine tool, Int. J. Mach. Tool. Manu., 43, 1561–1569, 2003.

Goodwin, G. C., Middleton, R. H., and Poor, H. V.: High-speed digital signal processing and control, Proceedings of the IEEE, 80, 240–259, 1992.

Ibrahim, O. and Khalil, W.: Kinematic and Dynamic Modeling of the 3-PRS Parallel Manipulator, in: Proceedings of the 12th IFToMM World congress, France, 18–21 June 20017, 1–6, 2007.

Khalil, W. and Ibrahim, O.: General Solution for the Dynamic Modeling of Parallel Robots, J. Intell. Robot. Syst., 49, 19–37, 2007.

Li, Y. M. and Xu, Q. S.: Kinematics and inverse dynamics analysis for a general 3-PRS spatial parallel manipulator, Robotica, 22, 219–229, 2004.

Li, Y. M. and Xu, Q. S.: Kinematic analysis of a 3-PRS parallel manipulator, Robot. CIM. Int. Manuf., 23, 395–408, 2007.

Liu, X. J. and Bonev, I. A.: Orientation Capability, Error Analysis, and Dimensional Optimization of Two Articulated Tool Heads With Parallel Kinematics, J. Manuf. Sci. Eng., 130, 1–9, 2008.

Mata, V., Provenzano, S., Valero, F., and Cuadrado, J. I.: Serial-robot dynamics algorithms for moderately large numbers of joints, Mech. Mach. Theory, 37, 739–755, 2002.

Mata, V., Farhat, N., Díaz-Rodríguez, M., Valera, A., and Page, A.: Dynamic parameters identification for parallel manipulator, Tech Education and Publishing, Vienna, Austria, 21–44, 2008.

Merlet, J. P.: Micro parallel robot MIPS for medical applications, in: Proceedings of the 8th international conference on emerging technologies and factory automation (ETFA 2001), France, 15–18 October 2001, 611–619, 2001.

Murray, J. and Lovell, G.: Dynamic modeling of closed-chain robotic manipulators and implications for trajectory control, IEEE T. Robotic. Autom., 5, 522–528, 1989.

Parsa, S. S., Carretero, J. A., and Boudreau, R.: Internal redundancy: an approach to improve the dynamic parameters around sharp corners, Mech. Sci., 4, 233–242, doi:10.5194/ms-4-233-2013, 2013.

Staicu, S.: Matrix modeling of inverse dynamics of spatial and planar parallel robots, Multibody Syst. Dyn., 27, 239–265, 2012.

Tsai, M. S. and Yuan, W. H.: Inverse dynamics analysis for a 3-PRS parallel mechanism based on a special decomposition of the reaction forces, Mech. Mach. Theory, 45, 1491–1508, 2010.

Tsai, M. S., Shiau, T. N., Tsai, Y. J., and Chang, T. H.: Direct kinematic analysis of a 3-PRS parallel manipulator, Mech. Mach. Theory, 38, 71–83, 2002.

Williamson, D.: Digital Control and Implementation: Finite Wordlength considerations, Prentice Hall, 1991.

Yiu, Y., Cheng, H., Xiong, Z. H., Liu, G., and Li, Z.: On the dynamics of parallel manipulators, in: Robotics and Automation, 2001, Proceedings 2001 ICRA, IEEE International Conference on, 4, 3766–3771, 2001.

Yuan, W. H. and Tsai, M. S.: A novel approach for forward dynamic analysis of 3-PRS parallel manipulator with consideration of friction effect, Robot. CIM. Int. Manuf., 30, 215–325, 2014.

13

A parametric investigation of a PCM-based pin fin heat sink

R. Pakrouh[1], M. J. Hosseini[2], and A. A. Ranjbar[1]

[1]Department of Mechanical Engineering, Babol University of Technology, P.O. Box 484, Babol, Iran
[2]Department of Mechanical Engineering, Golestan University, P.O. Box 155, Gorgan, Iran

Correspondence to: M. J. Hosseini (mj.hosseini@gu.ac.ir)

Abstract. This paper presents a numerical investigation in which thermal performance characteristics of pin fin heat sinks enhanced with phase-change materials (PCMs) designed for cooling of electronic devices are studied. The paraffin RT44 HC is poured into the aluminum pin fin heat sink container, which is chosen for its high thermal conductivity. The effects of different geometrical parameters, including number, thickness and height of fins, on performance are analyzed. Different aspects for heat transfer calculation, including the volume expansion in phase transition as well as natural convection in a fluid zone, are considered in the study. In order to validate the numerical model, previous experimental data and the present results are compared, and an acceptable agreement between these two is observed. Results show that increasing the number, thickness and height of fins leads to a significant decrease in the base temperature as well as operating time of the heat sink.

1 Introduction

Because of the fast development of modern technology and miniaturization of electronic packaging, new electronic devices generate a large amount of heat, which threatens their performance and efficiency. Therefore, thermal management of effective cooling systems has become one of the most important considerations in design of different pieces of electronic equipment. Considering the effects of temperature on the performance of electronic devices, an effective thermal design should be able to keep the working temperature of devices below their allowable maximum temperature during the entirety of normal operation.

There are common techniques, including air cooling, liquid cooling, piezoelectric pump and heat pipes, to remove high heat flux effectively from heat-generating electronic devices. Thermal energy storage, as a cooling method for electronic applications, is one of the techniques that have been widely researched in recent years. The thermal properties of some phase-change materials (PCMs), including melting- and solidification-related specifications, are listed by Abhat (1983). High latent heat of fusion, the capability of being a heat source at a constant temperature, and chemical stability are three of the favorite properties of PCMs. However these materials have the undesirable property of low thermal conductivity, which brings about a serious challenge in design application of PCM-based electronic cooling systems. In order to overcome this drawback, different techniques of enhancement have been proposed, including fins (Eftekhar et al., 1984; Henze and Humphrey, 1981), metal matrices (Ettouni et al., 2006) and nanoparticles (Ranjbar et al., 2011; Hosseini et al., 2013).

Several previous papers have studied the advantages of PCM utilization in electronic cooling systems. As reported by Pal and Joshi (2001), use of a PCM-based heat sink is an effective method for cooling of electronic devices. Kandasamy et al. (2007) in a combined experimental and numerical work studied the effects of various parameters such as power input level and the system configuration on PCM melting rate. Results showed that the geometrical shape of the package does not have a significant effect on the performance, while the input power influences the melting rate noticeably. Akhilesh et al. (2005) directed a numerical investigation to develop a thermal design procedure that maximizes the operating time of a composite heat sink which is made up of an elemental heat sink, PCM and high-conductivity base material. Hosseinizadeh et al. (2011) reported a parametric study on a plate fin PCM-based heat sink and found that increasing the number of fins and their height results in a considerable increase in overall thermal performance, whereas increasing the fins' thickness only brings about a slight improvement. Baby and Balaji (2012) compared the heat transfer performance of a plate fin heat sink with that of a pin fin heat sink in an experimental study. Their results showed that the pin fin heat sink exhibits a longer operation duration for electronic devices in comparison with similar plate fin heat sinks. Baby and Balaji (2013) experimentally developed an artificial neural network–genetic hybrid algorithm to determine the optimum configuration of a PCM-based pin fin heat sink in which the critical time is maximized. Shatikian et al. (2008) recommended a correlation among Nusselt, Stefan and Fourier numbers for a constant heat flux system. In their simulation, a complete formulation was endeavored which considers convection in the fluid zone and volume expansion of the PCM during phase transition. Nayak et al. (2006) studied a numerical model for PCM-based heat sinks with some different types of thermal conductivity enhancer (TCE) distribution. Their results showed that the use of TCE leads to a lower chip temperature. They also concluded that the convection in the melted PCM improves the temperature uniformity by increasing the effective heat transfer coefficient. Saha et al. (2010) presented a numerical investigation on melting of n-eicosane in a PCM-based plate fin heat sink to find the influence of the enclosure's aspect ratio on the performance of the system. They identified different non-dimensional numbers as effective parameters. They also have found a suitable length scale for each type of enclosure. Fok et al. (2010) experimentally investigated a PCM-based heat sink for cooling of portable hand-held electronic devices. They found that the orientation of the heat sink does not affect the heat sink performance meaningfully. Kozak et al. (2013) studied heat transfer as well as heat accumulation in a hybrid PCM–air heat sink both experimentally and numerically. The constant heat flux was applied to the bottom surface and the PCM subjected to a fan-driven forced convection. They have found that as the power input level increases, so too does the rate of sensible heat accumulation.

The present work evaluates the melting process of the PCM in the presence of pin-fin-type thermal conductivity enhancers (TCEs) for electronics cooling applications. The main purpose of this article is to investigate different effective parameters on the heat sink base temperature and melt fraction. The studied geometric parameters are the number of fins (for three values), fin thickness (for three values) and fin height (for two values).

2 Numerical model

2.1 Governing equations

The schematic diagram of the PCM-based pin fin heat sink is shown in Fig. 1. The heat sink is made of aluminum with a different number of fins as well as different fin thicknesses, fin heights and base thicknesses. The dimensions of heat sink base are 70 mm × 70 mm. The base thickness is kept constant at 2 mm. To simulate the heat revealed by the device, some elements are designated to generate 10.204 kW m^{-2}. This heat-generating device, which is inserted under the bottom of the sink, requires 50 W of power. The sidewalls of the heat sink are merely there as preservative walls of the melted PCM and have no role in the heat transfer process. Thus the sidewalls are assumed to be adiabatic and it is supposed that the heat supplied to the system is transferred between the PCM and the heat transfer component of the device including the base, the fins and finally the surrounding air. A total of 10 different simulations, summarized in Table 1, with different geometrical dimensions are carried out in the present work. The commercial paraffin RT44 HC (Rubitherm GmbH, 2012) was used as the PCM. Table 2 shows the thermophysical properties of the material.

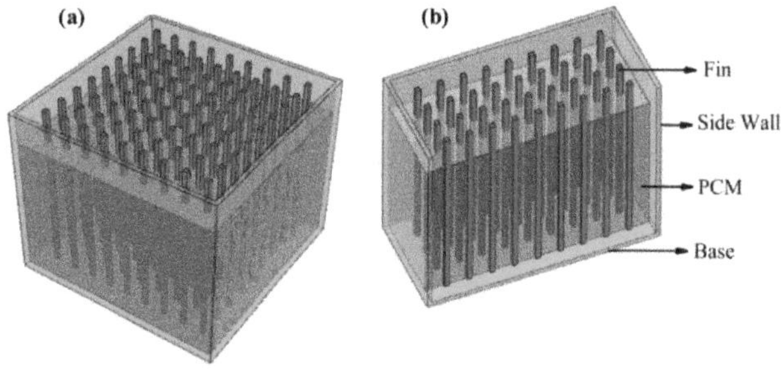

Figure 1. Example of a PCM-based pin fin heat sink: **(a)** actual geometry and **(b)** sectioned view.

Table 1. Numerical cases explored in this work.

Case	Fin height, H_F (mm)	Number of fins, N	Fin thickness, w_F (mm)
Case 1	15	25	2
Case 2	15	25	4
Case 3	15	25	6
Case 4	15	49	6
Case 5	15	100	6
Case 6	25	25	2
Case 7	25	25	4
Case 8	25	25	6
Case 9	25	49	6
Case 10	25	100	6

In order to simulate the melting process of the PCM, the enthalpy–porosity approach (Brent et al., 1988) is employed, wherein the porosity in each cell is set equal to the liquid fraction in that cell. In order to allow the PCM expansion during melting, the enclosure is filled to only 90 % of the fin height and the remaining volume is occupied by air. Numerical approaches, previously applied by other authors to heat sinks enhanced with plate fins (Shatikian et al., 2005;

Table 2. Properties of materials employed in the present study.

Material	ρ (kg m^{-3})	k (W m^{-1} K^{-1})	C_p (J kg^{-1} K^{-1})	T_m (°C)	ΔH (J kg^{-1})
Paraffin RT44	780 (solid) 760 (liquid)	0.2	2000	41–45	255 000
Aluminum	2719	202.4	871	660.4	–
Air	$1.2 \times 10^{-5}T^2 - 0.01134T + 3.498$	0.0242	1006.4	–	–

Kandasamy et al., 2007), considered a similar assumption to the expansion. A "volume-of-fluid" (VOF) model is used to solve the PCM–air system in which a moving internal interface is considered and interpenetration of the two fluids is disregarded (Hirts and Nichols, 1981). In this model, if the nth fluid's volume fraction in the computational cell is denoted as α_n, then the following three conditions are possible:

- $\alpha_n = 0$ the cell is empty of the nth fluid,
- $\alpha_n = 1$ the cell is full of the nth fluid,
- $0 < \alpha_n < 1$ the cell contains the interface between the fluids.

Thus, the variables and properties in any given cell are either purely representative of one of the fluids or are a mixture of fluids, depending on their volume fraction values. For the air phase, the density depends on its temperature as shown in Table 2. Thus due to the compressibility of the air during melting, a compressible model is utilized.

For the aluminum base and fins, only a conduction mechanism is considered. Accordingly, the energy equation is

$$\frac{\partial}{\partial t}(\rho_{\mathrm{Al}}\hbar) = \frac{\partial}{\partial x_i}\left(k_{\mathrm{Al}}\frac{\partial T}{\partial x_i}\right), \tag{1}$$

where $\hbar$ (J kg^{-1}) is the specific enthalpy and k_{Al} (W m^{-1} K^{-1}) is the thermal conductivity of aluminum. Also, the governing equations used here for modeling the PCM–air system are

(a) continuity equation,

$$\frac{D\alpha_n}{Dt} = 0; \tag{2}$$

(b) momentum equation,

$$\rho\frac{D\boldsymbol{v}}{Dt} = -\nabla P + \mu\nabla^2\boldsymbol{v} + \rho\boldsymbol{g} + \boldsymbol{S}; \tag{3}$$

(c) energy equation,

$$\rho\frac{D\hbar}{Dt} = k\nabla^2 T; \tag{4}$$

where α_n is the nth fluid's volume fraction in computational cells. As shown in Table 2, a density–temperature relation is assumed for air. For the PCM portion, the density in liquid phase can be expressed as

$$\rho = \frac{\rho_l}{\beta(T - T_l) + 1}, \tag{5}$$

where β is the thermal expansion coefficient and is set to 0.001, which is given in the study done by Humphries and Griggs (1977). The dynamic viscosity of the liquid PCM has been expressed as

$$\mu = 0.001 \times \exp\left(A + \frac{B}{T}\right), \tag{6}$$

where $A = -4.25$ and $B = 1790$ are reported by Reid et al. (1987). $S_{n,i}$ is the momentum sink due to the reduced porosity in the mushy zone that takes the following form:

$$\boldsymbol{S} = -\frac{C(1-\gamma)^2}{(\gamma^3 + \varepsilon)}\boldsymbol{v}, \tag{7}$$

where $C(1-\gamma)^2/(\gamma^3 + \varepsilon)$ is the mimic "porosity function" defined by Brent et al. (1988). The value of C is the mushy zone constant, which measures the amplitude of the velocity damping, and is considered to be $C = 10^5$ in the present study.

2.2 Boundary and initial conditions

In all simulations the initial temperature of the whole system is $T_0 = 27$ °C. Heat transfer from the fin tips to the ambient is considered negligible. Considering the symmetry of the geometry and following Dubovsky et al. (2009), only one-quarter of the fin horizontal cross section is included in the computation (Fig. 2). It is assumed that the distance between the exterior fins the and heat sink walls is half of the space between the fins; this assumption leads to a similar simulation for all the included fins. We note that the PCM layer's width (w_{P}) varies as the thickness (w_{F}) and number (N_{F}) of fins change. Table 3 presents the geometric variation of fin thickness and inter-fin distance.

The boundary conditions applied to the computational domain, in Fig. 2, are

(a) heat flux to the bottom,

$$-k_{\mathrm{Al}}\left.\frac{\partial T}{\partial y}\right|_{y=0} = q''; \tag{8}$$

Table 3. Fin thickness and inter-fin distance for different fin numbers.

	w_F (mm)	w_P (mm)	w_F (mm)	w_P (mm)	w_F (mm)	w_P (mm)
$N_F = 25$	2	12	4	10	6	8
$N_F = 49$	2	8	4	6	6	4
$N_F = 100$	2	5	4	3	6	1

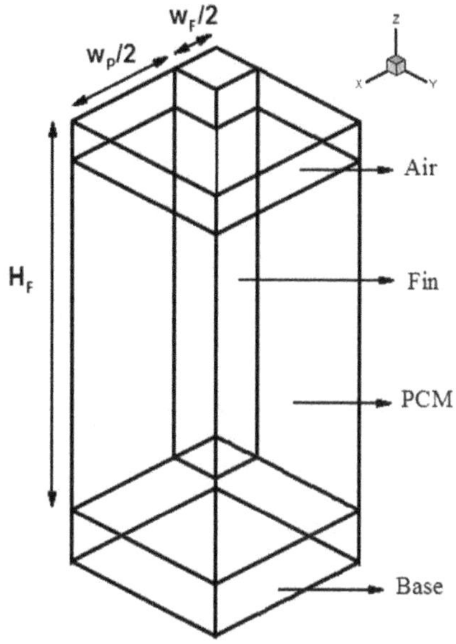

Figure 2. Schematic of the computational domain.

(b) symmetry boundary conditions at all enclosing sides,

$$\frac{\partial T}{\partial x} = \frac{\partial T}{\partial y} = 0, \ \frac{\partial u_i}{\partial x} = \frac{\partial u_i}{\partial y} = 0; \tag{9}$$

(c) adiabatic boundary condition at fins' tip,

$$-k_{\mathrm{Al}} \frac{\partial T}{\partial y} = 0. \tag{10}$$

2.3 Numerical procedure and validation

The SIMPLE algorithm within a 3-D in-house code (Rahimi et al., 2012) developed by authors was utilized for pressure–velocity coupling. The QUICK differencing scheme was adopted for solving the momentum and energy equations, whereas the PRESTO scheme was used for the pressure correction equation. The effects of time step and grid size on the solution were carefully examined in preliminary simulations. Therefore, the time step in the simulations is set to 0.005 s. Different grid sizes are used in the simulations, varying from 9500 to 34 000 cells for the fin height of 15 mm, 15 500 to 56 000 cells for the height of 25 mm, and 18 500 to 74 000 cells for the height of 35 mm. It is noticeable that, for the same fin height, the population of denser fins requires a larger number of grid cells. The convergence is checked at each time step, with the convergence criterion of 10^{-6} for the continuity equation, 10^{-5} for the momentum equation and 10^{-8} for the energy equation. In order to validate the melting computational model in our finite-volume CFD code, initial runs were performed and compared with experimental data of Hosseinizadeh et al. (2011). As can be seen from Fig. 3, there is a good agreement between the present calculation and those calculations of Hosseinizadeh et al. (2011).

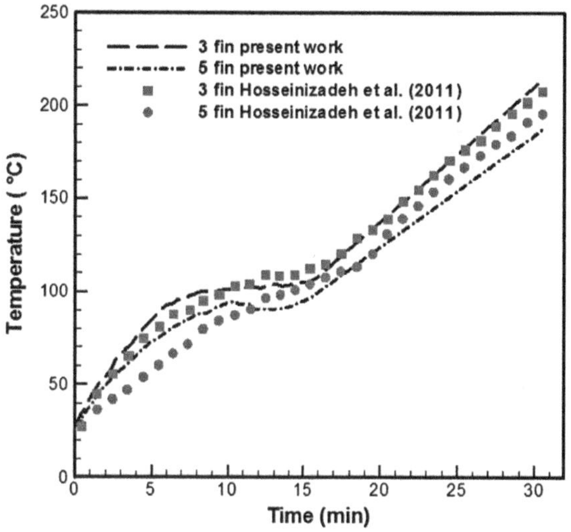

Figure 3. Comparison of base temperature profile between the present work and Hosseinizadeh et al. (2011).

3 Numerical results

In this section, the results of the parametric study are presented and discussed. The analysis involves the PCM melt fractions and the base temperatures as functions of time, which is intended to explain the effect of the geometry variation on the thermal performance of the heat sink.

3.1 Effect of number of fins

Figures 4 and 5 present the melt fraction and base temperature as functions of time for different number of fins at various fin heights. As expected, for both fin heights of 15 and 25 mm, the total melting time of the PCM decreases as the number of fins increases, as shown in Fig. 4a and b. This is due to the fact that the PCM volume decreases as the number

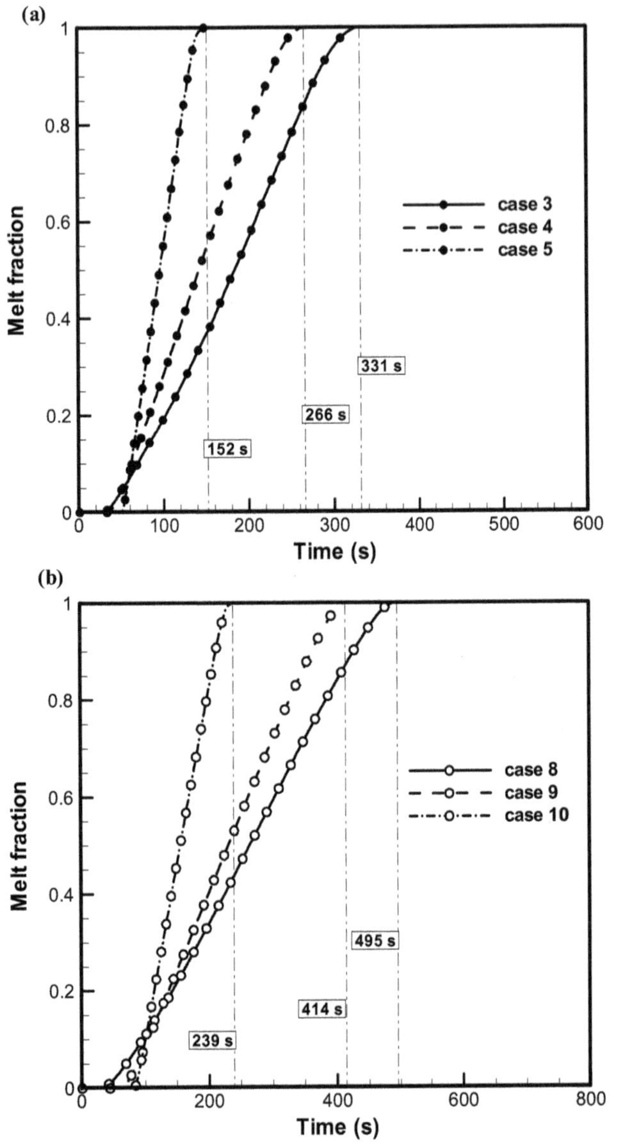

Figure 4. Melt fraction evolution with time for different numbers of fins: **(a)** various cases for fin height of 15 mm and **(b)** various cases for fin height of 25 mm.

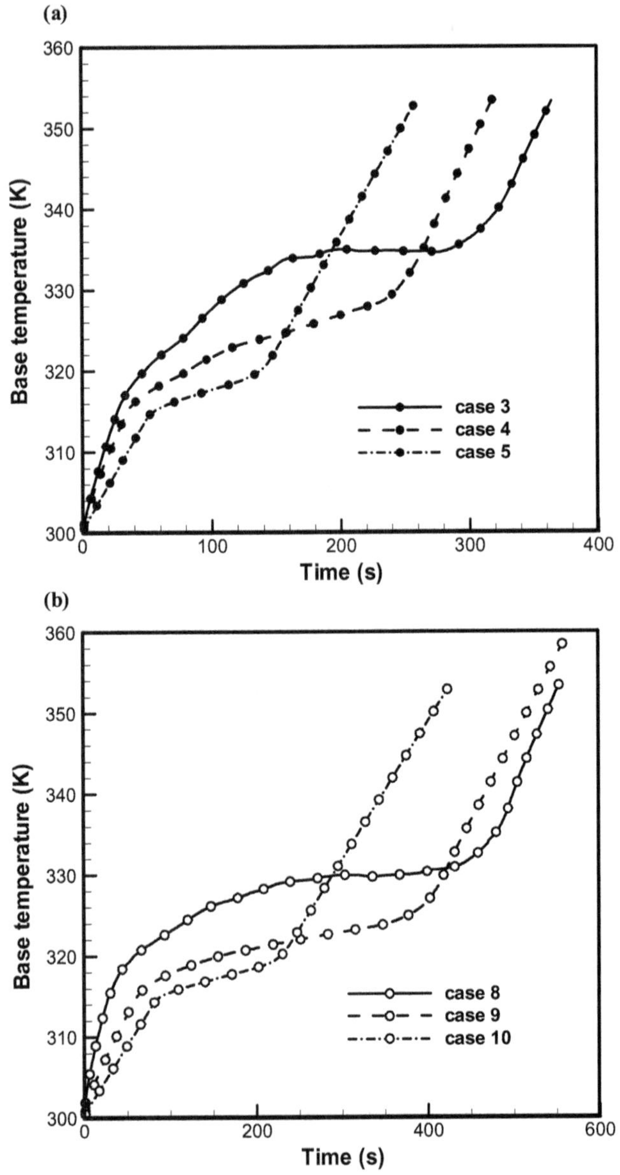

Figure 5. Base temperature evolution with time for different numbers of fins: **(a)** various cases for fin height of 15 mm and **(b)** various cases for fin height of 25 mm.

of fins rises. The lesser volume of the PCM leads to a latent heat potential reduction and thus a shorter melting period.

The effect of the number of fins is examined by checking the base temperature profiles in Fig. 5. It can be seen that the curves have similar forms and trends for both fin heights of 15 and 25 mm. Because of the extra surface, 100 fins provide more heat storage capacity compared to 25 fins, so a lower base temperature is observed for a greater number of fins, i.e., a 100-fin unit has a lower base temperature compared to units with 49 and 25 fins. It should be noted that although the base temperature decreases as the number of fins increases (which is significantly effective in electronics cooling systems), the latent heat period (which maintains the base temperature within a specified and almost constant range) decreases. For instance, in Fig. 5b, the latent heat period for heat sinks with 25, 49 and 100 fins is about 392, 310 and 138 s, respectively. As mentioned earlier, this is due to the decrease in the latent heat capacity of the system while the number of fins increases.

3.2 Effect of fin thickness

Figure 6 shows the melt fraction versus time for different pin fin thicknesses. As can be seen, the total melting time of PCM decreases as fin thickness increases. As mentioned, this is because the smaller PCM volume provides a lower

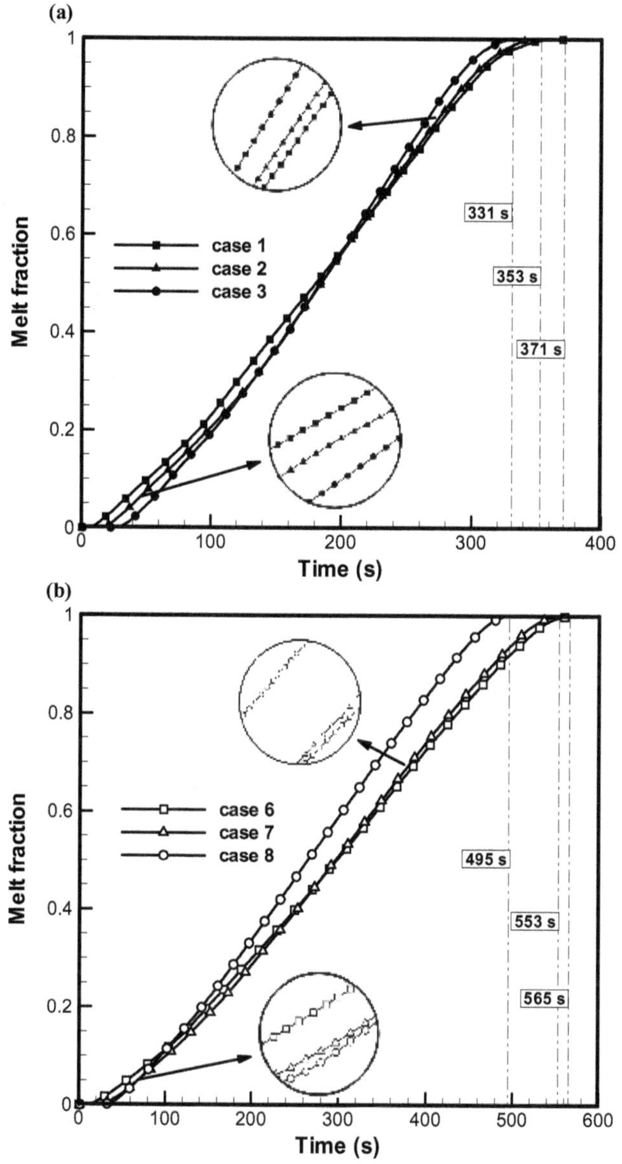

Figure 6. Melt fraction evolution with time for different fin thicknesses: **(a)** various cases for fin height of 15 mm and **(b)** various cases for fin height of 25 mm.

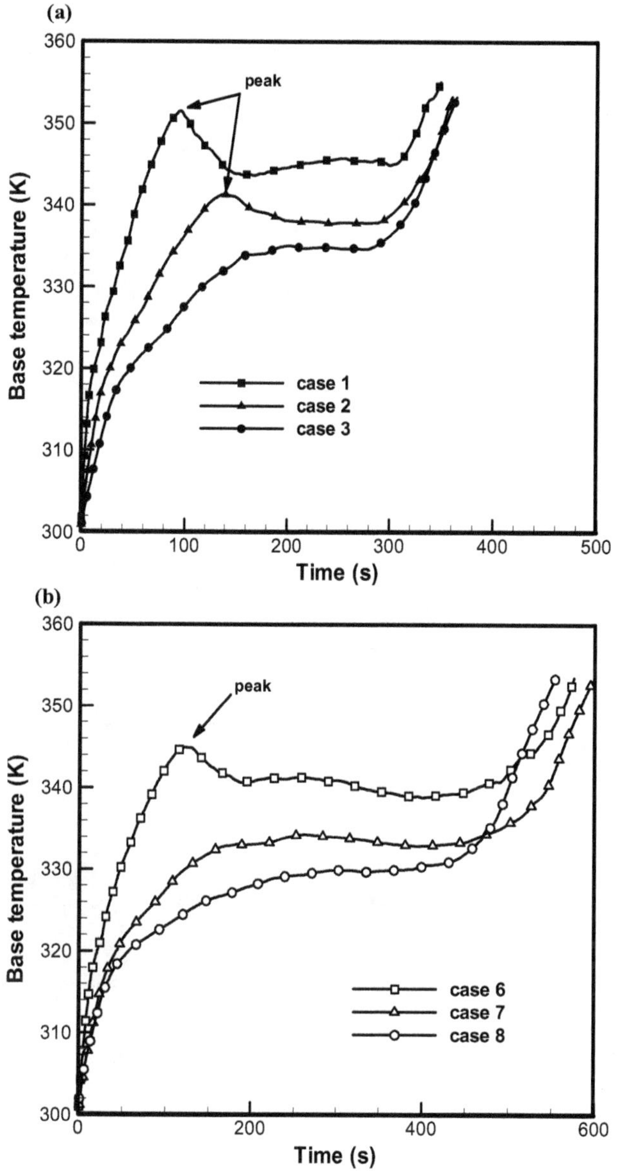

Figure 7. Base temperature evolution with time for different fin thicknesses: **(a)** various cases for fin height of 15 mm and **(b)** various cases for fin height of 25 mm.

latent heat absorption capacity. Second, the initiation of the PCM melting occurs later as the fins become thicker. This behavior was first observed by Hosseinizadeh et al. (2011) for PCM-based plate fin heat sinks. Their results showed that this property is related to the mode of heat transfer in the PCM, and heat is mainly transferred by conduction. According to Fourier's law ($q'' = -k\nabla T$), the conduction is related to the temperature gradient. This means that a higher temperature gradient results in greater heat conduction. In the present study the fins with 2 mm thickness give the higher temperature gradient compared to those of 4 and 6 mm thickness, and as we can see, a faster onset of the melting process is achieved in Fig. 6a and b.

Figure 7 compares the base temperature profiles for various fin thicknesses. It can be seen that, with increasing fin thickness, the base temperature decreases considerably during the latent heat period. For instance, case 1 exhibits a lower temperature than cases 2 and 3 by about 8 and 10 K respectively.

From Fig. 7, it can be observed that fast growth of the base temperature causes a peak in the profiles of thinner fins. In other words, due to the low heat capacity of the thinner fins, the base temperature rises rapidly at the early stages of melting. As the effect of natural convection increases, the flow circulation results in a local decrease in base temperature.

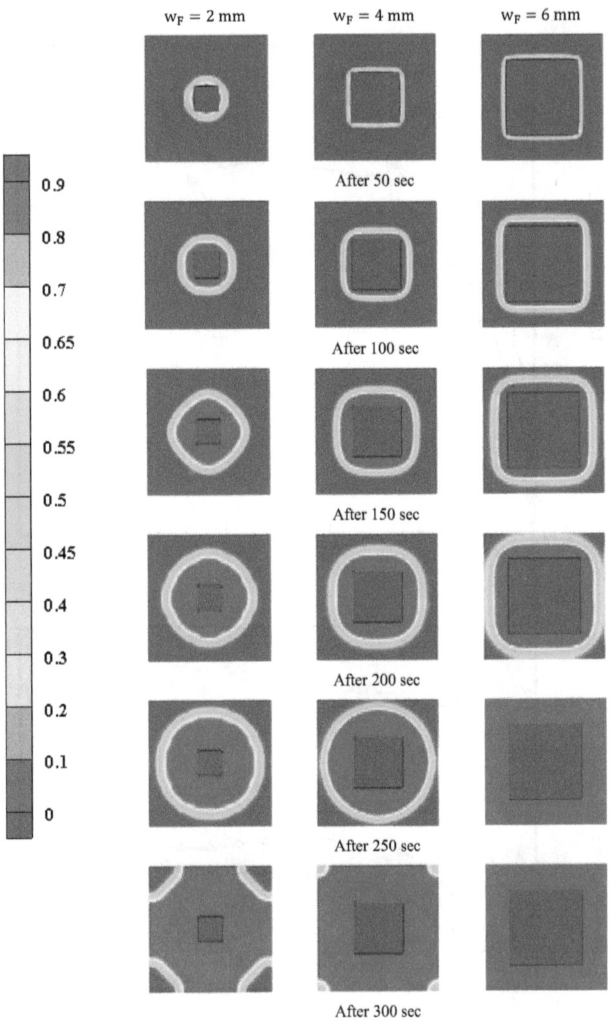

Figure 8. Melt fraction distribution for 49 fins and various fin thicknesses.

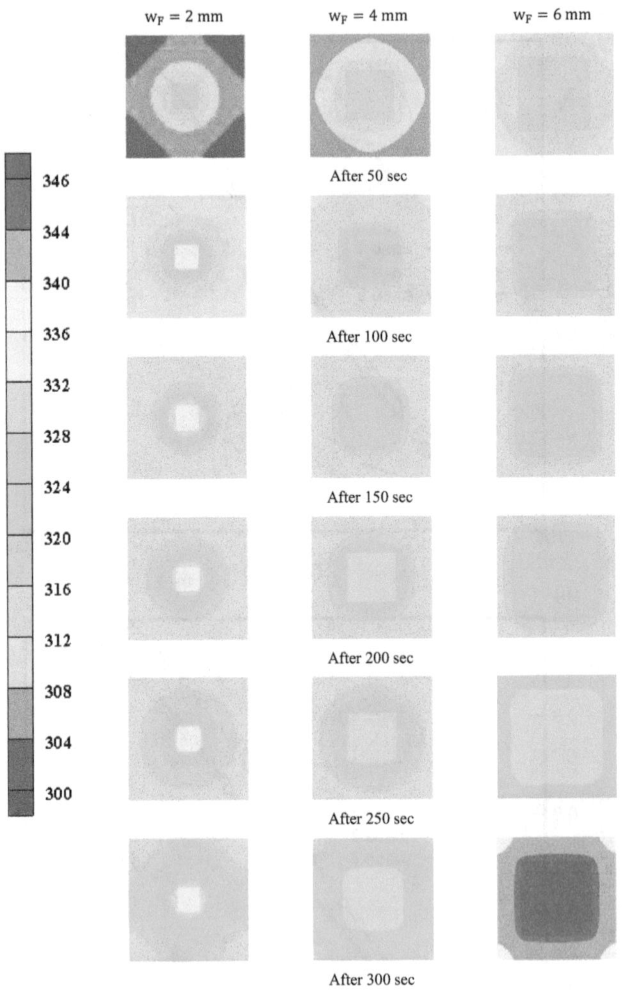

Figure 9. Temperature evolution for 49 fins and various fin thicknesses.

But for thicker fins, due to the high heat capacity of the system, heat is distributed uniformly and thus the base temperature curves are flattened. We can see that various fin thicknesses in Fig. 7b have smoother profiles compared to those in Fig. 7a. For example, the 4 mm thick fin of 15 mm fin height (case 2) includes the peak, but no peak is observed for that of 25 mm fin height (case 7).

Figure 8 shows plan view of the melt front evaluation with time for different fin thicknesses at half-height of the fin. As can be seen, in all cases the melting process starts at the four sidewalls of the fins. Over time, more of the PCM melts, and at final stages of melting processes, the liquid regions formed around the fins connect to each other. Temperature contours corresponding to Fig. 8 are given in Fig. 9. It is worth mentioning that, before 250 s, the thicker fin (w_F =6 mm) leads to a lower temperature compared to thinner ones (w_F = 4, 2 mm). After 250 s the temperature of the thicker fins rises to a large value. This is due to the fact that all of the PCM is liquid after 250 s, i.e., all the latent heat capacity of the system has been actualized, and hence the fin temperature increases more quickly.

As previously mentioned, the present work considers the effect of natural convection in the liquid PCM. Figure 10 deals with this subject for the wide case, case 6, and the narrow case, case 9. The right- and left-hand slices show the temperature evolution and velocity fields, respectively. As can be seen, in case 6, which contains a thick layer of PCM, the flow is trivial at a melt fraction of 0.1 but it becomes powerful at a melt fraction of 0.6. However, in case 9, which has a relatively thinner layer of PCM, the velocity field is weak even at a melt fraction of 0.6. This figure implies that the natural convection in the molten PCM depends on both the melt fraction and the geometrical characteristics of the heat sink. It should be noted that the strongest flow field is observed in the air region due to the lower density of air compared to that of liquid PCM. Figure 10 also shows that the use of

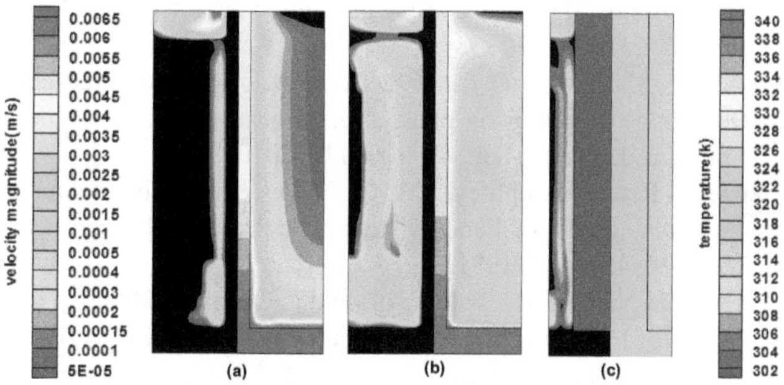

Figure 10. Temperature evolution and velocity fields: (**a**) case 6 and melt fraction of 0.1, (**b**) case 6 and melt fraction of 0.6, and (**c**) case 9 and melt fraction of 0.6.

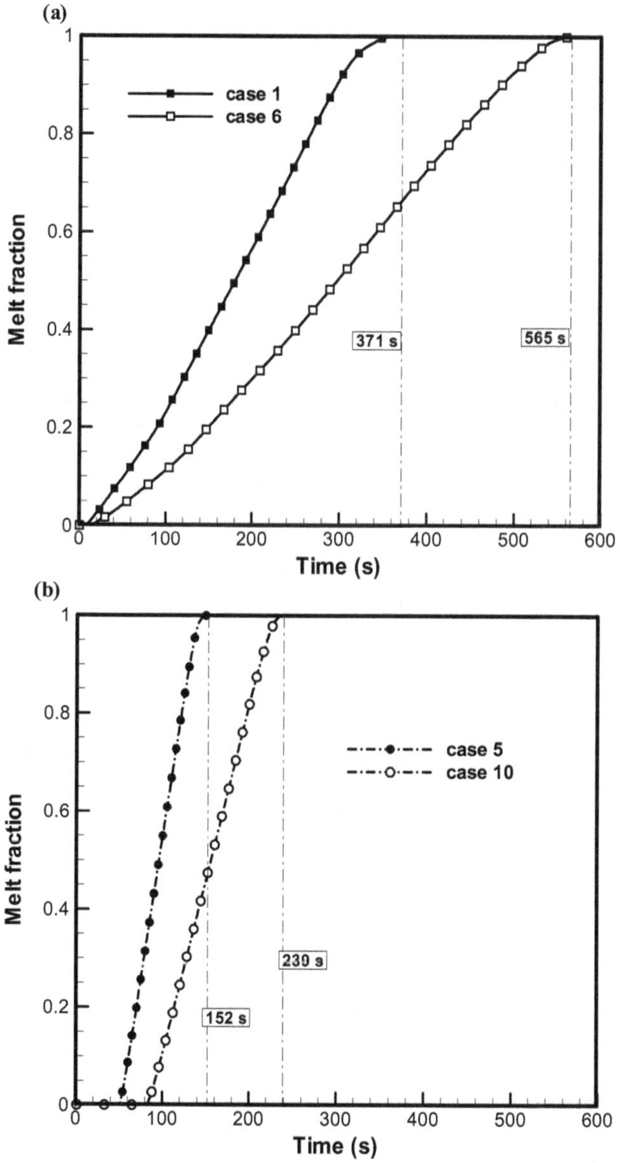

Figure 11. Melt fraction evolution with time for different fin heights: (**a**) various cases for 25 fins and $w_F = 2$ mm and (**b**) various cases for 100 fins and $w_F = 6$ mm.

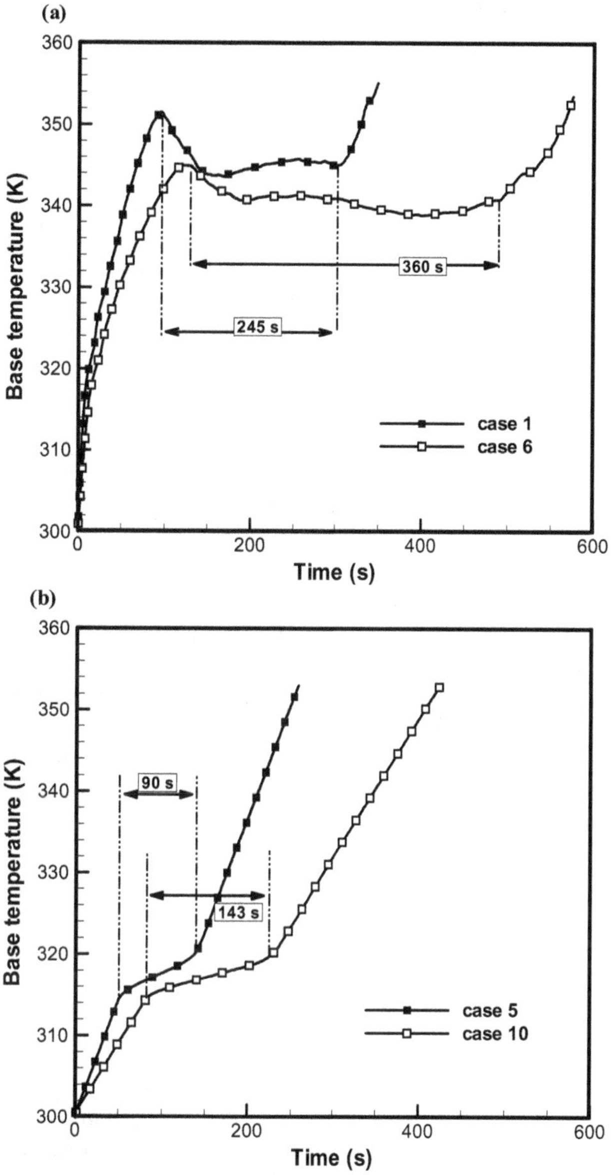

Figure 12. Base temperature evolution with time for different fin heights: (**a**) various cases for 25 fins and $w_F = 2$ mm and (**b**) various cases for 100 fins and $w_F = 6$ mm.

thick fins causes a more uniform temperature distribution in the system.

3.3 Effect of fin height

Figure 11 shows the melt fraction versus time for two fin heights. As expected, the total melting time increases with increasing fin height, e.g., from 371 to 565 s for cases 1 and 6. This is because of the large amount of PCM in the enclosure designed for the higher fins of the heat sinks. This observation is contrary to the results obtained by Hosseinizadeh et al. (2011) for plate fin heat sinks. This is due to the fact that, in their study, the height of the heat sink and consequently the amount of PCM content is assumed fixed, whereas in this study the heights of the heat sink and the fins are equal to each other. Therefore the PCM content increases as the fin height increases.

Figure 12 presents the temperature evaluation for the cases in Fig. 11. From this figure it can be seen that, for longer fins, the temperature growth is slower than that of the shorter fins, and thus the average temperatures of the heat sink are less. Furthermore, the latent heat period of the system in which the base temperature remains almost constant increases.

4 Conclusions

The use of PCMs in electronics cooling systems has been of increasing interest to researchers in recent years. Increasing the number of fins leads to a lower melting time. This is because of the fact that, by increasing the number of fins, the heat storage capacity of the system increases while the PCM volume decreases.

The total melting time of the PCM reduces when initiation of the PCM melting processes is delayed due to increased fin thickness. Finally, taller fins result in a lower melting time and lower base temperature. It can be concluded that increasing the number of fins as well as the fin thickness and height results in a lower base temperature, which means a lower chip temperature.

References

Abhat, A.: Low temperature latent heat thermal energy storage: heat storage materials, Sol. Energy, 30, 313–332, 1983.

Akhilesh, R., Narasimhan, A., and Balaji, C.: Method to improve geometry for heat transfer enhancement in PCM composite heat sinks, Int. J. Heat Mass Tran., 48, 2759–2770, 2005.

Baby, R. and Balaji, C.: Experimental investigations on phase change material based finned heat sinks for electronic equipment cooling, Int. J. Heat Mass Tran., 55, 1642–1649, 2012.

Baby, R. and Balaji, C.: Thermal optimization of PCM based pin fin heat sinks: An experimental study, Appl. Therm. Eng., 54, 65–77, 2013.

Brent, A. D., Voller, V. R., and Reid, K. J.: Enthalpy-porosity technique for modeling convection–diffusion phase change: application to the melting of a pure metal, Numer. Heat Tr. B-Fund., 13, 297–318, 1988.

Dubovsky, V., Barzilay, G., Granot, G., Ziskind, G., and Letan, R.: Study of PCM-based pin-fin heat sinks, Proceedings of the ASME Heat Transfer summer conference, San Francisco, California, USA, 2009.

Eftekhar, J., Haji-Sheikh, A., and Lou, D. Y. S.: Heat Transfer Enhancement in a Paraffin Wax Thermal Storage System, J. Sol. Energ.-T. ASME, 106, 299–306, 1984.

Ettouney, H., Alatiqi, I., Al-Sahali, M., and Al-Hajirie, K.: Heat transfer enhancement in energy storage in spherical capsules-filled with paraffin wax and metal beads, Energ. Convers. Manage., 47, 211–228, 2006.

Fok, S. C., Shen, W., and Tan, F. L.: Cooling of portable hand-held electronic devices using phase change materials in finned heat sinks, Int. J. Therm. Sci., 49, 109–117, 2010.

Henze, H. R. and Humphrey, J. A. C.: Enhanced heat conduction in phase-change thermal energy storage devices, Int. J. Heat Mass Tran., 24, 459–474, 1981.

Hirt, C. and Nichols, B.: Volume of fluid (VOF) method for the dynamics of free boundaries, J. Comput. Phys., 39, 201–225, 1981.

Hosseini, M. J., Ranjbar, A. A., Sedighi, K., and Rahimi, M.: Melting of Nanoprticle-Enhanced Phase Change Material inside Shell and Tube Heat Exchanger, Hindawi Publishing Corporation, Journal of Engineering, 2013, 784681, doi:10.1155/2013/784681, 2013.

Hosseinizadeh, S. F., Tan, F. L., and Moosania, S. M.: Experimental and numerical studies on performance of PCM-based heat sink with different configurations of internal fins, Appl. Therm. Eng., 31, 3827–3838, 2011.

Humphries, W. and Griggs, E.: A design handbook for phase change thermal control and energy storage devices, Tech. rep., 1074NASA Scientific and Technical Information Office, 1977.

Kandasamy, R., Wang, X., and Mujumdar, S.: Application of phase change materials in thermal management of electronics, Applied Thermal Energy, 27, 2822–2832, 2007.

Kozak, Y., Abramzom, B., and Ziskind, G.: Experimental and numerical investigation of a hybrid PCM-AIR heat sink, Appl. Therm. Eng., 59, 142–152, 2013.

Nayak, K. C., Saha, S. K., Srinivasan, K., and Dutta, P.: A numerical model for heat sinks with phase change materials and thermal conductivity enhancers, Int. J. Heat Mass Tran., 49, 1833–1844, 2006.

Pal, D. and Joshi, Y. K.: Melting in a side heated tall enclosure by a uniformly dissipating heat source, Int. J. Heat Mass Tran., 44, 375–387, 2001.

Rahimi, M., Ranjbar, A. A., Hosseini, M. J., and Abdollahzadeh, M.: Natural convection of nanoparticle-water mixture near its density inversion in a rectangular enclosure, Int. Commun. Heat Mass, 39, 131–137, 2012.

Ranjbar, A. A, Kashani, S., Hosseinizadeh, S. F., and Ghanbarpour, M.: Numerical heat transfer studies of a latent heat storage system containing nano-enhanced phase change material, Therm. Sci., 15, 169–181, 2011.

Reid, R., Prausnitz, J., and Poling, B.: The Properties of Gases and Liquids, McGraw–Hill, New York, 1987.

RUBITHERM: http://www.rubitherm.de/english/index.htm (last access: 18 February 2014), 2012.

Saha, S. K. and Dutta, P.: Heat transfer correlations for PCM-based heat sinks with plate fins, Appl. Therm. Eng., 30, 2485–2491, 2010.

Shatikian, V., Ziskind, G., and Letan, R.: Numerical investigation of a PCM-based heat sink with internal fins, Int. J. Heat Mass Tran., 48, 3689–3706, 2005.

Shatikian, V., Ziskind, G., and Letan, R.: Numerical investigation of a PCM-based heat sink with internal fins: Constant heat flux, Int. J. Heat Mass Tran., 51, 1488–1493, 2008.

14

Dynamic synthesis of machine with slider-crank mechanism

A. A. Jomartov[1], S. U. Joldasbekov[1], and Yu. M. Drakunov[2]

[1]Institute Mechanics and Mechanical Engineering, Almaty, Kazakhstan

[2]al-Farabi Kazakh National University, Almaty, Kazakhstan

Correspondence to: A. A. Jomartov (legsert@mail.ru)

Abstract. In this paper we consider the formulation and solution of the task of a dynamic synthesis machine with an asynchronous electric motor and a slider-crank mechanism. The constant parameters of the slider-crank mechanism (mass and moments of inertia and centers of gravity of links) and the parameters of the electrical motor are defined. The laws of motion of the machine and kinematic parameters of the mechanism are considered as given. We have developed the method of optimal dynamic synthesis of the machine, which consists of an asynchronous electric motor and a slider-crank mechanism. The criterion of optimization of the dynamic synthesis of a machine is the root mean square sum of the moments of driving forces, the forces of resistance and inertia forces which are reduced to the axis of rotation of the crank. The method of optimal dynamic synthesis of a machine can be used in the design of new and the improvement of known mechanisms and machines.

1 Introduction

The dynamic synthesis of mechanisms is among the most important and difficult problems encountered in the design of machines (Dresig and Vulfson, 1990; Dresig and Holzweißig, 2010; Wittenbauer, 1923; Homer and Eckhardt, 1998; Browne, 1965; Helmi and Hassan, 2008; Shabana, 1998; Erdman and Sandor, 1984). The main task of the dynamic synthesis of mechanisms is to determine the optimal values of the parameters and their combinations. The simplest task of the dynamic synthesis in the field of mechanics of machines is the Wittenbauer method (Wittenbauer, 1923) for determining the moment of inertia of the flywheel and the constant drive torque on the motor shaft. This method is a graph-analytical method and is based on plotting the energy-mass. Kolovsky et al. (2000) shows the method of dynamic analysis and synthesis of the machine. Shchepetilnikov (1975) has researched the static and dynamic balancing of different mechanisms. The solution of problems in the dynamics synthesis of machines depends significantly on the correct representation of the allowable levels of dynamic errors and dynamic forces. In many cases it is possible to allow significant dynamic errors in systems by introducing the necessary adjustment units during design. This is particularly obvious for cyclic machines working under steady state conditions; in such machines dynamic errors in the equations of motion of the working components can be compensated by adjustment mechanisms within defined limits (Astashev, Babitsky, and Kolovsky, 2000). The general tasks of dynamic synthesis have been set by various criteria: the minimization of reactions in the bearings, minimum deviation of the angular acceleration of the driving link, etc. (Penun uri, et al., 2011; Guangping and Zhiyong, 2011; Volkert and Herder, 2012; Wunderlich, 1968; Kim and Hee, 2012). While solving the dynamic problems, the analysis and synthesis are usually closely connected. In particular, many of the problems of synthesis, which establish the rational values of the system's parameters, often base on preliminarily solved problem of analysis. One of the most important and urgent tasks, for consideration, is the development of the optimization criteria. These criteria should be based on the most significant factors of the considered problem and at the same time have the foreseeable shape, to retain the role of active tool for the dynamics synthesis, when developing variants of a machine. When taking into account the elasticity of links, the question of criteria, without losing its importance, is further complicated. In this case, in addition to geometric and kinematic character-

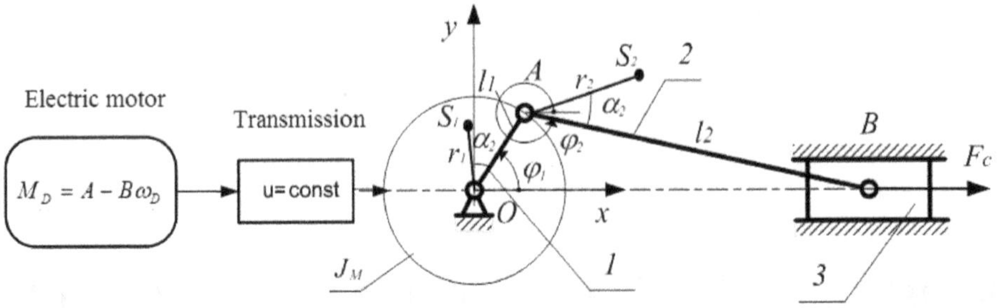

Figure 1. Diagram of the machine.

istics, other factors, characterizing the frequency features of the system, level of proximity of working models to the dynamically unstable modes, level of additional dynamic load, caused by oscillations and many other factors appear as dynamic criteria (Vulfson, 2015). The aim of this paper is to develop the method of optimal dynamic synthesis of the machine, which consists of a motor and a mechanism, using the methods of synthesis and choosing the optimization criteria which were proposed in the work (Vulfson, 2015).

2 Dynamic synthesis of a machine

Problem statement. Machine motion with rigid links is described by the following equations:

$$J_n\ddot{\varphi}_1 + \frac{1}{2}\frac{dJ_n}{d\varphi_1}\dot{\varphi}_1^2 = Q = M_D + M_C(\varphi_1, \dot{\varphi}_1) \tag{1}$$

$$\tau\dot{M}_D + M_D = A - B\dot{\varphi}_1,\ \text{at}\ \tau \cong 0,\ M_D = A - B\dot{\varphi}_1. \tag{2}$$

Where φ_1 – generalized coordinate, J_n – the reduced moment of inertia, Q – the generalized force, M_D – the torque of asynchronous electric motor, M_C $(\varphi_1, \dot{\varphi}_1)$ – the resistance moment, A, B – the parameters of the static characteristics of asynchronous electric motor, τ – the electrical time constant of the asynchronous electric motor.

Let's solve the problem of dynamic synthesis of machine, which is to determine the parameters of its mechanisms (mass and moments of inertia and the centre of gravity of links) and motor parameters, the law of motion of the machine set $\phi_1^* = \phi_1^*(t)$ or $\omega_1^* = \omega_1^*(t)$ and the kinematic parameters of the mechanism are considered as given. As the objective function for the dynamic synthesis of machine, we construct a functional on the basis of Eq. (1), with taking into account the characteristics of the motor (Eq. 2).

$$G = \sum_{i=1}^{n}\left[J_n\ddot{\varphi}_{1i} + \frac{1}{2}\frac{\partial J_n}{\partial\varphi_1}\dot{\varphi}_{1i}^2 - Q_i\right]^2. \tag{3}$$

The required parameters of the machine are determined by minimizing the root mean square sum of the moments of driving forces, the forces of resistance and inertia forces which are reduced to the axis of rotation of the crank, for the n positions of the mechanism.

Let us consider the optimal dynamic synthesis of a machine, which consists of an asynchronous electric motor and a slider-crank mechanism. As a criterion for the optimization of the dynamic synthesis machine we choose the root mean square sum of the moments of driving forces, the forces of resistance and inertia forces which are reduced to the axis of rotation of the crank, for the n positions of the slider-crank mechanism. A diagram of the machine is shown in Fig. 1.

By solving the task of kinematic analysis of the slider-crank mechanism (see. Fig. 1), taking into account, we have the formulas

$$\begin{cases} \varphi_2 = 2\pi - \arcsin(\lambda\sin\varphi_1), \\ x_C = l_1\cos\varphi_1 + l_2\cos\varphi_2 \\ x_{S_2} = l_1\cos\varphi_1 + r_2\cos(\varphi_2 + \alpha_2), \\ y_{S_2} = l_1\sin\varphi_1 + r_2\sin(\varphi_2 + \alpha_2) \end{cases} \tag{4}$$

After differentiation of Eq. (4) by the generalized coordinates, we obtain the first transfer functions and second transfer functions

$$\begin{cases} \varphi'_2 = -\lambda\dfrac{\cos\varphi_1}{\cos\varphi_2}, \\ x'_C = l_1\sin\varphi_1(\varphi'_2 - 1) \\ x'_{S_2} = -l_1\sin\varphi_1 - r_2\sin(\varphi_2 + \alpha_2)\cdot\varphi'_2, \\ y'_{S_2} = l_1\cos\varphi_1 + r_2\cos(\varphi_2 + \alpha_2)\cdot\varphi'_2 \\ \varphi''_2 = \lambda\dfrac{\sin(\varphi_1 - \varphi_2)}{\cos^2\varphi_2}, \\ x''_C = l_1[\cos\varphi_1(\varphi'_2 - 1) + \sin\varphi_1\cdot\varphi''_2] \\ x''_{S_2} = -l_1\cos\varphi_1 - r_2[\cos(\varphi_2 + \alpha_2)\cdot\varphi'^2_2 \\ y''_{S_2} = -l_1\sin\varphi_1 - r_2[\sin(\varphi_2 + \alpha_2)\cdot\varphi'^2_2 \\ \quad + \sin(\varphi_2 + \alpha_2)\cdot\varphi''_2], \\ \quad - \cos(\varphi_2 + \alpha_2)\cdot\varphi''_2] \end{cases} \tag{5}$$

The input parameters of the dynamic synthesis of a machine are: F_C – the force of technological resistance; $\phi_1, \dot{\phi}_1, \ddot{\phi}_1$ – the generalized coordinates and the velocities and accelerations of the crank 1; l_1 and l_2 – the geometric dimensions of the coupler and crank; $u = \omega_D/\omega_1$ – the transmission ratio; ω_c – the average angular velocity of the crank in a steady motion, and δ – the coefficient of unevenness of movement.

The output parameters of dynamic synthesis are: m_1, m_2, m_3 – the mass of the crank, coupler and slider; $a_1 = r_1\cos\alpha_1$, $b_1 = r_1\sin\alpha_1$, $a_2 = r_2\cos\alpha_2$, $b_2 = r_2\sin\alpha_2$ – the coordinates

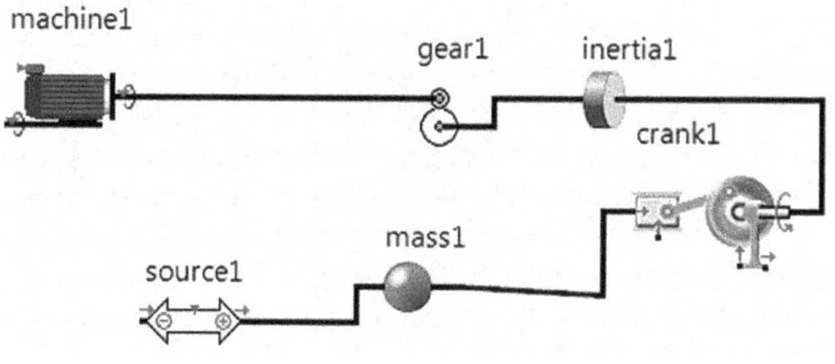

Figure 2. Dynamic model of machine with slider-crank mechanism on SimulationX.

of the centers of mass of the crank and coupler; J_D – the moment of inertia of the rotor of the electric motor; J_M – the moment of inertia of the flywheel; J_1, J_2 – the moments of inertia of the crank and coupler.

The parameters of the machine are determined from the condition of the minimum root mean square sum of the moments of driving forces, the forces of resistance and inertia forces which are reduced to the axis of rotation of the crank, for the n positions of the slider-crank mechanism.

This optimization criterion is given by

$$G = \sum_{i=1}^{n}\left[J_n\ddot{\varphi}_{1i} + \frac{1}{2}\frac{\partial J_n}{\partial\varphi_1}\dot{\varphi}_{1i}^2 - Q_i\right]^2. \qquad (6)$$

In Eq. (6) the reduced moment of inertia of the machine is given by

$$J_n = J_D u^2 + J_M + J_1 + m_3 x'^2_C + m_2(x'^2_{S_2} + y'^2_{S_2}) + J_2\varphi'^2_2 \quad (7)$$

and the remaining parameters are determined by the expressions:

$$\frac{1}{2}\frac{\partial J_n}{\partial\phi_1} = m_3 x'_C x''_C + m_2(x'_{S_2}x''_{S_2} + y'_{S_2}y''_{S_2}) + J_2\phi'_2\phi''_2$$

$$Q = M_D u + F_C x'_C - m_1 g(a_1\cos\varphi_1 - b_1\sin\varphi_1) - m_2 g y'_{S_2}$$

Let us write the expression M_D through the parameters A and B the static characteristics of the asynchronous electric motor.

Then

$$Q = (A - Bu\dot{\phi}_1)u + F_C x'_C - m_1 g(a_1\cos\phi_1 - b_1\sin\phi_1) - m_2 g y'_{S_2}.$$

We substitute the expressions of first transfer functions (Eq. 5) in Eq. (7)

$$J_n = J_0 + m_3 l_1^2\sin^2\phi_1(\phi'_2 - 1)^2 + J_B\phi'^2_2 + 2m_2 l_1 a_2\cos(\phi_2 - \phi_1)\phi'_2 - 2m_2 l_1 b_2\sin(\phi_2 - \phi_1)\phi'_2. \qquad (8)$$

In Eq. (8) we introduce the notation

$$J_0 = J_D u^2 + J_M + J_1 + m_2 l_1^2,\ J_B = J_2 + m_2(a_2^2 + b_2^2).$$

The partial derivative of the reduced moment of inertia is

$$\frac{1}{2}\frac{\partial J_n}{\partial\phi_1} = m_3 l_1^2\sin\phi_1[\cos\phi_1(\phi'_2 - 1) + \sin\phi_1\phi''_2](\phi'_2 - 1) + J_B\phi'_2\phi''_2 + m_2 l_1 a_2[\cos(\phi_2 - \phi_1)\phi''_2 - \sin(\phi_2 - \phi_1)(\phi'_2 - 1)\phi'_2] - m_2 l_1 b_2 [\sin(\phi_2 - \phi_1)\phi''_2 + \cos(\phi_2 - \phi_1)(\phi'_2 - 1)\phi'_2]. \qquad (9)$$

The final expression for the generalized force can be written

$$Q = (A - Bu\dot{\phi}_1)u + F_C l_1\sin\phi_1(\phi'_2 - 1) - m_1 g a_1\cos\phi_1 + m_1 g b_1\sin\phi_1 - m_2 g l_1\cos\phi_1 - m_2 g a_2\cos\phi_2\cdot\phi'_2 + m_2 g b_2\sin\phi_2\cdot\phi'_2. \qquad (10)$$

The expression enclosed in square brackets of Eq. (6) is called a function of deviation Δ. It is based on the relations (Eqs. 8–10) and is the form

$$\Delta = J_0\ddot{\phi}_1 + m_3 l_1^2\sin^2\phi_1(\phi'_2 - 1)^2\ddot{\phi}_1 + J_B\phi'^2_2\ddot{\phi}_1 + 2m_2 l_1 a_2\cos(\phi_2 - \phi_1)\phi'_2\ddot{\phi}_1 - 2m_2 l_1 b_2\sin(\phi_2 - \phi_1)\phi'_2\ddot{\phi}_1 + m_3 l_1^2\sin\phi_1[\cos\phi_1(\phi'_2 - 1) + \sin\phi_1\phi''_2](\phi'_2 - 1)\dot{\phi}_1^2 + J_B\phi'_2\phi''_2\dot{\phi}_1^2 + m_2 l_1 a_2 [\cos(\phi_2 - \phi_1)\phi''_2 - \sin(\phi_2 - \phi_1)(\phi'_2 - 1)\phi'_2]\dot{\phi}_1^2 - m_2 l_1 b_2[\sin(\phi_2 - \phi_1)\phi''_2 + \cos(\phi_2 - \phi_1)(\phi'_2 - 1)\phi'_2]\dot{\phi}_1^2 - (A - Bu\dot{\phi}_1)u - F_C l_1\sin\phi_1(\phi'_2 - 1) + m_1 g a_1\cos\phi_1 - m_1 g b_1\sin\phi_1 + m_2 g l_1\cos\phi_1 + m_2 g a_2\cos\phi_2\cdot\phi'_2 - m_2 g b_2\sin\phi_2\cdot\phi'_2. \qquad (11)$$

Let us simplify and group the expression Eq. (11)

$$\Delta = J_0\ddot{\phi}_1 + m_3 l_1^2\{\sin^2\phi_1(\phi'_2 - 1)^2\ddot{\phi}_1 + [\sin 2\phi_1(\phi'_2 - 1)/2 + \sin^2\phi_1\phi''_2](\phi'_2 - 1)\dot{\phi}_1^2\} + Bu^2\dot{\phi}_1 + m_2 a_2\{2l_1\cos(\phi_2 - \phi_1)\phi'_2\ddot{\phi}_1 + l_1[\cos(\phi_2 - \phi_1)\phi''_2 - \sin(\phi_2 - \phi_1)(\phi'_2 - 1)\phi'_2]\dot{\phi}_1^2 + g\cos\phi_2\cdot\phi'_2\} - m_2 b_2\{2l_1\sin(\phi_2 - \phi_1)\phi'_2 \ddot{\phi}_1 + l_1[\sin(\phi_2 - \phi_1)\phi''_2 + \cos(\phi_2 - \phi_1)(\phi'_2 - 1)\phi'_2]\dot{\phi}_1^2 + g\sin\phi_2\cdot\phi'_2\} + (m_1 a_1 + m_2 l_1)g\cos\phi_1 - m_1 g b_1\sin\phi_1 + J_B(\phi'^2_2\ddot{\phi}_1 + \phi'_2\phi''_2\dot{\phi}_1^2) - Au - F_C l_1\sin\phi_1(\phi'_2 - 1).$$

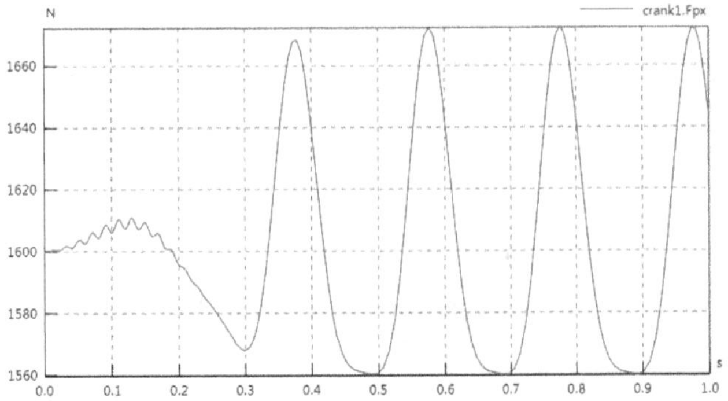

Figure 3. A graph of the longitudinal force acting on the slider for a typical machine.

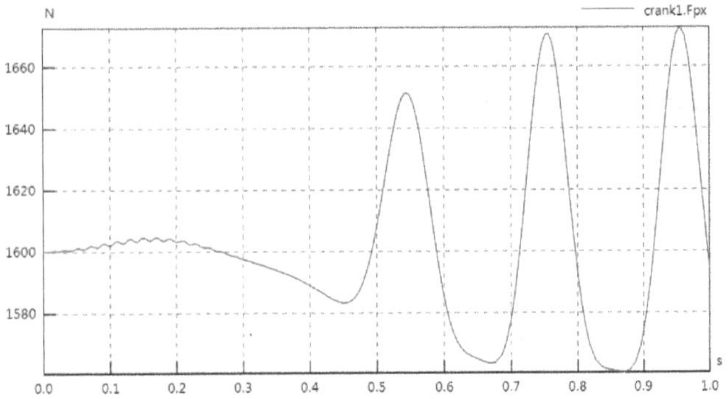

Figure 4. A graph of the longitudinal force acting on the slider for the synthesized machine with a slider-crank mechanism.

The function of deviation Δ is represented as a generalized polynomial

$$\Delta_i = [P_1 f_1(\phi_{1i}) + P_2 f_2(\phi_{1i}) + \ldots + P_9 f_9(\phi_{1i}) - F(\phi_{1i})]^2.$$

Then the optimality criterion (Eq. 6) of the machine takes the form

$$G = \sum_{i=1}^{n} [P_1 f_1(\varphi_{1i}) + P_2 f_2(\varphi_{1i}) + \ldots + P_9 f_9(\varphi_{1i}) - F(\varphi_{1i})]^2. \quad (12)$$

Here, the following notation

$$P_1 = J_O,\ P_2 = m_1 b_1,\ P_3 = J_B,\ P_4 = m_2 a_2,\ P_5 = m_2 b_2,$$
$$P_6 = Au,\ P_7 = Bu^2,\ P_8 = m_1 a_1 + m_2 l_1,\ P_9 = m_3 l_1^2. \quad (13)$$

$$\begin{aligned}
&F(\varphi_1) = F_C l_1 \sin\varphi_1(\varphi'_2 - 1)\\
&f_1(\varphi_1) = \ddot{\varphi}_1,\\
&f_2(\varphi_1) = -g\sin\varphi_1,\\
&f_3(\varphi_1) = \varphi'^2_2\ddot{\varphi}_1 + \varphi'_2\varphi''_2\dot{\varphi}_1^2,\\
&f_4(\varphi_1) = 2l_1\cos(\varphi_2-\varphi_1)\varphi'_2\ddot{\varphi}_1 + l_1[\cos(\varphi_2-\varphi_1)\varphi''_2 -\\
&-\sin(\varphi_2-\varphi_1)(\varphi'_2-1)\varphi'_2]\dot{\varphi}_1^2 + g\cos\varphi_2\cdot\varphi'_2,\\
&f_5(\varphi_1) = -2l_1\sin(\varphi_2-\varphi_1)\varphi'_2\ddot{\varphi}_1 +\\
&+ l_1[\sin(\varphi_2-\varphi_1)\varphi''_2 + \cos(\varphi_2-\varphi_1)(\varphi'_2-1)\varphi'_2]\dot{\varphi}_1^2\\
&+ g\sin\varphi_2\cdot\varphi'_2,\\
&f_6(\varphi_1) = -1,\\
&f_7(\varphi_1) = \dot{\varphi}_1,\\
&f_8(\varphi_1) = g\cos\varphi_1,\\
&f_9(\varphi_1) = \sin^2\varphi_1(\varphi'_2-1)^2\ddot{\varphi}_1 + [\sin 2\varphi_1(\varphi'_2-1)/2\\
&+ \sin^2\varphi_1\varphi''_2](\varphi'_2-1)\dot{\varphi}_1^2.
\end{aligned}$$

From the minimum of Eq. (12), we obtain a system of equations from which coefficients P_1, P_2,…,P_9 are determined

$$\frac{\partial G}{\partial P_k} = 0, \quad k = 1, 2, \ldots, 9$$

This system of equations in expanded form is:

$$\begin{cases}
C_{11}P_1 + C_{12}P_2 + \ldots + C_{19}P_9 = \sigma_1\\
C_{21}P_1 + C_{22}P_2 + \ldots + C_{29}P_9 = \sigma_2\\
\ldots\ldots\ldots\ldots\ldots\ldots\ldots\ldots\ldots\ldots\\
C_{91}P_1 + C_{92}P_2 + \ldots + C_{99}P_9 = \sigma_9
\end{cases} \quad (14)$$

where

$$C_{jk} = C_{kj} = \sum_{i=1}^{n} f_j(\phi_{1i}) f_k(\phi_{1i}),$$

$$\sigma_k = \sum_{i=1}^{n} F(\phi_{1i}) f_k(\phi_{1i}), \quad j,k = 1, 2, \ldots, 9 \quad (15)$$

After finding the coefficients P_1, P_2, …, P_9, we define the physical parameters of the relations (Eq. 13).

An algorithm for solving the above task is as follows:

1. The segment $\phi_1 \in [0, 2\pi]$ is split into n equal parts.
2. Let us take as a first approximation for the angular velocity of the following law.

   $$\omega_1 = \dot{\varphi}_1 = \omega_c + 0.5\delta\omega_c\cos(\varphi_1).$$

3. Then, the crank angular acceleration is determined by the formula

   $$\varepsilon_1 = \ddot{\varphi}_1 = -0.5\delta\omega_c\sin(\varphi_1)\cdot\dot{\varphi}_1.$$

4. The coefficients C_{jk}, σ_k are defined by Eq. (15);
5. We solve the system of linear equations (Eq. 14) and define the parameters P_k.

The required physical parameters are determined from Eq. (13)

Typically, the mass m_3 of the slider is set on the basis of technological requirements. In this case, the order of the system of linear equations is reduced by 1, i.e. $k = 1, 2, \ldots, 8$ and the final system becomes

$$\sum_{j=1}^{8} C_{kj} P_j = \sigma_k \qquad k = 1, 2, \ldots, 8,$$

where

$$C_{jk} = C_{kj} = \sum_{i=1}^{n} f_j(\phi_{1i}) f_k(\phi_{1i}), \sigma_k = \sum_{i=1}^{n} F(\phi_{1i}) f_k(\phi_{1i}),$$
$$j, k = 1, 2, \ldots, 8$$

Expression for the function $F(\phi_1)$ will be in the form

$$F(\phi_1) = F_C l_1 \sin\phi_1 (\phi'_2 - 1) - m_3 l_1^2 \{\sin^2\phi_1 (\phi'_2 - 1)^2 \ddot{\phi}_1 + [\sin 2\phi_1 (\phi'_2 - 1)/2 + \sin^2\phi_1 \phi''_2](\phi'_2 - 1)\dot{\phi}_1^2\}.$$

The formation of the coefficients C_{jk}, σ_k and the solution of () were carried out in the computing software Maple.

3 Example

Synthesis of the machine was carried out for the following initial data:

$$m_3 = 1.4\,\mathrm{kg}, l_1 = 0.04\,\mathrm{m}, l_2 = 0.16\,\mathrm{m}, F_C = 160\,\mathrm{N},$$
$$u = 5, \omega_c = 20\,\mathrm{s}^{-1}.$$

The coefficients of the polynomial (Eq. 10), are: $P_1 = 2.1022$, $P_2 = 1.2619$, $P_3 = 0.02478$, $P_4 = 0.1263$, $P_5 = 0.2023$, $P_6 = 1522.8$, $P_7 = 50.627$, $P_8 = 8.1963$, The physical parameters of the machine are: $J_0 = 2.1022\,\mathrm{kg\,m^2}$, $J_B = 0.02478\,\mathrm{kg\,m^2}$, $m_1 = 3.1263\,\mathrm{kg}$, $m_2 = 2.2023\,\mathrm{kg}$, $A = 304.56\,\mathrm{Nm}$, $B = 2.015\,\mathrm{Nms}$.

The angular velocity of the electric motor idling is $\omega_o = \frac{1500 \cdot \pi}{30} = 151.15\,\mathrm{rad\,s^{-1}}$, the number of revolutions of the electric motor is $n_o = 1443\,\mathrm{rpm}$. Let us accept $n_o = 1500\,\mathrm{rpm}$, $\omega_H = \frac{1500 \cdot \pi}{30} = 157\,\mathrm{rad\,s^{-1}}$. The nominal angular velocity electric motor shaft is $\omega_H = 145.5\,\mathrm{rad\,s^{-1}}$. The nominal torque of the electric motor is $M_H = A \cdot B \cdot \omega_H = 11.38\,\mathrm{Nm}$. The nominal power of the electric motor is 1.655 kW.

For dynamic analysis of the machine with a slider-crank mechanism let us use the software package SimulationX. Figure 2 shows the dynamic model of the machine with a slider-crank mechanism on SimulationX.

The calculation model was carried out for a typical machine with a slider-crank mechanism and an electric motor of 3 kW power, and for a synthesized machine with slider-crank mechanism and an electric motor of 1.655 kW power. Fig. 3 shows a graph of the longitudinal force acting on the slider for a typical machine with a slider-crank mechanism. Fig. 4 shows a graph of the longitudinal force acting on the slider for the synthesized machine with a slider-crank mechanism.

As can be seen from a comparison of Fig. 3 and Fig. 4 the synthesized machine unit with a crankshaft-slide mechanism provides the required technological loading, with less power from the electric motor.

4 Conclusions

The method of optimal dynamic synthesis of a machine, which consists of an asynchronous electric motor and a slider-crank mechanism, has been developed. The criterion of optimization of the dynamic synthesis of the machine is built on the basis of the equations of the machine motion with rigid links. The criterion of optimization of the dynamic synthesis of the machine is the root mean square sum of the moments of driving forces, the forces of resistance and inertia forces which are reduced to the axis of rotation of the crank.

The program in Maple for solving this task of the dynamic synthesis of a machine with an asynchronous electric motor and a slider-crank mechanism has been compiled.

The method of optimal dynamic synthesis of a machine, which consists of an asynchronous electric motor and a slider-crank mechanism, has been validated using a numerical simulation of the synthesized machine with a slider-crank mechanism.

In future work, to validate of the new method of optimal dynamic synthesis of a machine, which consists of an asynchronous electric motor and a slider-crank mechanism, we expected to conduct of experimental tests of the synthesized machine.

References

Astashev, V. K., Babitsky, V. I., and Kolovsky, M. Z.: Dynamics and control of machines, Springer Berlin Heidelberg, 235 pp., 2000.

Browne, J. W.: The Theory of Machine Tools, Cassell and Co. Ltd., 374 pp., 1965.

Dresig, H. and Holzweißig, F.: Dynamics of Machinery, Springer Berlin Heidelberg, 328 pp., 2010.

Dresig, H. and Vulfson, I. I.: Dynamik der Mechanismen, Wien, New York, Springer-Verlag, 621 pp., 1990.

Homer, D. E.: Design of machines and mechanisms, 621 pp., 1998.

Erdman, A. G. and Sandor, G. N.: Mechanism Design: Analysis and Synthesis, Prentice Hall, New Jersey, 688 pp., 1984.

Guangping, H. and Zhiyong, G.: Dynamics synthesis and control for a hopping robot with articulated leg, Mech. Mach. Theory, 46, 1669–1688, 2011.

Helmi, A. Y. and Hassan, E.-H.: Machining technology: machine tools and operations, Taylor & Francis Group, LLC, 672 pp., 2008.

Kim, B. S. and Yoo, H. H.: Unified synthesis of a planar four-bar mechanism for function generation using a spring-connected arbitrarily sized block model, Mechanism and Machine Theory, 49, 141–156, 2012.

Kolovsky, M. Z., Evgrafov, A. N., Semenov, Yu. A., and Slousch, A. V.: Advanced theory of mechanisms and machines, Springer Berlin Heidelberg, 396 pp., 2000.

Penunuri, F., Peon-Escalante, R., Villanueva, C., and Pech-Oy, D.: Synthesis of mechanisms for single and hybrid tasks using differential evolution, Mechanism and Machine Theory, 46, 1335–1349, 2011.

Shabana, A. A.: Dynamic of Multibody Systems, Cambridge University Press, 2nd ed., 384 pp., 1998.

Shchepetil'nikov, V. A.: The balancing of bar mechanisms with unsymmetrical links, Mechanism and Machine Theory, 10, 461–466, 1975.

Van der Wijk, V. and Herder, J. L.: Synthesis method for linkages with center of mass at invariant link point – Pantograph based mechanisms, Mech. Mach. Theory, 48, 15–28, 2012.

Vulfson, I. I.: Dynamics of cyclic machines, Springer Berlin Heidelberg. 2015.

Wittenbauer, F.: Graphische Dynamik, Julius Springer, Berlin, 1923.

Wunderlich, W.: Concerning the trajectory of the center of mass of the four-bar linkage and the slider-crank mechanism, J. Mech., 3, 391–396, 1968.

15

Modeling and control of piezoelectric inertia–friction actuators: review and future research directions

Y. F. Liu[1], J. Li[1], X. H. Hu[1], Z. M. Zhang[2], L. Cheng[3], Y. Lin[4], and W. J. Zhang[1,2]

[1]Complex and Intelligent System Laboratory, School of Mechanical and Power Engineering, East China University of Science and Technology, Shanghai, China

[2]Department of Mechanical Engineering, University of Saskatchewan, Saskatoon, Canada

[3]Institute of Automation, Chinese Academy of Sciences, Beijing, China

[4]Department of Mechanical and Industrial Engineering, Northeastern University, Boston, USA

Correspondence to: W. J. Zhang (chris.zhang@usask.ca) and J. Li (lijinme@ecust.edu.cn)

Abstract. This paper provides a comprehensive review of the literature regarding the modeling and control of piezoelectric inertia–friction actuators (PIFAs). Examples of PIFAs are impact drive mechanisms (IDMs) and friction-driving actuators (FDAs). In this paper, the critical challenges are first identified in modeling and control of PIFAs. Second, a general architecture of PIFAs is proposed to facilitate the analysis and classification of the literature regarding modeling and control of PIFAs. This general architecture covers all types of PIFAs (e.g., FDAs, IDMs) and thus serves as a general conceptual model of PIFAs. There is an additional benefit with this general architecture of PIFAs, namely that it is conducive to innovation in PIFAs, as new specific PIFAs may be designed in order to tailor to a specific application (for example, both FDAs and IDMs are viewed as specific PIFAs). Finally, the paper presents future directions in modeling and control for further improvement of the performance of PIFAs.

1 Introduction

Given the rapid development of nanotechnology, nanopositioning is becoming increasingly important in devices such as aerospace positioning systems and scanning probe microscopes (Croft et al., 2000; Zhang et al., 2012; Higuchi et al., 1990). One generic requirement with these devices is the realization of an accurate long-range motion (accuracy: nanometers) or accurate motion along the full length of the range that the mechanical system permits.

Piezoelectric actuators (PAs) are widely used in devices to realize an extremely accurate motion because of their simple mechanical structure, small dimension, and ability to generate a motion with high frequency, large force, and high resolution (Ouyang et al., 2008). However, PAs are not suitable for meeting the aforementioned generic requirement due to their limited motion range (approximately 10–100 μm with a common piezoelectric stack). For instance, positioning in an electro-discharge machine (Furutani et al.,1993, 1997) needs extremely high accuracy as well as a long moving range (which PAs alone could not achieve). A combination of the piezoelectric actuation principle and the inertia–friction actuation principle, namely the piezoelectric inertia–friction actuation (PIFA) principle, promises to meet the aforementioned generic requirement (i.e., high accuracy and long drive range). The underlying reason is that this combination can nicely meet the engineering hybridization criterion (Zhang et al., 2010) – that is, that the two principles are indeed complementary to each other.

One example of the PIFA system is the well-known FDA (friction-driving actuator) (Zhang et al., 2012) (Fig. 1), which consists of the stage, the end effector, the ground (which is at rest with respect to the other components), and the PA which connects to the ground at one end and to the stage at the other end. The working principle of the FDA (Fig. 1) consists of three steps:

1. The initial state: there is no input voltage applied to the PA. The PA remains at its initial length, and the stage and end effector stay at the initial position.

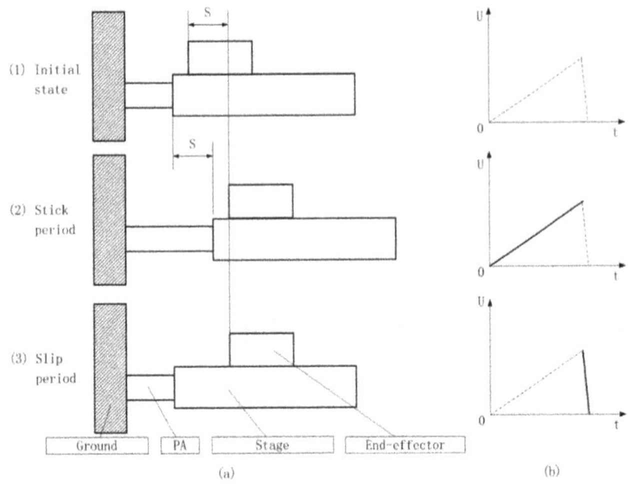

Figure 1. The principle of the FDA: **(a)** actuation steps and **(b)** input voltage.

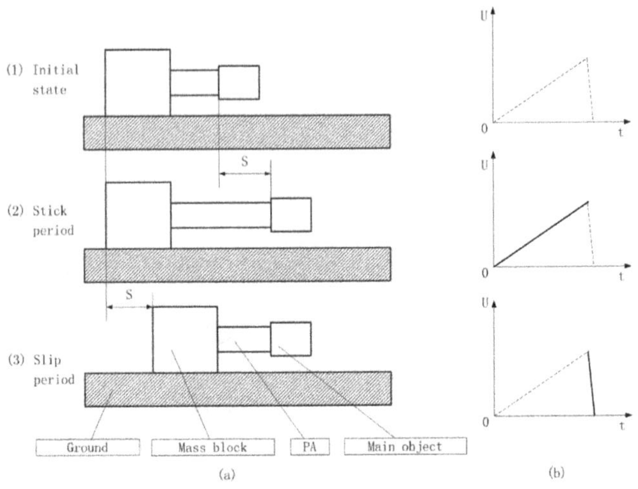

Figure 2. The principle of the IDM: **(a)** actuation steps and **(b)** input voltage.

2. The stick period: during this period, the input voltage increases very slowly. The PA will extend to a distance S at a very low velocity. During this period, the friction force between the end effector and the stage is large enough to overcome the inertial force of the end effector with respect to the stage. Therefore, the end effector sticks on the stage and moves forward, together with the stage, the distance S.

3. The slip period: during this period, the input voltage decreases very quickly. The PA will contract back to the initial position in a very short time. During this period, the friction force between the end effector and the stage is not sufficiently large to overcome the inertial force of the end effector with respect to the stage. Therefore, the end effector slips on the stage and remains at its position with respect to the ground while the stage goes back to the initial position.

It is clear after the above three steps that the end effector advances S without consideration of any backlash. By repeating the foregoing three steps, the end effector will keep moving forward. The driving range is only limited by the length of the stage, and it is theoretically infinite. If one changes the waveform of input voltage, the end effector will change the direction of movement.

Another well-known example of the PIFA system is an IDM (impact drive mechanism) (Soderqvist, 1976) (Fig. 2). An IDM consists of the main object, the ground, the mass block or inertial mass, and the PA that further connects to the main object at one end and to the mass block at the other. The working principle of the IDM (Fig. 2) consists of three steps.

1. The initial state: there is no input voltage applied to the PA. The PA remains at its initial length, and the main object and the mass block stay at the initial position.

2. The stick period: during this period, the input voltage increases very slowly. The PA will slowly extend to a distance S, and the mass block accordingly moves forward a distance S at a very low velocity. During this period, the friction force between the main object and the ground is large enough to overcome the inertial force of the mass block. As such, the main object will stay at the initial position.

3. The slip period: during this period, the input voltage decreases very quickly. The PA will contract back in a very short time. During this period, the friction force between the main object and the ground is not sufficiently large to overcome the inertial force of the mass block, and therefore the main object will slip over a distance S on the ground.

Throughout the three steps, the main object generates a displacement S without consideration of any backlash. If we keep repeating the foregoing three steps, the main object will keep moving forward. The driving range is only limited by the length of the ground. It should be clear that the main object will change the direction of movement if we change the waveform of input voltage.

The common elements in FDAs and IDMs are the coupling of the frictional and inertial effects. In fact, the general architecture of such an actuation, namely piezoelectric inertia–friction actuation (PIFA), is illustrated in Fig. 3. The PIFA system consists of four objects: A, B, C, and D. The connectivity among the four objects is such that one connection is present between A and C, which is a PA in this case, and the other connection is present between A and B, which is a frictional contact.

To prove the generality of this architecture, FDAs and IDMs, as discussed before, are revisited. In the FDA, A is the stage, B is the end effector, C is the ground (which is at

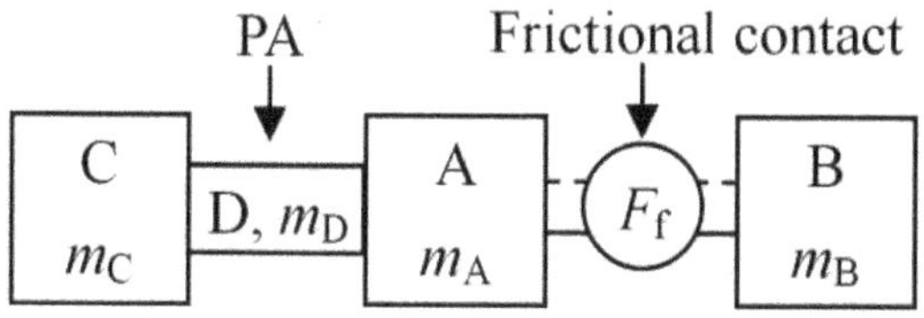

Figure 3. The general architecture of the PIFA system. A, B, C, D (PA): the objects in the system. m_A, m_B, m_C: the masses of the objects. F_f: the frictional force between object A and object B. PA: piezoelectric actuator.

rest with respect to the other components D), and D is the PA which connects to the ground object C at one end and to the stage A at the other. In the IDM, A is the main object; B is the ground; C is the mass block or inertial mass; and D is the PA which connects the main object at one end and to the mass block at the other end.

It should be noted that the inchworm actuator (Shamoto et al., 2000; Shamoto and Moriwaki, 1997) is different from the PIFA, so the present paper does not cover the inchworm. A detailed elaboration on the difference between the PIFA and the inchworm can be found in Zhang et al. (2012) and Ouyang et al. (2008).

On a general note, an important benefit from the generalization of FDAs and IDMs into the general architecture of the PIFA is the opening of a potential source of innovation for new devices that combine the piezoelectric and inertia–friction principles. For instance, both FDAs and IDMs can be viewed as specific PIFAs with respect to the general architecture of PIFAs. The discussion of this benefit is not within the scope of this paper, and for details the reader is directed to Zhang et al. (2012). The present paper will derive another benefit from this generalization, i.e., its facility for classifying the knowledge of modeling and control of the specific types of PIFAs such as FDAs and IDMs in such a way that the general architecture serves as a common platform (i.e., the PIFA architecture) for comparison of different theories or technologies. Indeed, with this general architecture, the knowledge for PIFAs such as FDAs and IDMs can be analyzed and compared, and this will also have benefits for the further development of both FDAs and IDMs.

It is well known that accurate models and feedback control methods are very important in order to achieve high performance with dynamic systems. The models also provide a tool for the optimization of the system design. The modeling and control of PIFAs are difficult due to effects such as material hysteresis, creep behavior, and friction hysteresis coupled with thermal effect (if the PIFA is operated over sufficiently long periods of time) (Li et al., 2008). In addition, the PIFA system may concern both step movement (similar to a step motor, concerning only a step length called the normal mode) and fine movement within a single step (called the fine mode or scanning mode) (Spiller and Hurak, 2011).

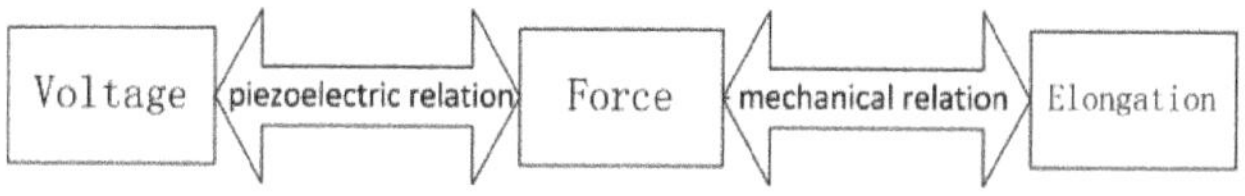

Figure 4. Internal relations of the piezoelectric actuator (no sequence between these two relations).

Not many review papers on the problem of modeling and control are available regarding specific types of PIFAs, namely IDMs and FDAs. Nguyen et al. (2013) presented a review of the modeling of a FDA (a specific type of PIFA). However, they did not consider the dynamics of PAs, nor the dynamics of the other two moving components (except the ground object). They did, however, consider friction, but their discussion is not comprehensive in that the discussion overlooked many interesting friction models. The present paper attempts to overcome this shortcoming, and a discussion of future work will also be presented in the final section.

2 Modeling of PIFA systems

The dynamics of the whole system of a PIFA depend on the dynamics of both the PA and those of the frictional contact. Therefore, a dynamic model of the whole system must capture the dynamics of the PA and the frictional contact (Fig. 3).

2.1 Modeling of PA

A PA converts an electrical signal into a physical displacement. In particular, when a voltage is applied to a piezoelectric material (e.g., a stack), the material will produce a significantly large deformation. By controlling the voltage, the displacement can be controlled. Therefore, the piezoelectric actuator is a coupled electrical–mechanical system. There are two basic processes (thus relations) in a PA, as shown in Fig. 4: (1) the piezoelectric relation and (2) the mechanical relation.

2.1.1 The piezoelectric relation

The piezoelectric relation is where a voltage is applied to the two ends of a piezoelectric material and the voltage generates an internal force within the piezoelectric material, causing the material to deform. Both the hysteresis and creep effect occur during this process. Ideally, the PA model should cover these two effects. However, in the existing literature, the creep effect is mostly ignored (because the PA works with high frequency in the PIFA systems and the creep effect is very small). The present paper does not cover the creep effect. The challenge in modeling of the piezoelectric relation is thus how to account for the hysteresis behavior.

The following model is widely used in the literature for describing the hysteresis behavior in the V–F relation (Croft et al., 2000; Ha et al., 2005, 2006; Fung et al., 2008a, b; Peng

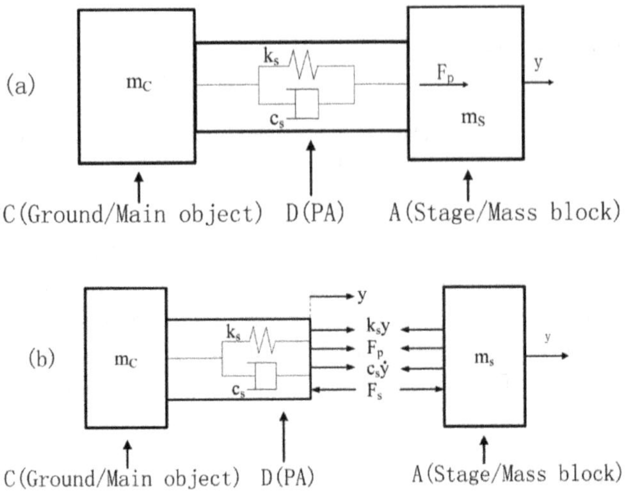

Figure 5. The spring-mass-damping system: **(a)** total system of the PA and A (stage/mass block) and **(b)** force analysis of the PA and A (stage/mass block) (Adriaens et al., 2000).

and Chen, 2011), i.e.,

$$F = H(V), \tag{1}$$

where H represents the hysteresis effect or behavior. The relation H usually uses the Bouc–Wen (Bouc, 1967; Wen, 1976) or Preisach model (Mayergoyz, 1991). It is noted that the inclusion of the hysteresis effect in the model of the V–F relation may lead to a significant computational overhead, which could then compromise the performance of PIFAs if a model-based feedback controller is used (Cheng et al., 2012). Therefore, there are also some studies (Furutani et al., 1997) that simply ignore the hysteresis effect. However, this may cause a large error in understanding the behavior of PAs.

2.1.2 The mechanical relation

The mechanical relation is where a force is applied to the piezoelectric material to cause a change in the length (y) of the material along the force or voltage direction. Usually, in the modeling of PIFAs, the model for the $F-y$ relation uses a lumped model. Figure 5 shows a spring-mass-damping (lump) system. In Fig. 5, y represents the displacement of A (stage/mass block), m_s represents the mass of A (stage/mass block), c_s represents the damping coefficient of A (stage/mass block), k_s represents the stiffness of A (stage/mass block), F_S represent the interaction force between the PA and A (stage/mass block), and F_P represents the drive force generated by the PA.

The main differences in the mechanical relation among the existing PA models lie in how the inertia (e.g., mass) of the PA is represented. To date, there are three methods to deal with this issue in literature. The first method completely neglects the mass of the PA (Ha et al., 2005, 2006; Chang and Li, 1999; Jiang et al., 2000; Lambert et al., 2003; Edeler et al., 2011). The mechanical relation with this method is represented by

$$F_\mathrm{P} = ky + c\dot{y} + m\ddot{y}, \tag{2}$$

$$\begin{cases} k = k_\mathrm{s} + k_\mathrm{PA} \\ c = c_\mathrm{s} + c_\mathrm{PA} \\ m = m_\mathrm{s}, \end{cases} \tag{3}$$

where y represents the displacement of A (stage/mass block), k represents the stiffness of the PIFA system, c represents the damping of the PIFA system, m represents the mass of the PIFA system, m_s represents the mass of A (stage/mass block), c_s represents the damping coefficient of A (stage/mass block), k_s represents the stiffness of (stage/mass block), c_PA represents the PA damping coefficient, k_PA represents the stiffness of A (stage/mass block), and F_P represents the drive force generated by the PA. It is noted that k, c, and m are at the system level, meaning that they are an aggregated property of the stiffness, damping, and mass of each component in the system (PA and stage/mass block in this case). In Kang (2007), some reasons for neglecting the mass of the PA are listed. Basically, if the mass of the PA is much less than the mass of the other PIFA components, the mass of the PA can be neglected. This situation may occur with a single PA system but not in the PIFA. In a PIFA, the sizes of the PA and the other neighboring components are comparable, and this is true in particular when a micro-PIFA is considered. Therefore, for PIFAs, in most cases, the mass of the PA can be a significant factor, and it cannot be ignored.

The second method is to treat the mass of the PA with a lumped model (Yakimov, 1997; Breguet and Clavel, 1998). Particularly, the mass of the PA is considered as concentrated at one point (usually the midpoint), and the mechanical relation is then represented by

$$F_\mathrm{P} = ky + c\dot{y} + m\ddot{y} \tag{4}$$

$$\begin{cases} k = k_\mathrm{s} + k_\mathrm{PA} \\ c = c_\mathrm{s} + c_\mathrm{PA} \\ m = m_\mathrm{s} + \frac{m_\mathrm{PA}}{2}, \end{cases} \tag{5}$$

where the parameters are the same as those in the first method except that m_PA represents the mass of the PA.

The third method is to treat the mass of the PA as a distributed lumped system (Pozzi and King, 2003). In particular, the method considers that the PA is composed of n pieces. Every piece is treated the same way as in the second method. Thus, there are $2n$ equations to describe the dynamics of the PA.

Of the three methods, the third method is the most accurate one, as it captures more of the inertia of the PA. However, it suffers from computational overhead, which is a detrimental factor for the real-time performance of PIFAs with a feedback control strategy. A trade-off thus needs to be made when the model of the PIFA is used for control of the PIFA, which is the accuracy of the dynamic model of PIFAs versus the real-time information acquisition.

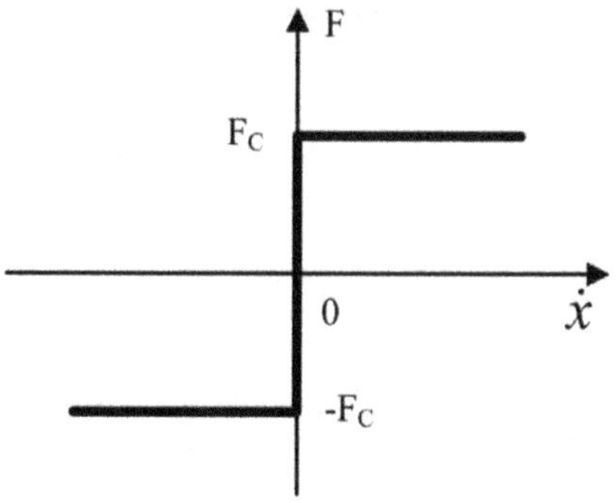

Figure 6. Coulomb friction model.

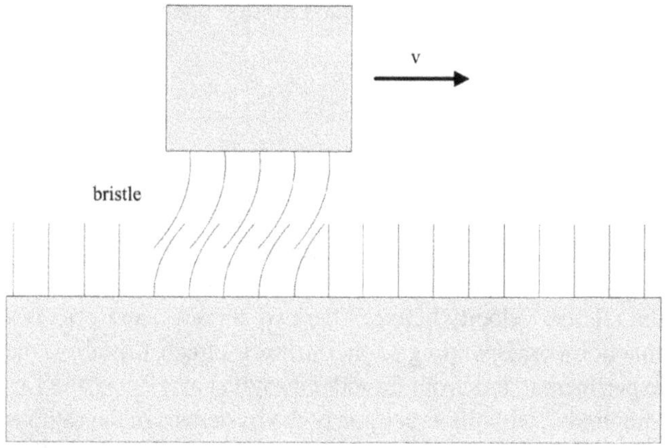

Figure 7. Bristle theory (Canudas de Wit et al., 1995).

2.2 Modeling of friction

There are six main models of friction used in modeling the friction in the PIFA system: the Coulomb model, the reset-integrator model, the LuGre model, the elastoplastic model, the Leuven model, and the Dahl model. Among these, the Coulomb, LuGre, and elastoplastic models are the most popular ones. The literature regarding these models in PIFAs is discussed in the following.

2.2.1 Coulomb model

The expression for this model is as follows:

$$F = \begin{cases} F_c \cdot \mathrm{Sgn}(\dot{x}) & \text{if } \dot{x} \neq 0, \\ F_{\mathrm{app}} & \text{if } \dot{x} = 0 \text{ and } F_{\mathrm{app}} < F_c \end{cases} \quad (6)$$

$$F_c = \mu F_N, \quad (7)$$

where F is the friction force, $\dot{x}$ is the sliding speed, F_{app} is the applied force, F_C is the Coulomb friction force, μ is the Coulomb friction coefficient (or the dynamic friction coefficient), and F_N is the normal force between the two contact surfaces.

It can be observed (from the model) that the friction force only depends on the applied force F_{app} and the direction of the sliding speed $\dot{x}$. It has only two values, F_c and $-F_c$, as shown in Fig. 6

This model cannot describe the influence of the sliding speed (viscous friction) and the transition between static friction and dynamic friction (the pre-sliding friction and the Stribeck effect (Canudas de Wit et al., 1995) and the zero amplitude phenomenon (Edeler et al., 2011)). Unfortunately, these factors may significantly impact the behavior and performance of PIFAs. Furthermore, there is also some difficulty in determining the direction of the velocity when the velocity is zero; when the velocity approaches zero, a high-frequency oscillating motion may occur.

This model was used in Chang and Li (1999), Jiang et al. (2000), Pohl (1987), Darby and Pellegrino (1997), and Okamoto and Yoshida (1998). In Pohl (1987), Darby and Pellegrino (1997), and Okamoto and Yoshida (1998), a very simple model for modeling of a PA (e.g., ignoring the hysteresis of a PA) and inertia of the entire PIFA was presented. The Coulomb model was used to model the friction in these works; additionally, the simulation and experimental results were compared, and they showed the same trend between the experimental and simulation results but with a large error. It was further found that these models completely failed to predict the performance of their experimental system when the frequency of the input voltage was high. It is clear that the result from these works suggests that the Coulomb model for friction in PIFAs is not adequate. Other works such as Furutani et al. (1998), Chang and Li (1999), and Jiang et al. (2000), which also used the Coulomb model for friction, did not discuss the adequacy of the Coulomb model for frictional contact in PIFAs.

Some modifications to the Coulomb model have been made in the existing literature. For instance, Patrascu and Stramigioli (2007) used a model for friction that was based on the Coulomb model with an empirical Stribeck effect. The entire system in Patrascu and Stramigioli (2007) was modeled as a simple mass-spring system. They showed that the model predicted results quite close to the experimental results. This suggests the need to include the Stribeck effect in the friction model for PIFAs.

2.2.2 LuGre model

It is known from the literature that the LuGre model can usually provide a relatively good result with acceptable complexity. This model was developed based on bristle theory (Canudas de Wit et al., 1995). In this theory, a frictional surface is composed of numerous elastic bristles. The frictional force arises from the interaction of the elastic bristles of the two contact surfaces, as shown in Fig. 7.

The model further assumes that all of the bristles on the two contact surfaces are the same in terms of their bending stiffness. The expression for the LuGre model is as follows:

$$\begin{cases} F = \sigma_0 z + \sigma_1 \dfrac{dz}{dt} + \sigma_2 \mathrm{v} \\ \dfrac{dz}{dt} = v - \dfrac{|v|}{g(v)} z, \end{cases} \tag{8}$$

where F is the friction force, σ_0 is the average stiffness of the bristles, σ_1 is the damping coefficient, σ_2 is the viscous coefficient, z is the average deflection of the bristles, v is the relative velocity between the two surfaces, and $g(v)$ is a function corresponding to the Stribeck effect. Based on the experimental data from Canudas de Wit et al. (1995), the LuGre model, with the exception of the hysteresis in the regions that include the pre-sliding period, has captured almost all of the important frictional characteristics such as the Stribeck effect, static friction, viscous friction, frictional lag, and pre-sliding.

The LuGre model was used in the work of Kang (2007), Breguet and Clavel (1998), Canudas de Wit et al. (1995), Bergander and Breguet (2003), Zhang (2008), and Li et al. (2009). In Bergander and Breguet (2003), a dynamic model of PIFAs was established, in which the LuGre model was used to model the friction. The goal of their study was to examine how to attenuate vibration in the PIFA so that the velocity of the PIFA could be increased (note: the velocity is restricted by the vibration). There is no direct evidence in their work to show the adequacy of the LuGre model; however, the success of their method to attenuate the vibration and consequently double the velocity of the PIFA may provide indirect evidence of the adequacy of the LuGre model. In Kang (2007), a comparison was made between the LuGre, elastoplastic model, and reset-integrator model, based on the simulation. The comparison showed that the LuGre model was the most suitable one for their system in terms of the usability of the model.

One problem with the LuGre model is that the accuracy of the model changes with an increase in the number of operations of the system. Such a phenomenon is called drift. The underlying reason for the drift is due to the plastic deformation of the asperity. However, the LuGre model is unable to capture this kind of deformation.

Some modifications to the LuGre model have been made in the existing literature. In our group, Li et al. (2009) added the thermal effect into the LuGre model, and two approaches to integrate the thermal effect into the model were proposed. The first approach was to consider the parameters in the LuGre model as functions of temperature. The second approach was to consider the model as having two parts: (i) the LuGre model, which corresponds to the friction force without the thermal effect, and (ii) the thermal effect, in which the temperature is the only variable. The two approaches were compared with a conclusion that the second approach is better than the first one in terms of both accuracy and computational cost. This was perhaps the first time in which the thermal effect was considered in a friction model. However, the method of Li et al. (2009) took a black-box approach to establishing a model; see (Li et al., 2009) for a more detailed discussion of the black-box model. Therefore, the model-building process is more complex, and the model accuracy is limited.

It should be noted that, in Zhang (2008), the viscous term in the LuGre model was abandoned for PIFAs. This is because the author found that the viscous term has little effect on PIFA behavior, which is also due to the limited velocity with the PIFA. In short, it remains to be a future work to examine the LuGre model for its suitability for PIFAs, especially to determine a coupling relation between the friction and the temperature rise at the contact surface in PIFAs.

2.2.3 Elastoplastic model

The elastoplastic model is an improvement of the LuGre model. The expression for the model is as follows:

$$\begin{cases} F = \sigma_0 z + \sigma_1 \dfrac{dz}{dt} + \sigma_2 v \\ \dfrac{dz}{dt} = v\left(1 - \alpha(z, v) \dfrac{\sigma_0}{g(v)} z\right)^{\mathrm{i}}, \end{cases} \tag{9}$$

where F is the friction force; σ_0 is the average stiffness of the bristles; σ_1 is the damping coefficient; σ_2 is the viscous coefficient, z is the average deflection of the bristles; v is the relative velocity between the two surfaces (which is calculated from the relative displacement between the two surfaces, x, using the relation $v = \dot{x}$); x is divided into two parts: $x = z + w$, where z is the elastic part (also the deformation of the bristles) and w is the plastic part; and $g(v)$ is a function corresponding to the Stribeck effect. $\alpha(z, v)$ is a parameter which is defined by

$$\alpha(z, v) = \begin{cases} 0 & |z| < z_{\mathrm{ba}} \\ \dfrac{1}{2}\sin\left(\pi \dfrac{z - \left(\frac{z_{\max} + z_{\mathrm{ba}}}{2}\right)}{z_{\max} - z_{\mathrm{ba}}}\right) + \dfrac{1}{2} & z_{\mathrm{ba}} < |z| < z_{\max} \\ 1 & |z| > z_{\max}, \end{cases} \tag{10}$$

where z_{ba} is the breakaway average deflection and $z_{\max}$ is the maximum average deflection or the steady-state deflection.

The improvement of the elastoplastic model over the LuGre model is that elastoplastic model divides the relative displacement in the pre-sliding stage into two parts – the elastic part z and the plastic part w – by introducing the parameter $\alpha(z, v)$. The elastoplastic model provides a more detailed description of the pre-sliding. The introduction of the notion of the breakaway average deflection z_{ba} may overcome the drift problem in the LuGre model (Edeler et al., 2011). However, the elastoplastic model is much more complex than the LuGre model due to the introduction of the parameters w, z, and $\alpha(z, v)$.

In the works of Edeler et al. (2011), Peng and Chen (2011), Dupont et al. (2000, 2002), Chen et al. (2008), and Rakotondrabe et al. (2009), the elastoplastic model was used to model friction. In Dupont et al. (2000), the elastoplastic model was compared with the Coulomb and LuGre models. The results showed that only elastoplastic model can cover both presliding displacement and stiction in the friction process. In Chen et al. (2008), both the Dahl and elastoplastic model were used. According to them, there was no large difference between the two models. The authors came to the conclusion that both the Dahl and elastoplastic model are effective in representing the friction in the PIFA system. This may be because the test bed they used to identify the model parameters was too coarse. They used a supporting cylinder to support the driving object. There was a friction between the supporting cylinder and the driving object, which may contribute to the dynamics of the entire system; however, this friction was ignored in their work.

In Rakotondrabe et al. (2009), a state-space model of PIFAs was presented. This model is only for one period of motion. In Kang (2007), two models (the LuGre and elastoplastic model) were carefully compared. They concluded that the elastoplastic model is not a good choice for PIFA systems unless a sound reason is given.

Some modifications of the elastoplastic model have been made in literature. For instance, in Edeler et al. (2011), the authors performed a comprehensive investigation of the "zero-amplitude phenomenon", and they developed a model called the "CEIM" model based on the elastoplastic model. In the CEIM model, they made an empirical modification by introducing the preload to the parameters z_{ba}, z_{max} and σ_0. This modification enabled the model to cover the influence of the preload at zero amplitude.

2.2.4 Leuven, reset-integrator, and Dahl models

In Ha et al. (2005), they used a combined Leuven model and Bouc–Wen model to describe the friction and friction-induced hysteresis. They showed that the friction-induced hysteresis can be captured. However, how effective the Leuven model is at describing the friction behavior was not shown. In Fung et al. (2008a), hysteresis was considered as a characteristic of friction and not only as a characteristic of the piezoelectric materials. Certainly, both friction and the piezoelectric materials have hysteresis, and so their work is a pioneering work in the area of modeling of the PIFA. However, there is no direct evidence (in their work) of the adequacy of the Leuven model for friction. The reset-integrator model was presented in Chao et al. (2006). They developed a feedback controller for the PIFA system. However, the performance of the model, in the aspect of friction modeling, was not discussed. In Chen et al. (2008), the Dahl and elastoplastic model were compared as discussed previously.

2.2.5 Discussion

In short, the Coulomb, LuGre, elastoplastic, rest-integrator, Leuven, and Dahl models have been applied to PIFAs in the existing literature. The first three models have been widely used, and the LuGre model appears to be the most promising one. However, not much attention has been to thermal effects on the friction behavior of PIFAs. Friction can cause a significant temperature rise, which may significantly affect the performance of PIFAs (Li et al., 2008).

Table 1. General model of a PIFA.

Following Newton's second law	
For object A	$m_{\mathrm{A}}\ddot{x}_{\mathrm{A}} = {}^{D}_{A}F_{\mathrm{a}} - {}^{\mathrm{B}}_{\mathrm{A}}F_{\mathrm{f}}$
For object B	$m_{\mathrm{B}}\ddot{x}_{\mathrm{B}} = {}^{\mathrm{A}}_{\mathrm{B}}F_{\mathrm{f}}$
For object C	$m_{\mathrm{C}}\ddot{x}_{\mathrm{C}} = -{}^{\mathrm{D}}_{\mathrm{C}}F_{\mathrm{a}} P\left({}^{\mathrm{C}}_{\mathrm{D}}F_{\mathrm{a}}\right.$
For object D	$\left.{}^{\mathrm{A}}_{\mathrm{D}}F_{\mathrm{a}},\ V\right) = 0$
Following Newton's third law	
${}^{\mathrm{B}}_{\mathrm{A}}F_{\mathrm{f}} = {}^{\mathrm{A}}_{\mathrm{B}}F_{\mathrm{f}} = F(N);\ {}^{\mathrm{C}}_{\mathrm{D}}F_{\mathrm{a}} = {}^{\mathrm{D}}_{\mathrm{C}}F_{\mathrm{a}};\ {}^{\mathrm{A}}_{\mathrm{D}}F_{\mathrm{a}} = {}^{\mathrm{D}}_{\mathrm{A}}F_{\mathrm{a}}$	

2.3 Model integration

Model integration is used to integrate the PA model and the friction model for an entire PIFA system. The integration is based mainly on Newton's law. In the following, we first propose a general model for any PIFA system based on the proposed general architecture (Fig. 3), and then the literature will be discussed in the context of this general model.

Fig. 8 shows a separate force diagram for each component. In Fig. 8, there is a friction force ${}^{\mathrm{B}}_{\mathrm{A}}F_{\mathrm{f}}$ on A from B, and an actuation force ${}^{\mathrm{D}}_{\mathrm{A}}F_{\mathrm{a}}$ on A from D. There is a friction force ${}^{\mathrm{A}}_{\mathrm{B}}F_{\mathrm{f}}$ on B from A. There is an actuation force ${}^{\mathrm{D}}_{\mathrm{C}}F_{\mathrm{a}}$ on C from D. There is an actuation force ${}^{\mathrm{A}}_{\mathrm{D}}F_{\mathrm{a}}$ on D from A, and another actuation force ${}^{\mathrm{C}}_{\mathrm{D}}F_{\mathrm{a}}$ on D from C. With respect to the origin O, the positions of A, B, and C are x_{A}, x_{B}, and x_{C}, respectively. One end of the D has the same position as A, and the other end of D has the same position as C. The masses of the objects are m_{A}, m_{B}, m_{C}, and m_{D}. Thus, the following equations for a general PIFA system can be derived, as shown in Table 1.

There are eight equations in Table 1, with two general functions (P, F), where P is a model for the PA and F is a model for friction. A specific model of a specific PIFA is dependent on a specific P and a specific F. Specific F refers to the different friction models that were reviewed earlier in this paper. Specific P refers to the different models that were reviewed earlier in this paper.

In the literature, most of the models for the PIFA consider the components (except the PA) to be rigid. Further, in Adriaens et al. (2000), both inertia and damping of a stage/mass block system, driven by the PA, were considered. The same approach can be found in Kang (2007), Zhang (2008), and Chen et al. (2008). By considering the damping, the accuracy of the model has been improved.

It is noted that most PIFA models, except for the one in Yakimov (1997) and Breguet and Clavel (1998), have not considered gravitational effects. As a result, they are re-

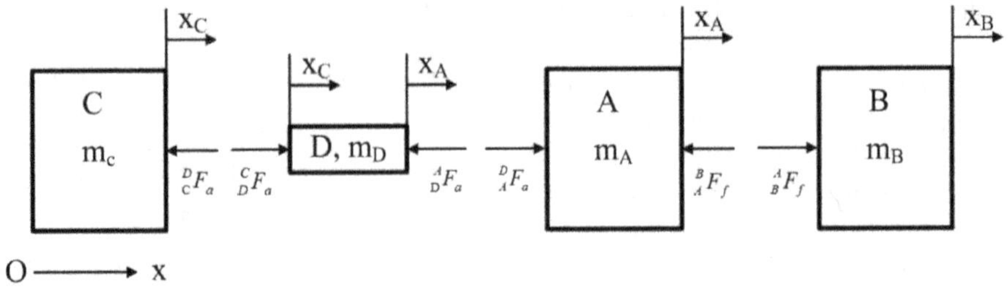

Figure 8. Free-body diagram of the PIFA components.

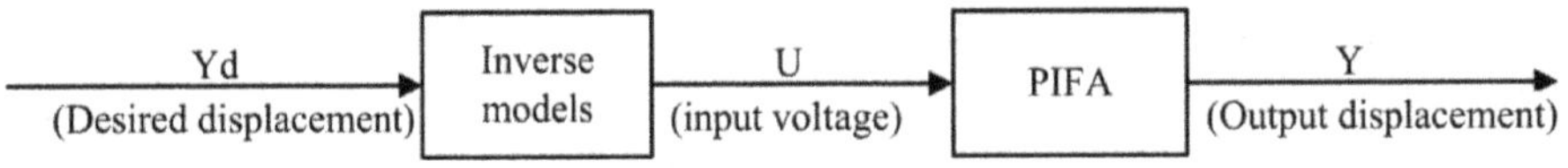

Figure 9. Principle of feed-forward control.

stricted to actuations in the horizontal direction only. Furthermore, the previous models for PIFAs are for PIFAs with one degree of freedom and with a rigid connection between the PA and the other components. In fact, a so-called soft actuation concept may be applicable to PIFAs, in which several components may be quite compliant or "soft".

2.4 Utility of the PIFA model

The models for the PIFAs were used for the optimization of PIFA performance. For example, in the work of Darby and Pellegrino (1997), the model was used to optimize the input voltage waveform. In the work of Ha et al. (2005) and Jiang et al. (2000), the models were used for the optimization of the design of the whole PIFA system, including the structure of the components and the input waveform. Additionally, the model was used to test the feasibility of new PIFA designs in Lambert et al. (2003). In addition to the optimization of PIFA systems, a partial PIFA model (e.g., model of the PA alone) was used to develop a feed-forward controller or compensator for PIFAs. For instance, in the work of Ha et al. (2005), the PA model was used to compensate for the PA hysteresis in PIFAs.

3 Control of PIFA systems

3.1 Feed-forward control

The feed-forward control is commonly used to improve the quality of the output motion of a PIFA system. This is done by compensating for the hysteresis of some components in the PIFA (e.g., PA) by designing the structure of the PIFA and/or modifying the input voltage wave to attenuate vibration in the output (Holub et al., 2006). The principle of feed-forward control is shown in Fig. 9.

There are different strategies available for compensation for the hysteresis of the PA. One approach is to adjust the phase lag of the end effector (Chang and Du, 1998). Another strategy is to invert the hysteresis model (which represents the relation between the driving force and the output displacement of the PA; Ha et al., 2005).

The feed-forward control of PIFAs is challenging because it is difficult to generate an accurate model for the PIFA. This difficulty is further due to many factors in the system, including material hysteresis, friction hysteresis, creep, vibration during the stick and slip points, and temperature rise (due to the friction, and which further changes the friction). Another difficulty arises from the inherent uncertainty in such a system, e.g., friction-induced wear of the material (Bergander et al., 2000) and thermally induced degradation of the material.

There is a mechatronic technique that can be used to compensate for hysteresis and which uses charge control instead of voltage control (Newcomb and Flinn, 1982; Fleming et al., 2006). As such, the information flow of the system is from charge to voltage. The hysteresis occurs in the path from the voltage to the deformation. By designing the flow path from the charge to voltage, it is possible to eliminate the hysteresis from the charge to the deformation. More recently, a hybrid charge control drive was developed for PIFAs to achieve two goals (Spiller and Hurak, 2011): (1) to compensate for the hysteresis in the piezoelectric material and (2) to increase the slew rate (a fast return in voltage to achieve slip). The hybrid charge control drive is an excellent idea in that it actually provides a supplemental means to compensate for the hysteresis, and since the whole system has greater means to cope with non-linear properties, the whole system gains the ability to focus on overcoming other problems, e.g., the vibration of the end effector.

3.2 Feedback control

Feedback control (of a PIFA system) is usually used to address the dynamic uncertainty due to friction or outside disturbances. Feedback control requires a sensor in order to measure the output (the displacement in the case of PIFAs)

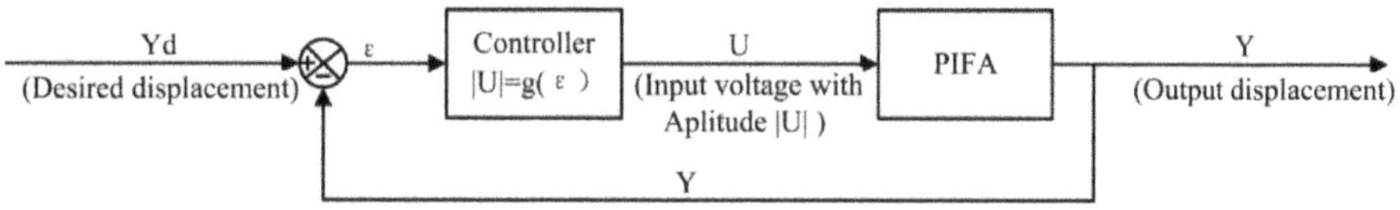

Figure 10. Position feedback control with amplitudes of input voltage.

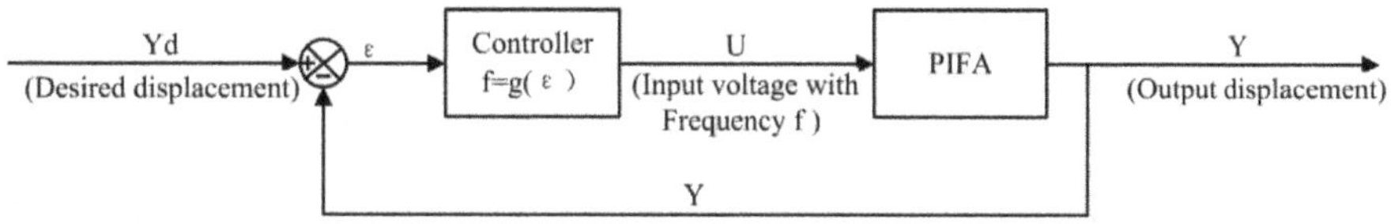

Figure 11. Position feedback control with frequency of input voltage.

for the subsequent evaluation of the output. The outcome of the evaluation is then used to adjust the control input to the (plant) system or controller itself (i.e., adaptive control).

There are two types of feedback control (of the PIFA) in the literature in terms of the output signals, namely positioning feedback and velocity feedback. In the positioning feedback control method, the feedback signal is the displacement of the end effector. The control input is the voltage or the charge shown in Fig. 10.

For instance, in Fahlbusch et al. (1999), the authors presented a study where the output displacement information is acquired using a CCD camera with image analysis. The controller is a fuzzy logic controller. Their approach was, however, not verified by either experiment or simulation. The motivation for their work was to examine the feasibility of using an image as a position sensor, as it is indeed difficult to build a sensor in PIFAs due to the limited space available in the applications. In contrast, an image-based sensor is not intrusive to the system at all, which is clearly an advantage. In Shim and Gweon (2001), a laser interferometric sensor was used to obtain the displacement of the end effector, and the control input was the voltage. However, they did not describe their control law. The PIFA had three degrees of freedom, which were coupled. It is unfortunate that the experimental verification of their control system did not show much promise due to the poor construction of the test bed.

In the positioning feedback control method, the control signal could also be the frequency of the voltage (or switching frequency for the PIFA in particular); see the work described in Breguet and Clavel (1998). The control strategy is shown in Fig. 11.

In Breguet and Clavel (1998), an interferometric sensor was used to measure the output displacement. The control law is a proportional one, i.e., the frequency is proportional to the displacement error. The control system enables the system to reach steady state within 2 ms without any overshoot (there is a ± 5 nm noise due to the resolution of the interferometric sensor).

A control scheme with two or more inputs has also been studied in the literature. For instance, in Rakotondrabe et al. (2008), the voltage and frequency were taken as two control inputs. The output was the displacement. The control law for the two controls was a proportional law – i.e., both the amplitude and the frequency of the voltage are proportional to the displacement error.

The control strategy is shown in Fig. 12.

This control system appears to have a superior performance in the normal or stepping mode of the operation of the PIFA. However, it is not clear whether there is performance improvement in the scanning mode. The definition of stepping mode and scanning mode is mentioned in Sect. 1.

In addition to the positioning feedback control method, a velocity feedback control method has also been proposed in the literature (Chao et al., 2006). The challenge with this method was how to accurately measure the velocity information. Fortunately, there is an effect called the double piezoelectric effect (DPE), in which the PAs can provide both the position and velocity information at the same time. It is noted that the velocity information can be directly used to compensate for the vibration of the end effector. However, there are only a few studies in the literature that discuss the control of PIFAs using velocity information. The control strategy of velocity feedback control is shown in Fig. 13.

In Zou et al. (2005), the authors proposed a more sophisticated learning control method. The control law was composed of two parts: forward dynamics (i.e., inversion-based) and a feedback controller (proportional and iterative learning). The forward dynamics were obtained using the transfer function of the system. Their approach is promising because the iterative learning controller is simple and yet provides greater enhancement of control accuracy (Chen et al., 2011; Ouyang et al., 2006). However, the PIFA does not seem to be repetitive on its own, as each step is slightly different, and the motion at the current step has some dependence on the motion at the previous steps.

In summary, the control for PIFAs is a challenging problem due to the presence of many uncertain factors, such as

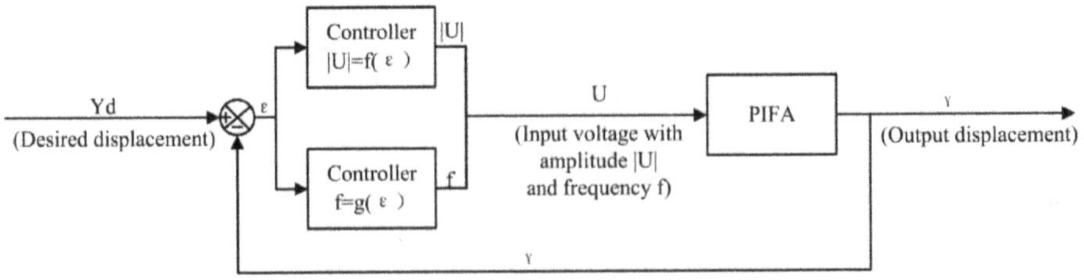

Figure 12. Position feedback control with both amplitudes and frequency.

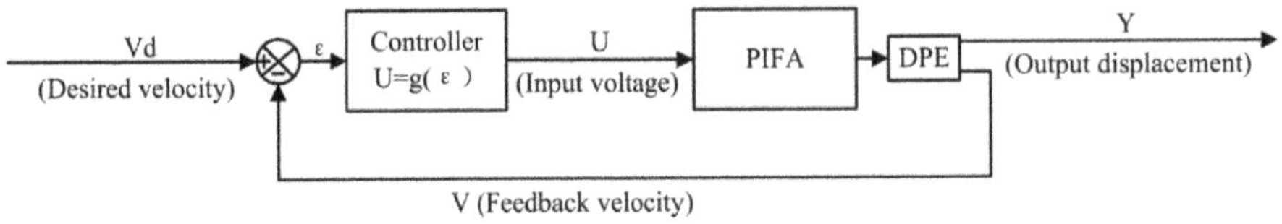

Figure 13. Velocity feedback control.

friction and hysteresis. The challenge is also due to the principle of its actuation, which consists of both the scanning mode (stick period) and step mode (slipping period). It is known that conventional feedback control is usually not suitable for the step movement. However, the PIFA requires good feedback control for achieving reasonable accuracy and reliability. In the current literature, the feedback control for PIFAs is mostly model-free with a proportional control law.

4 Concluding remarks with future directions for research

4.1 Concluding remarks

This paper has presented a critical review of the work on the modeling and control of PIFAs. Modeling can be separated from control, and the model can be very useful for optimizing the design of PIFAs. In this paper, modeling was mainly considered useful to control, in particular for feed-forward control and compensation. To generalize the review results, this paper has also proposed a general architecture for PIFA systems. This architecture consists of a generic structure suitable for any PIFA, that is, four components with the relations among them, including the inertia-varying components, such as piezoelectric materials and friction. These components serve as templates, and an individual PIFA can be built by instantiating the template. This general architecture was useful for the systematic generation of the dynamic model of any PIFA system, as demonstrated in this paper in Sect. 1. This implies that the design of a PIFA system (including its structure and controller) can be automated by the computer. Several concluding remarks are further made in the following:

1. A dynamic model that captures all of the dynamics including friction, thermal effects, hysteresis, and vibration is not currently available in the literature. Such a model could be called a comprehensive model. It is noted that a comprehensive model would be very helpful for feed-forward control or compensation to further improve the performance of PIFA systems (note: the comprehensive model is supposed to capture all the dynamics of PIFAs).

2. The experimental studies of PIFAs are generally weak in the current literature, as most test beds are not well designed and the experimental data are thus not convincing enough; for instance the support means have changed the dynamics of the PIFA in Li et al. (2009). This has compromised the reliability of the validation of the control methods.

3. Feedback control laws are usually simple and of a proportional type. The dynamic model is used mostly for the purpose of compensation, especially compensating for the hysteresis of the PA material. Such a feedback system is not quite robust and does not adequately address the highly uncertain dynamics of PIFAs.

4.2 Directions for future research

First, it would be of interest to develop a computer-aided design system for PIFAs. Such a system, equipped with a computer user interface, would greatly facilitate the design of PIFA systems. The proposed general architecture of PIFAs is a possible starting point for this computer-aided design system, as a general model for the plant is a necessity for any effective computer aiding for design and control of an underlying plant system (Zhang, 1994; Li and Zhang, 1998).

Second, a comprehensive dynamic model needs to be developed to provide more accurate compensation for the hysteresis of PAs as well as the hysteresis arising from friction. Such a model can also be useful as a part of the entire feedback control system.

Third, the problem of degradation requires some attention. It is known that a PIFA is subject to a high rate of degradation due to friction, which always plays an active role. Friction can consume energy in a reversible manner and can cause surface degradation. With the degradation in mind, the dynamic properties of the PIFA system are time-varying. We propose that a model-updating technique be used to update the model in response to degradation. Furthermore, the model-updating technique may also be expanded to the controller by updating the parameters in the controller. This can then be called an offline adaptive controller.

Fourth, the compensation is expanded to the whole system instead of to the PA component alone. In particular, the hysteresis due to the friction (intertwined with the temperature rise) needs to be compensated for. New friction models can be developed to take into account the coupling effect of friction and temperature variation (a rise in the continuous operation).

Fifth, PIFAs with different orientations and different degrees of freedom and different configurations have wide applications, such as in atom force microscopy, 3-D printing, and micro-robots. This calls for the study of control methods for such PIFAs. For instance, when the direction of the output of a PIFA is vertical, the effects of gravity need to be considered as well.

Last, a relatively new concept called resilient PIFAs may be worth investigation. It is well known that PIFAs are very sensitive to disturbances, especially friction-induced degradation in the interface among these composing objects (A, B, C, D) (Fig. 3). This implies that PIFAs may easily suffer from malfunctioning. As such, system recovery from the dysfunctions of PIFAs (i.e., resilience (Zhang and Luttervelt, 2011; Zhang and Lin, 2010) is an interesting problem worthy of future research. In the recovery process, the system needs to be reconfigured and, consequently, the dynamic model for the system needs to be updated or regenerated. In this context, computer generation or automatic generation of the dynamic model for PIFAs is an essential requirement.

Acknowledgements. This work was supported by the National Natural Science Foundation of China (grant no. 51375166), National Natural Science Foundation of China (grant no. 61422310), the China Scholarship Council (CSC), and the Fundamental Research Funds for the Central Universities of East China University of Science and Technology. W. J. Zhang also wants to thank the NSERC for partial support for this research through a Discovery Grant.

References

Adriaens, H., De Koning, W. L., and Banning, R.: Modeling piezoelectric actuators, Mechatronics, IEEE/ASME Trans., 5, 331–341, 2000.

Bergander, A. and Breguet, J.-M.: Performance improvements for stick-slip positioners, MHS 2003: Proceeding of 2003, International Symposium on Micromechatronics and Human Science, Nagoya, Japan, 2003, 59–66, 2003.

Bergander, A., Breguet, J. M., Schmitt, C., and Clavel, R.: Micropositioners for microscopy applications based on the stick-slip effect, Mhs 2000: Proceedings of the 2000 International Symposium on Micromechatronics and Human Science, 213–216, doi:10.1109/Mhs.2000.903315, 2000.

Bouc, R.: Forced vibration of mechanical systems with hysteresis, Preceedings of the 4th International Conference on Nonlinear Oscillations, Prague, Czechoslovakia, p. 315, 1967.

Breguet, J. M. and Clavel, R.: Stick and slip actuators: design, control,performances and applications, MHS 1998, Proceedings of the 1998, International Symposium on Micromechatronics and HumanScience, Nagoya, Japan, 1998, 89–95, 1998.

Canudas de Wit, C., Olsson, H., Astrom, K. J., and Lischinsky, P.: A new model for control of systems with friction, Automatic Control, IEEE Trans., 40, 419–425, doi:10.1109/9.376053, 1995.

Chang, S. and Li, S.: A high resolution long travel friction-drive micropositioner with programmable step size, Rev. Sci. Instr., 70, 2776–2782, 1999.

Chang, S. H. and Du, B. C.: A precision piezodriven micropositioner mechanism with large travel range, Rev. Sci. Instr., 69, 1785–1791, doi:10.1063/1.1148842, 1998.

Chao, S. H., Garbini, J. L., Dougherty, W. M., and Sidles, J. A.: The design and control of a three-dimensional piezoceramic tube scanner with an inertial slider, Rev. Sci. Inst., 77, 063710–063717, 2006.

Chen, X. B., Kong, D., and Zhang, Q. S.: On the dynamics of piezoelectric-driven stick-slip actuators, in: Advances in Machining and Manufacturing Technology Ix, edited by: Yao, Y., Xu, X., and Zuo, D., Key Eng. Mat., 648–652, 2008.

Chen, Z., Wang, Y., Ouyang, P., Huang, J., and Zhang, W.: A novel iteration-based controller for hybrid machine systems for trajectory tracking at the end-effector level, Robotica, 29, 317–324, 2011.

Cheng, L., Lin, Y. Z., Hou, Z. G., Tan, M., Huang, J., and Zhang, W. J.: Integrated Design of Machine Body and Control Algorithm for Improving the Robustness of a Closed-Chain Five-Bar Machine, IEEE Asme. T. Mech., 17, 587–591, doi:10.1109/Tmech.2012.2183378, 2012.

Croft, D., Shedd, G., and Devasia, S.: Creep, hysteresis, and vibration compensation for piezoactuators: Atomic force microscopy application, P. Amer. Contr. Conf., 123, 2123–2128, 2000.

Darby, A. and Pellegrino, S.: Inertial stick-slip actuator for active control of shape and vibration, J. Intel. Mat. Syst. Str., 8, 1001–1011, 1997.

Dupont, P., Armstrong, B., and Hayward, V.: Elasto-plastic friction model: contact compliance and stiction, American Control Conference 2000, Proceedings of the 2000, 1072–1077, 2000.

Dupont, P., Hayward, V., Armstrong, B., and Altpeter, F.: Single state elastoplastic friction models, Automatic Control, IEEE Trans. Autom. Control, 47, 787–792, 2002.

Edeler, C., Meyer, I., and Fatikow, S.: Modeling of stick-slip micro-drives, J. Micro-Nano Mechatr., 6, 65–87, 2011.

Fahlbusch, S., Fatikow, S., Seyfried, J., and Buerkle, A.: Flexible microrobotic system MINIMAN: design, actuation principle and control. Proceeding of the 1999 IEEE/ASME, International Conference on Advanced Intelligent Mechatronics, Atlanta, USA, 19–23 September 1999, 156–161, 1999.

Fleming, A. J., Behrens, S., and Moheimani, S. O. R.: Inertial vibration control using a shunted electromagnetic transducer (vol 11, pg 84, 2006), IEEE-Asme. T. Mech., 11, 367–367, doi:10.1109/Tmech.2006.878790, 2006.

Fung, R.-F., Han, C.-F., and Chang, J.-R.: Dynamic modeling of a high-precision self-moving stage with various frictional models, Appl. Math. Model., 32, 1769–1780, doi:10.1016/j.apm.2007.06.012, 2008a.

Fung, R. F., Han, C. F., and Ha, J. L.: Dynamic responses of the impact drive mechanism modeled by the distributed parameter system, Appl. Math. Model., 32, 1734–1743, 2008b.

Furutani, K., Mohri, N., Higuchi, and T., and Saito, N.: Development of Pocket-Size Electro-Discharge Machine with Multiple Degrees of Freedom, Proceedings of the 1993 JSME International Conference on Advanced Mechatronics, Tokyo, Japan, 2–4 August 1993, 561–566, 1993.

Furutani, K., Mohri, N., and Higuchi, T.: Self-running type electrical discharge machine using impact drive mechanism, Proc. Adv. Int. Mechatron. AIM, 97, 88–93, 1997.

Furutani, K., Higuchi, T., Yamagata, Y., and Mohri, N.: Effect of lubrication on impact drive mechanism, Prec. Eng., 22, 78–86, 1998.

Ha, J. L., Fung, R. F., and Yang, C. S.: Hysteresis identification and dynamic responses of the impact drive mechanism, J. Sound Vibr., 283, 943–956, 2005.

Ha, J.-L., Fung, R.-F., Han, C.-F., and Chang, J.-R.: Effects of frictional models on the dynamic response of the impact drive mechanism, J. Vibr. Acoust., 128, 88–96, 2006.

Higuchi, T., Yamagata, Y., Furutani, K., and Kudoh, K.: Precise Positioning Mechanism Utilizing Rapid Deformations of Piezoelectric Elements, Micro Electro Mech. Syst., 222–226, 1990.

Holub, O., Spiller, M., and Hurak, Z.: Stick-slip based micropositioning stage for transmission electron microscope, 9th IEEE International Workshop on Advanced Motion Control, 484–487, doi:10.1109/amc.2006.1631707, 2006.

Jiang, T., Ng, T., and Lam, K.: Optimization of a piezoelectric ceramic actuator, Sens. Act. A: Phys., 84, 81–94, 2000.

Kang, D.: Modeling of the piezoelectric-driven stick-slip actuators, Master, University of Saskatchewan, 1–111, 2007.

Lambert, P., Valentini, A., Lagrange, B., De Lit, P., and Delchambre, A.: Design and performances of a one-degree-of-freedom guided nano-actuator, Robot. Comp.-Int. Manufact., 19, 89–98, doi:10.1016/S0736-5845(02)00065-0, 2003.

Li, J., Chen, X., An, Q., Tu, S., and Zhang, W.: Friction models incorporating thermal effects in highly precision actuators, Rev. Sci. Instr., 80, 045104–045106, 2009.

Li, J. W., Yang, G. S., Zhang, W. J., Tu, S. D., and Chen, X. B.: Thermal effect on piezoelectric stick-slip actuator systems, Rev. Sci. Instr., 79, 046108, doi:10.1063/1.2908162, 2008.

Li, Q. and Zhang W. J.: On methodology of using model-based reasoning approach to intelligent CAE systems development, Eng. Appl. Artif. Intel., 11, 327–336, 1998.

Mayergoyz, I. D.: The Classical Preisach Model of Hysteresis, in: Mathematical Models of Hysteresis, Springer New York, 1–63, 1991.

Newcomb, C. and Flinn, I.: Improving the linearity of piezoelectric ceramic actuators, Electr. Lett., 18, 442–444, 1982.

Nguyen, H. X., Edeler, C., and Fatikow, S.: Modeling of Piezo-Actuated Stick-Slip Micro-Drives: An Overview, Adv. Sci. Technol., 81, 39–48, 2013.

Okamoto, Y. and Yoshida, R.: Development of linear actuators using piezoelectric elements, Electronics and Communications in Japan, Part III: Fundamental Electronic Science (English translation of Denshi Tsushin Gakkai Ronbunshi), 81, 11–17, doi:10.1002/(SICI)1520-6440(199811)81:11<11::AID-ECJC2>3.0.CO;2-U, 1998.

Ouyang, P., Tjiptoprodjo, R., Zhang, W., and Yang, G.: Micro-motion devices technology: The state of arts review, Int. J. Adv. Manufact. Technol., 38, 463–478, 2008.

Ouyang, P. R., Zhang, W. J., and Gupta, M. M.: An adaptive switching learning control method for trajectory tracking of robot manipulators, Mechatronics, 16, 51–61, doi:10.1016/j.mechatronics.2005.08.002, 2006.

Patrascu, M. and Stramigioli, S.: Modeling and simulating the stick–slip motion of the μ Walker, a MEMS-based device for μ SPAM, Microsyst. Technol., 13, 181–188, 2007.

Peng, J. Y. and Chen, X. B.: Modeling of Piezoelectric-Driven Stick-Slip Actuators, IEEE/ASME Trans. Mechatr., 16, 394–399, doi:10.1109/tmech.2010.2043849, 2011.

Pohl, D.: Dynamic piezoelectric translation devices, Review of scientific instruments, 58, 54–57, 1987.

Pozzi, M. and King, T.: Piezoelectric modelling for an impact actuator, Mechatronics, 13, 553–570, doi:10.1016/S0957-4158(02)00004-1, 2003.

Rakotondrabe, M., Haddab, Y., and Lutz, P.: Voltage/frequency proportional control of stick-slip micropositioning systems, IEEE Trans. Control Syst. Technol., 16, 1316–1322, 2008.

Rakotondrabe, M., Haddab, Y., and Lutz, P.: Development, modeling, and control of a micro-/nanopositioning 2-dof stick–slip device, Trans. IEEE/ASME Mechatron., 14, 733–745, 2009.

Shamoto, E. and Moriwaki, T.: Development of a "walking drive" ultraprecision positioner, Precis. Eng., 20, 85–92, doi:10.1016/S0141-6359(97)00060-3, 1997.

Shamoto, E., Murase, H., and Moriwaki, T.: Ultraprecision 6-axis table driven by means of walking drive, Cirp Annals 2000: Manufacturing Technology, 299–302, 2000.

Shim, J. Y. and Gweon, D. G.: Piezo-driven metrological multiaxis nanopositioner, Rev. Sci. Instr., 72, 4183–4187, 2001.

Spiller, M., and Hurak, Z.: Hybrid charge control for stick-slip piezoelectric actuators, Mechatronics, 21, 100–108, doi:10.1016/j.mechatronics.2010.09.002, 2011.

Wen, Y.-K.: Method for random vibration of hysteretic systems, J. Eng. Mech. Div., 102, 249–263, 1976.

Yakimov, V.: Vertical ramp-actuated inertial micropositioner with a rolling-balls guide, Rev. Sci. Instr., 68, 136–139, 1997.

Zhang, Q. S.: Development and Characterization of a Novel Piezoelectric-Driven Stick – Slip Actuator with Anisotropic-Friction Surfaces, M. Sc. thesis, 1–91, 2008.

Zhang, W. J.: An integrated environment for CAD/CAM of mechanical systems, TU Delft, Delft University of Technology, 1–263, 1994.

Zhang, W. J., Ouyang, P., and Sun, Z.: A novel hybridization design principle for intelligent mechatronics systems, Proceedings of International Conference on Advanced Mechatronics (ICAM2010), 4–6, 2010.

Zhang, Z. M., An, Q., Li, J. W., and Zhang, W. J.: Piezoelectric friction-inertia actuator-a critical review and future perspective, Int. J. Adv. Manuf. Technol., 62, 669–685, doi:10.1007/s00170-011-3827-z, 2012.

Zhang, W. J. and van Luttervelt, C. A.: Towards a Resilient Manufacturing System, Annals of CIRP, 60, 469–472, 2011.

Zhang, W. J. and Lin Y.: Principles of Design of Resilient Systems and its Application to Enterprise Information Systems, Enterpr. Inf. Syst., 4, 99–110, 2010.

Zou, Q., Giessen, C. V., Garbini, J., and Devasia, S.: Precision tracking of driving wave forms for inertial reaction devices, Rev. Sci. Instr., 76, 023701–023709, 2005.

16

ECAP process improvement based on the design of rational inclined punch shapes for the acute-angled Segal 2θ-dies: CFD 2-D description of dead zone reduction

A. V. Perig[1] **and N. N. Golodenko**[2]

[1]Manufacturing Processes and Automation Engineering Department, Donbass State Engineering Academy, Shkadinova Str. 72, 84313 Kramatorsk, Ukraine

[2]Department of Water Supply, Water Disposal and Water Resources Protection, Donbass National Academy of Civil Engineering and Architecture, Derzhavin Str. 2, 86123 Makeyevka, Ukraine

Correspondence to: A. V. Perig (olexander.perig@gmail.com)

Abstract. This article is focused on a 2-D fluid dynamics description of punch shape geometry improvement for Equal Channel Angular Extrusion (ECAE) or Equal Channel Angular Pressing (ECAP) of viscous incompressible continuum through acute-angled Segal 2θ-dies with $2\theta < 90°$. It has been shown both experimentally with physical simulation and theoretically with computational fluid dynamics that for the best efficiency under the stated conditions, the geometric condition required is for the taper angle $2\theta_0$ of the inclined oblique punch to be equal to the 2θ angle between the inlet and outlet channels of the Segal 2θ-die. Experimentally and theoretically determined rational geometric condition for the ECAP punch shape is especially prominent and significant for ECAP through the acute angled Segal 2θ-dies. With the application of Navier-Stokes equations in curl transfer form it has been shown that for the stated conditions, the introduction of an oblique inclined $2\theta_0$-punch results in dead zone area downsizing and macroscopic rotation reduction during ECAP of a viscous incompressible continuum. The derived results can be significant when applied to the improvement of ECAP processing of both metal and polymer materials through Segal 2θ-dies.

1 Introduction

For the last 20 years a number of research efforts in materials science related fields have been focused on wider development, implementation, commercialization and improvement of new material forming methods known as Severe Plastic Deformation (SPD) schemes (Boulahia et al., 2009; Haghighi et al., 2012; Han et al., 2008; Laptev et al., 2014; Minakowski, 2014; Nagasekhar et al., 2006; Nejadseyfi et al., 2015; Perig et al., 2013a, b, 2015; Perig and Laptev, 2014; Perig, 2014; Rejaeian and Aghaie-Khafri, 2014). The classical SPD processing method is Segal's Equal Channel Angular Extrusion (ECAE) or Equal Channel Angular Pressing (ECAP) material forming technique (Segal, 2004). ECAE or ECAP realization is based on one or several extrusion passes of a lubricated metal or polymer material through a die with two intersecting channels of equal cross-section (Segal, 2004). Materials' processing by ECAP results in the accumulation of large shear strains and material structure refinement with physical properties enhancement (Boulahia et al., 2009; Nagasekhar et al., 2006; Nejadseyfi et al., 2015; Segal, 2004). The standard die geometry ABC-abc for ECAP processing is the so-called Segal 2θ-die geometry, where the inlet AB-ab and outlet BC-bc die channels have an intersection angle 2θ (Figs. 1–2). Moreover Segal 2θ-dies have neither external nor internal radii at the channel intersection points B; b (Figs. 1–2).

In recent years we have seen major research interest in the introduction of fluid mechanics techniques (Minakowski, 2014; Perig et al., 2010; Perig and Golodenko, 2014a, b; Re-

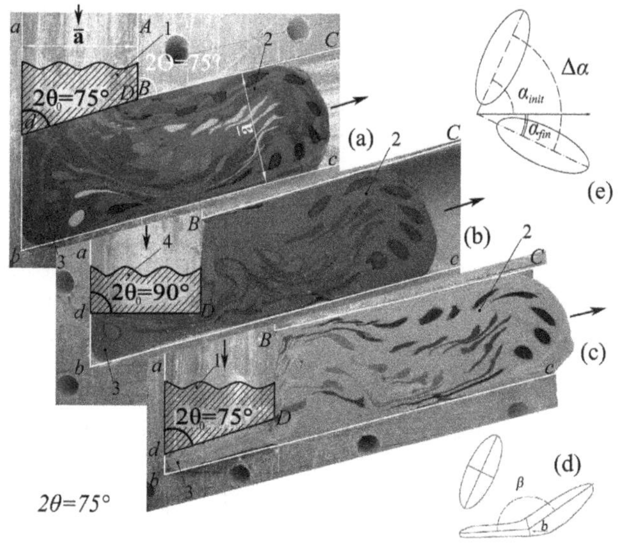

Figure 1. Physical simulation with soft model-based experiments of punch shape (1, 4) influence on ECAP flow of viscous continuum (2) through acute-angled Segal die ABC-abc with channel intersection angle $2\theta = 75° < 90°$: (4) classical punch of rectangular shape dD in **(b)**; (1) modified shape of inclined $2\theta_0$-punch dD in **(a)** and **(c)**, where $2\theta_0 = 2\theta$; (3) the experimentally derived shape of the dead zone for material flow during ECAP; the schematic diagrams of macroscopic rotation **(e)** and rotational inhomogeneity **(d)** formation during viscous continuum ECAP.

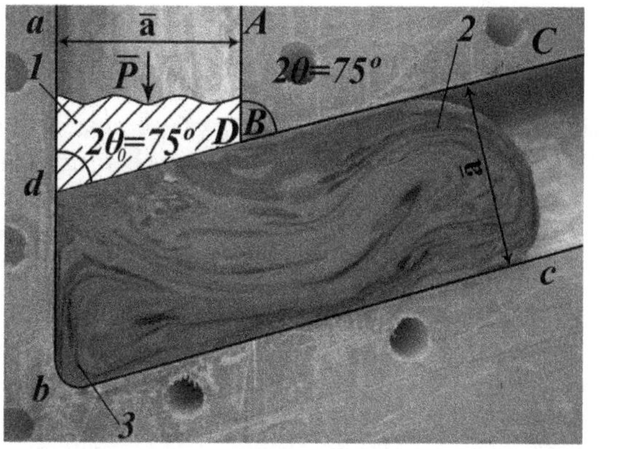

Figure 2. Soft physical model of the workpiece after 3 ECAP passes through Segal 2θ-die via route C with modified shape of $2\theta_0$-inclined or $2\theta_0$-beveled punch, where $2\theta = 2\theta_0 = 75°$.

jaeian and Aghaie-Khafri, 2014) to the solution of ECAP problems. This interest is results from growing application of ECAP SPD techniques to processing of polymers (Boulahia et al., 2009; Perig et al., 2010; Perig and Golodenko, 2014a, b) and powder materials (Haghighi et al., 2012; Nagasekhar et al., 2006) where viscosity effects become essential.

At the same time the phenomenological description of polymer materials flow through Segal 2θ-dies with Navier-Stokes equations has not been adequately addressed in previously known publications (Minakowski, 2014; Perig et al., 2010; Perig and Golodenko, 2014a, b; Rejaeian and Aghaie-Khafri, 2014). This underlines the importance of the present research, dealing with fluid dynamics 2-D simulation of material flow through the acute-angled Segal 2θ-dies with channel intersection angles of $2\theta > 0°$ and $2\theta < 90°$.

Another problem during ECAP material processing through the acute-angled Segal 2θ-dies with $2\theta < 90°$ is connected with the formation of large dead zones (3) in the viscous material flow in Fig. 1b as well as enormous and dangerous mixing $\Delta\alpha$ of viscous material (2) in Fig. 1b and e during viscous continuum ECAP through acute-angled dies with channel intersection angles of $2\theta < 90°$ when standard classical rectangular punches (4) are applied (Fig. 1b). So simple physical simulation experiments in Fig. 1b for viscous continuum ECAP through the die ABC-abc with $2\theta = 75°$ confirm the disadvantages of using a standard punch (4) with rectangular shape AD-ad ($2\theta_0 = 90°$) in Fig. 1b. It is very important to note that known approaches in published articles (Boulahia et al., 2009; Haghighi et al., 2012; Han et al., 2008; Laptev et al., 2014; Minakowski, 2014; Nagasekhar et al., 2006; Nejadseyfi et al., 2015; Perig et al., 2013a, b; Perig and Laptev, 2014; Perig, 2014; Rejaeian and Aghaie-Khafri, 2014; Segal, 2004; Wu and Baker, 1997) have never addressed the possibility of changing the standard rectangular punch shape AD-ad in Fig. 1b for material ECAP through acute-angled Segal dies with $2\theta < 90°$.

This fact emphasizes the importance and underlines the prime novelty of the present article addressing the viscous fluid dynamics description of the influence of classical (Fig. 1b) and novel modified $2\theta_0$-inclined or $2\theta_0$-beveled (Figs. 1a, c and 2) punch shape AD-ad on viscous flow features of processed workpieces during ECAP SPD pressure forming through acute-angled Segal 2θ-dies with channel intersection angles of $2\theta > 0°$ and $2\theta < 90°$.

2 Aims and scopes of the article – prime novelty statement of research

The present article is focused on the experimental and theoretical description of viscous workpiece flow through 2θ acute-angled angular dies of Segal geometry during ECAP by a classical rectangular punch and a novel modified $2\theta_0$-inclined or $2\theta_0$-beveled punch.

The aim of the present research is the phenomenological continuum mechanics based description of viscous workpiece flow through the 2θ acute-angled angular dies of Segal geometry during ECAE with an application of classical rectangular and novel modified $2\theta_0$-inclined or $2\theta_0$-beveled punch shapes.

The subject of the present research is the process of ECAP working through the 2θ acute-angled angular dies of Segal geometry with viscous flow of polymeric workpiece models, forced by the external action of classical rectangular and novel modified $2\theta_0$-inclined or $2\theta_0$-beveled punch shapes.

The object of the present research is to establish the characteristics of the viscous flow of workpiece models through the 2θ acute-angled angular dies of Segal geometry with respect to workpiece material rheology and geometric parameters of different punch shapes on viscous ECAP process.

The experimental novelty of the present article is based on the introduction of initial circular gridlines to study the punch shape influence on viscous workpiece ECAP flow through the 2θ angular acute-angled dies of Segal geometry.

The prime novelty of the present research is the numerical finite-difference solution of Navier-Stokes equations in the curl transfer form for the viscous workpiece flow through 2θ acute-angled angular dies of Segal geometry during ECAP, taking into account the classical rectangular and novel modified $2\theta_0$-inclined or $2\theta_0$-beveled punch shapes.

3 Physical simulation study of punch shape influence on viscous flow

Physical simulation techniques using plasticine workpiece models are often used in material forming practice (Chijiwa et al., 1981; Han et al., 2008; Laptev et al., 2014; Perig et al., 2010, 2013a, b, 2015; Perig and Laptev, 2014; Perig and Golodenko, 2014a, b; Perig, 2014; Sofuoglu and Rasty, 2000; Wu and Baker, 1997).

In order to estimate the character of viscous flow during ECAP through a 2θ acute-angled angular die of Segal geometry ABC–abc under the action of a classical rectangular punch and a novel modified $2\theta_0$-inclined or $2\theta_0$-beveled punch shapes we have utilized physical simulation techniques in Figs. 1–2. The plasticine workpiece models in Figs. 1–2 have been extruded through a ECAP die ABC–abc with channel intersection angle $2\theta = 75°$ using a standard punch (4) with rectangular shape ($2\theta_0 = 90°$) in Fig. 1b and novel modified $2\theta_0 = 75°$-inclined or $2\theta_0 = 75°$-beveled punch (1) in Figs. 1a, c and 2 as the first experimental approach to polymeric materials flow (Figs. 1–2).

The aim of the physical simulation is an experimental study of dead zone abc formation and deformation zone abc location during viscous ECAP flow of workpiece plasticine models under the external action of rectangular and inclined punches. The physical simulation in Figs. 1–2 is also focused on the experimental visualization of rotary modes of SPD during ECAP of viscous polymer models for the different punch geometries. The experimental results in Figs. 1–2 are original experimental research results, obtained by the authors.

The plastic die model of ECAP die ABC-abc with channel intersection angle $<ABC = <abc = 2\theta = 75°$ and the width of inlet aA and outlet cC die channels 35 mm is shown in Figs. 1–2. Potato flour was used as the lubricator in Figs. 1–2.

The main experimental visualization technique in Figs. 1–2 is based on the manufacture of the initial plasticine physical models of the workpieces in the shapes of rectangular parallelepipeds, freezing of these rectangular parallelepipeds, marking the initial circular gridlines on the front sides of the frozen parallelepipeds, perforation of through-holes in the parallelepipeds at the centers of the initial circular gridlines, repeated freezing of the plasticine (Fig. 1) parallelepipeds, heating of the plasticine (Fig. 1) pieces with different colors to the half-solid state, and placing the half-solid multicolor plasticine (Fig. 1) into the through-holes of the frozen parallelepipeds using a squirt without needle technique.

In this way the initial plasticine-based (Fig. 1) circular gridlines were marked throughout the initial plasticine (Fig. 1) workpieces. The initial circular gridlines transform into deformed elliptical ones as workpieces flow from inlet to outlet die channels during ECAP (Figs. 1a, c and 2). The gridline-free dead zones (p. b) were visualized through the physical simulation techniques introduction in Figs. 1–2. It was found that dead zone (p. b) formation takes place in the vicinity of the external angle abc of channel intersection zone Bb. It was experimentally shown that the best reduction of dead zone size (3) for an ECAE die with $2\theta = 75°$ could be achieved through the replacement of the standard rectangular punch AD-ad with ($2\theta_0 = 90°$) in Fig. 1b with the new $2\theta_0$-inclined or $2\theta_0$-beveled punch AD-ad with $2\theta_0 = 75°$.

It was experimentally found in Figs. 1–2 that the deformation zone BCDc during ECAP of the viscous models is not located in the channel intersection zone Bb but is located in the beginning of the outlet die channel BC-bc. The relative location of the elliptical markers in outlet die channel BC-bc show the formation of two rotary modes of SPD during ECAP (Fig. 1).

Checking the successive locations of one color elliptical markers in Fig. 1, we see that the major axis of every elliptical marker rotates with respect to the axis of the outlet die channel bc. We define the term of macroscopic rotation as the relative rotation of the major axis of an elliptical marker with respect to the flow direction axis bc. The macroscopic rotation is the first visually observable rotary mode during ECAP forming of the viscous workpiece model.

Visual comparison of Fig. 1b with Figs. 1a, c and 2 shows that the macroscopic rotation is an unknown function of ECAP die channel intersection angle 2θ and $2\theta_0$-punch shape geometry. However under SPD ECAP treatment some deformed elliptical markers within the viscous material have additional bending points and have the form of "commas" or "tadpoles" in Figs. 1–2. If the elliptical marker has an additional bending point during ECAP, then we will call the vicinity of the marker with this "waist" as a zone of rotational inhomogeneity within the workpiece material, which is usually located at the beginning of the outlet die channel BC-bc in Figs. 1–2. The rotational inhomogeneity is the second visually observable rotary mode during ECAP forming of the viscous workpiece model, which strongly depends on

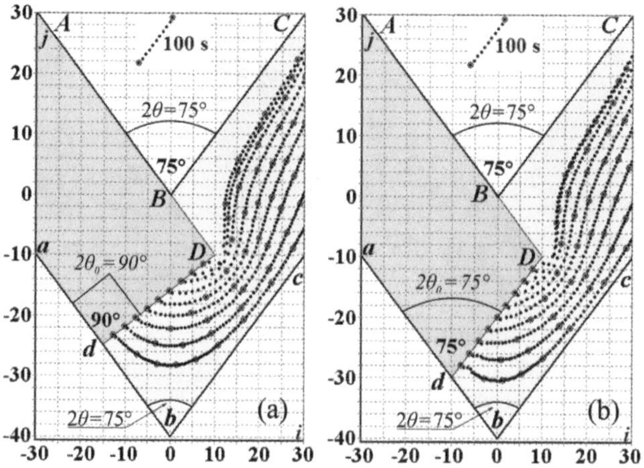

Figure 3. Computational flow lines for the Segal die with $2\theta = 75°$ for the coordinate steps $\bar{\xi} = 1.10$ mm; $\bar{\eta} = 1.44$ mm for the rectangular punch dD ($2\theta_0 = 90°$) **(a)** and for the inclined punch dD ($2\theta_0 = 75°$) **(b)**, where time iteration step is $\bar{t}_{\text{it}} = 610\,\mu$s, transition time is $\bar{t}_{\text{tr}} = 11.3$ s.

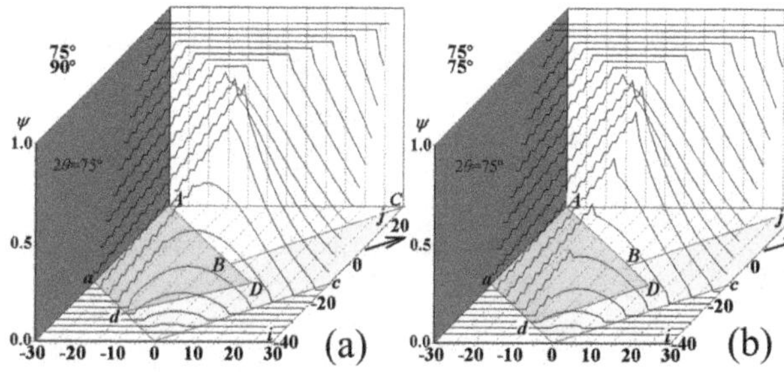

Figure 4. Computational dimensionless flow function ψ for the Segal die with $2\theta = 75°$ for the rectangular punch dD ($2\theta_0 = 90°$) **(a)** and for the inclined punch dD ($2\theta_0 = 75°$) **(b)**.

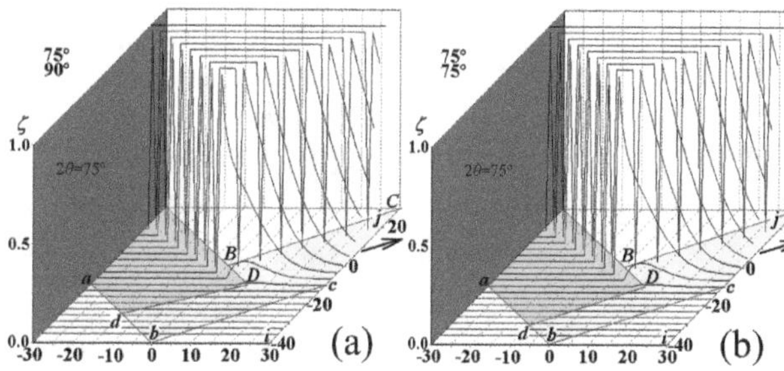

Figure 5. Computational dimensionless curl function ζ for the Segal die with $2\theta = 75°$ for the rectangular punch dD ($2\theta_0 = 90°$) **(a)** and for the inclined punch dD ($2\theta_0 = 75°$) **(b)**.

the ECAP die channel intersection angle 2θ and $2\theta_0$-punch shape geometry.

The experimental results in Figs. 1–2 have indicated the formation of the following zones within worked materials' volumes: (I) the dead zone (p. b); (II) the deformation zone BCDc; (III) the macroscopic rotation zone (BC-bc), and (IV) the zone of rotational inhomogeneity (BC-bc). The complex of physical simulation techniques in Figs. 1–2 introduces the initial circular gridlines technique with the application of plasticine workpieces with the initial circular colorful gridlines in the shape of initial colorful cylindrical plasticine inclusions (Fig. 1). The application of the initial circular gridlines experimental technique and the introduction of a novel modified $2\theta_0$-inclined or $2\theta_0$-beveled punch shapes has not been addressed in previous known ECAP research (Boulahia et al., 2009; Haghighi et al., 2012; Han et al., 2008; Laptev et al., 2014; Minakowski, 2014; Nagasekhar et al., 2006; Nejadseyfi et al., 2015; Perig et al., 2013a, b; Perig and Laptev, 2014; Perig, 2014; Rejaeian and Aghaie-Khafri, 2014; Segal, 2004; Wu and Baker, 1997).

The proposed complex of experimental techniques for physical simulation of SPD during ECAP in Figs. 1–2 will find the further applications in the study of viscous ECAP through the dies with more complex Iwahashi, Luis-Perez, Utyashev, Conform and equal radii geometries for the different punch shape geometries and different routes of multi-pass ECAP working.

4 Numerical simulation study of punch shape influence on flow lines, and punching pressure during viscous ecap flow through Segal 2θ-dies

In order to derive the mathematical model of the viscous material flow during ECAP through the acute-angled Segal 2θ-die taking into account the punch shape AD-ad effect on viscous flow dynamics we will apply the Navier-Stokes equations (Appendices A–D). The results of the numerical simulation study are shown in computational diagrams in Figs. 3–10.

Computational results in Figs. 3–10 illustrate the punch shape influence on geometry (Fig. 3), kinematics (Figs. 4–8) and dynamics (Figs. 9–10) of the viscous flow during ECAP. Computational plots in Figs. 3–10 are based on a finite-difference solution of the Navier–Stokes equations in curl transfer form Eqs. (A1)–(A2) with initial Eq. (B1) and boundary Eqs. (C1)–(C7) conditions.

Instabilities of the numerical solutions, which appear at the outlet frontiers cC (Figs. 3–10), propagate upstream.

CFD-derived computational flow lines in Fig. 3b directly show the reduction of dead zone area dDbc when we use the modified $2\theta_0$-inclined or $2\theta_0$-beveled punch shape, where $2\theta = 2\theta_0 = 75°$ (Fig. 3b). CFD-derived computational flow lines in Fig. 3a also outline the largest dead zone area dDbc when we use the standard punch (Fig. 3a) with rectangular shape ($2\theta_0 = 90°$). CFD-derived computational di-

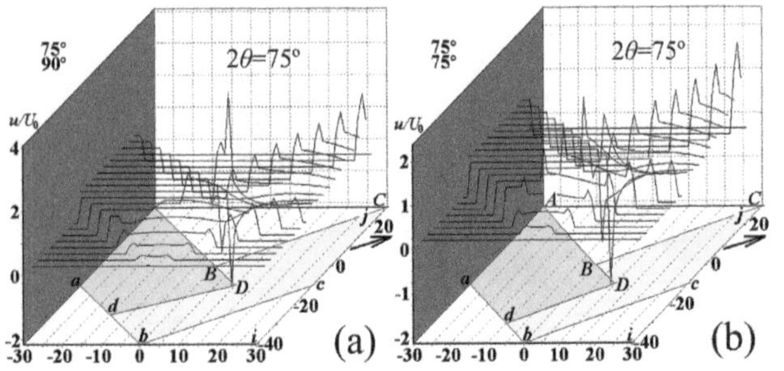

Figure 6. Computational u components of flow velocities for the Segal die with $2\theta = 75°$ for the rectangular punch dD ($2\theta_0 = 90°$) **(a)** and for the inclined punch dD ($2\theta_0 = 75°$) **(b)**.

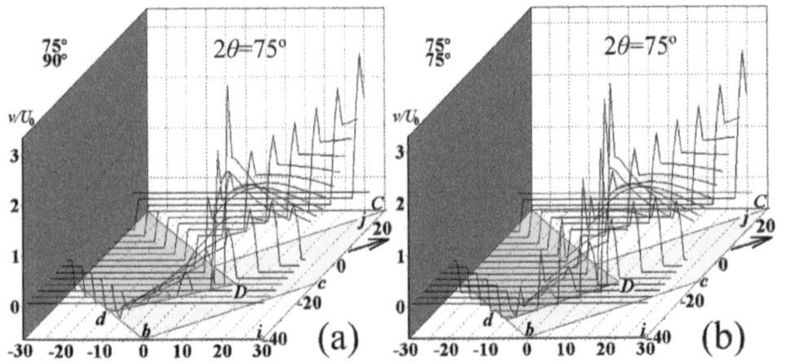

Figure 7. Computational v components of flow velocities for the Segal die with $2\theta = 75°$ for the rectangular punch dD ($2\theta_0 = 90°$) **(a)** and for the inclined punch dD ($2\theta_0 = 75°$) **(b)**.

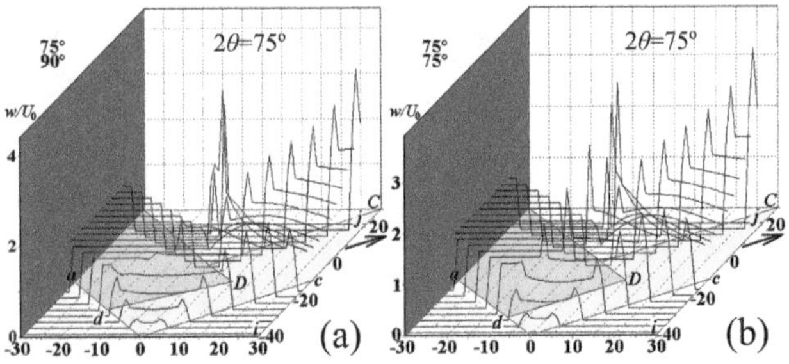

Figure 8. Computational dimensionless full flow velocities w for the Segal die with $2\theta = 75°$ for the rectangular punch dD ($2\theta_0 = 90°$) **(a)** and for the inclined punch dD ($2\theta_0 = 75°$) **(b)**.

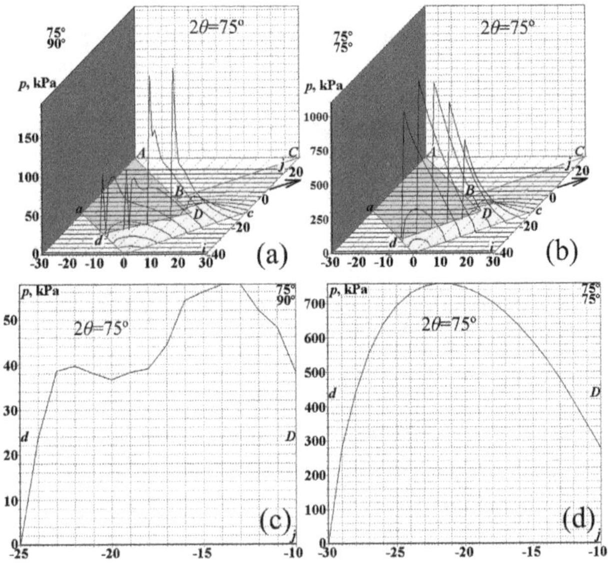

Figure 9. Computational dimension punching pressure for the Segal die with $2\theta = 75°$ for the rectangular punch dD ($2\theta_0 = 90°$) **(a, c)** and for the inclined punch dD ($2\theta_0 = 75°$) **(b, d)**.

agrams for ECAP punching pressure in Figs. 9–10 show that the application of the standard rectangular punch with $2\theta_0 = 90°$ requires lower punching pressures (Fig. 10). The CFD-based simulation in Figs. 9–10 indicates that the use of the modified $2\theta_0$-inclined or $2\theta_0$-beveled punch shapes requires higher punching pressures for ECAP of viscous incompressible continuum through the acute-angled Segal 2θ-dies with $2\theta < 90°$.

Higher values of punching pressure for modified $2\theta_0$-inclined or $2\theta_0$-beveled punch shapes in comparison with the standard rectangular punch with $2\theta_0 = 90°$ in Figs. 9–10 result from the fact that the compressive strains in such schemes are higher than shear strains.

So in order to force the plasticine model through the 2θ-die by the modified $2\theta_0$-inclined punch we have to apply higher punching force in order to reach the necessary shear stresses. This fact is shown in Figs. 9–10.

The increased punching pressure required for the modified $2\theta_0$-inclined punches and for the acute angled 2θ-dies with $2\theta < 90°$ results in decreased dead zone in angle b and a decreased shear stress component (Figs. 6b, 7b, 8b, 9b, d, and 10).

For the modified $2\theta_0$-inclined punches and the obtuse angled 2θ-dies with $2\theta > 90°$ the decreased punching pressure results from increased effective punch area dD and increased shear stress component (Fig. 10).

5 Discussion of derived results

The technological issue addressed in this article has direct industrial importance in material forming applications. The introduction of the fluid dynamics numerical simulation (Figs. 3–10) provides us with a better understanding of physical simulation results in Figs. 1–2.

Addressing Eqs. (A1)–(A2) in Appendix A again, the partial derivatives of dimensionless flow function ψ define the flow velocity components: $\partial\psi/\partial y = u$; $\partial\psi/\partial x = (-v)$. In the 3-D spatial diagrams for flow function ψ in Fig. 4 near the die corner b with rectangular Cartesian coordinates $(0, -40)$ we have the following effect of punch shape ad-AD on dead zone dDb size. With the application of a rectangular punch with $2\theta_0 = 90°$ in Figs. 1b, 3a, 4a, 5a, 6a, 7a, 8a, 9a, c, 10 we see a large dead zone dDb with zero flow function $\psi = 0$ (Fig. 4) and zero flow velocities $u = 0$ (Fig. 6); $v = 0$ (Fig. 7). But with the introduction of an inclined 2θ-

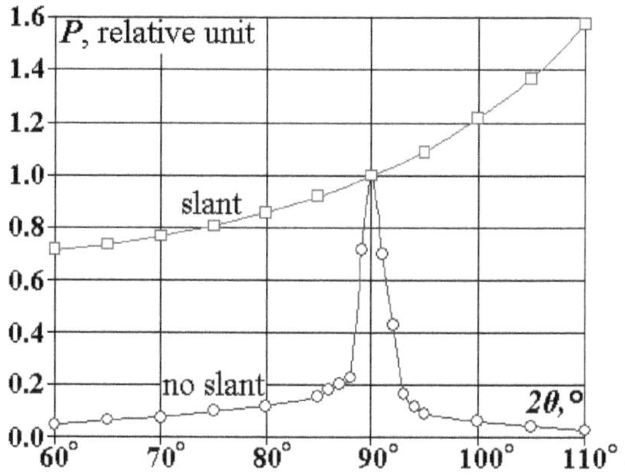

Figure 10. Computational dimensionless punching pressures for plasticine viscous liquid flow through Segal dies with $60° \leq 2\theta \leq 110°$ for the rectangular punch dD ($2\theta_0 = 90°$) (○) and for the modified $2\theta_0$-inclined or $2\theta_0$-beveled punch dD ($2\theta_0 = 2\theta$) (□).

punch with $2\theta_0 = 75°$ (inclined punch in Figs. 1a, c, 2, 3b, 4b, 5b, 6b, 7b, 8b, 9b, d, 10) we see a smaller dead zone size dDb. Computational flow lines (Fig. 3) are the lines near which flow function ψ (Fig. 4) is constant $\psi = \text{const}$. The computed effect in Fig. 4, which shows the absence of the "sawtooth" shape of the ψ-function over the die area dDb confirms that the dDb area is just the dead zone and not a vortex or eddy zone with circulating flow. Figure 5 show us that the curl function $\zeta = 0$ is also zero in the dead zone dDb.

Polycrystalline material is a natural composite, which contains ultra fine single crystals and amorphous viscous fluid between single crystals for fastening and connecting these single crystals among themselves. Laminar-flow layers of such amorphous fluid move with different velocities as well as single crystal sides, adjacent to laminar-flow layers. Curl ζ (Eq. A2) characterizes single crystal relative rotation during its linear displacement along the flow lines in Fig. 5. As a result of internal friction the contacting facets of single crystals become smooth like smoothing of river or sea pebbles under action of viscous flow.

This is the hydrodynamic explanation of the increase of the material plasticity during ECAP, which follows from the computational diagrams in Figs. 3–10. Under the action of mechanical loads at the boundaries of the contacting facets of single crystals, there appear no micro-cracks because of their flatness. The curl is zero in dead zone dDb. So in the material dead zone dDb no smoothing of single crystals facets takes place. As a result, material plasticity cannot be improved in the material dead zone dDb.

Such hydrodynamic illustrations (Figs. 3–10) directly confirm experimentally derived results (Figs. 1–2) with physical simulation of punch shape effect on material flow kinematics during ECAE through the acute-angled Segal 2θ-die.

6 Conclusions

In the present work we addressed the $2\theta_0$-punch shape effect on material flow dynamics during ECAP through the numerical solution of the boundary value problem Eqs. (A1)–(A2), (B1), (C1)–(C7) for Navier–Stokes equations in curl transfer form (Figs. 3–10), taking into account the standard rectangular and improved $2\theta_0$-inclined or $2\theta_0$-beveled punch shapes.

Both physical (Fig. 1b) and fluid dynamics (Figs. 3a, 4a, 5a, 6a, 7a, 8a, 9a, c, 10) simulations show that the application of a standard rectangular punch with $2\theta_0 = 90°$ for workpiece ECAP through acute-angled Segal 2θ-dies with $2\theta < 90°$ is highly undesirable because of the resulting large material dead zone areas dDb in the neighborhood of the external die angle $2\theta = \angle(abc)$.

Both physical (Figs. 1a, c and 2) and fluid dynamics (Figs. 3b, 4b, 5b, 6b, 7b, 8b, 9b, d, 10) simulations reveal that the introduction of $2\theta_0$-inclined or $2\theta_0$-beveled punch shapes with dDbc for material ECAP processing through the acute-angled Segal 2θ-dies with $2\theta < 90°$ and $2\theta_0 = 2\theta$ is a very promising technique because of minimal material dead zone areas dDb and the resulting minimal material waste in the neighborhood of external die angle $2\theta = \angle(abc)$, e.g. for $2\theta = 75°$.

Appendix A: Navier–Stokes equations in curl transfer form

The curl transfer equation in dimensionless variables will have the following form (Roache, 1976):

$$\frac{\partial \zeta}{\partial t} = -Re\left(\frac{\partial (u\zeta)}{\partial x} + \frac{\partial (v\zeta)}{\partial y}\right) + \left(\frac{\partial^2 \zeta}{\partial x^2} + \frac{\partial^2 \zeta}{\partial y^2}\right), \quad \text{(A1)}$$

where the dimensionless curl function will be defined as (Fig. 5):

$$\zeta = \frac{\partial u}{\partial y} - \frac{\partial v}{\partial x}. \quad \text{(A2)}$$

Appendix B: Initial conditions for curl transfer equation

We now study the steady-state regime of viscous flow for a physical model of polymer material (Figs. 3–10). So the initial conditions we will assume in the form of a rough approximation to the stationary solution (Figs. 3–10):

$$u^0_{i,j} = 0; \; v^0_{i,j} = 0; \; \zeta^0_{i,j} = 0; \; \psi^0_{i,j} = 0. \quad \text{(B1)}$$

Appendix C: Boundary conditions for curl transfer equation

The boundary conditions for the die walls we will define as the viscous material "sticking" to the walls of the die (Figs. 3–10).

At the inner upper boundary DBC (Figs. 3–10) we have

$$\psi_{i,j} = 1; \; \zeta_{i,j} = 1. \quad \text{(C1)}$$

At the external lower boundary dbc (Figs. 3–10) we have

$$\psi_{i,j} = 0; \; \zeta_{i,j} = 0. \quad \text{(C2)}$$

For the punch frontal edge dD (Figs. 3–10) we have

$$\psi_{10,-10} = 1; \; \psi_{9,-11} = 1 - 2/N;$$
$$\psi_{i,j} = \psi_{i+2,j+2} - 2/N, \quad \text{(C3)}$$

where N is the quantity of ordinate steps along the channel width.

For the angular points, which are located in the vertices of the concave angles b and B (Figs. 3–10) we have

$$\zeta_{i,j} = 0. \quad \text{(C4)}$$

For the angular point D (Figs. 3–10) of the convex angle in the finite-difference equation, written for the mesh point (10, −11) we have the following curl

$$\zeta_{10,10} = 2\psi_{10,-11}. \quad \text{(C5)}$$

For the angular point D (Figs. 3–10) of the convex angle in the finite-difference equation, written for the mesh point (11, 10) we have the curl

$$\zeta_{10,10} = 2\psi_{11,10}. \quad \text{(C6)}$$

At the outlet line cC we have

$$\psi_{N+1,j} = \psi_{N-3,j} - 2\psi_{N-2,j} + 2\psi_{N,j};$$
$$\zeta_{N+1,j} = \zeta_{N-3,j} - 2\zeta_{N-2,j} + 2\zeta_{N,j}. \quad \text{(C7)}$$

Appendix D: Numerical values of physical parameters for the problem

The numerical results of integration of curl transfer Eqs. (A1)–(A2) with initial Eq. (B1) and boundary Eqs. (C1)–(C7) conditions are outlined in Figs. 3–10 for the following numerical values:

- the dimensional width of inlet and outlet die channels is $\overline{a} = 35$ mm;
- the dimensional length of die channel is $\overline{L} = 16 \cdot \overline{a} = 16 \cdot 35 \times 10^{-3}$ m $= 0.56$ m;
- the dimensional average ECAP punching velocity is $\overline{U}_0 = 0.1 \times 10^{-3}$ m s^{-1};
- the dimensional time of processed workpiece material motion in die channel is $\overline{t}^* = \overline{L}/\overline{U}_0 = 0.56/(0.1 \times 10^{-3}) = 5600$ s;
- the maximum value of dimensionless curl is $\zeta = 1$;
- the dimensional curl is $\overline{\zeta} = \zeta \cdot \overline{U}_0/\overline{a} = (1 \times 0.1 \times 10^{-3}$ m s$^{-12})/(35 \times 10^{-3}$ m$) = 2.86 \times 10^{-3}$ s^{-1};
- the dimensional average angular velocity of rotation for viscous material layers is $\overline{\omega} = |\mathrm{rot}w|/2 = \overline{\zeta}/2 = 1.43 \times 10^{-3}$ s^{-1};
- the number of turns for viscous material layers during the time of workpiece material motion in die channel is $N^* = \overline{\omega}\overline{t}^*/2\pi = (1.43 \times 10^{-3} \times 5600)/2 \times 3.14) = 1.27$;
- the dimensional density of the viscous plasticine physical model of extruded polymer material is $\overline{\rho} = 1850$ kg m^{-3};
- the dimensional plasticine yield strength is $\overline{\sigma}_s = 217$ kPa (Sofuoglu and Rasty, 2000);
- the dimensional specific heat capacity of plasticine material is $\overline{c} = 1.004$ kJ/(kg K^{-1});
- the dimensional thermal conductivity is $\overline{\lambda} = 0.7$ J/(m s^{-1} K^{-1}) (Chijiwa et al., 1981);
- the dimensional punching temperature is $\overline{t}_{\mathrm{temp}} = 20$ °C;
- the dimensional dynamic viscosity for viscous Newtonian fluid model of plasticine workpiece during ECAE is $\overline{\eta}_{\mathrm{vis}} = 1200$ kPa s^{-1};

- the dimensional kinematic viscosity for viscous Newtonian fluid model of plasticine workpiece during ECAE is $\overline{\nu}_{vis} = \overline{\eta}_{vis}/\overline{\rho} = 1.2 \times 10^6/1850 = 648.648\ \mathrm{m^2\ s^{-1}}$;
- Reynolds number is $Re = \overline{U}_0\overline{a}\overline{\rho}/\overline{\eta}_{vis} = \overline{U}_0\overline{a}/\overline{\nu}_{vis} = 5.396 \times 10^{-9}$;
- the half number of coordinate steps along the x and y axes is $N = 40$;
- the number of coordinate steps along the x and y axes is $2 \times N = 80$;
- the relative error of iterations is $e = 1/1000$;
- the dimensional time moment for the first isochrone building is $t_1 = 100$ s;
- die channel intersection angle of Segal die is $2\theta = 75°$;
- punch shape inclination angles adD are $2\theta_0 = 90°$ (rectangular punch in Figs. 1b, 3a, 4a, 5a, 6a, 7a, 8a, 9a, c, 10) and $2\theta_0 = 75°$ (inclined punch in Figs. 1a, c, 3b, 4b, 5b, 6b, 7b, 8b, 9b, d, 10);
- the dimensional horizontal and vertical coordinate steps along the $x-$ and y axes are $\overline{\xi} = 1.10$ mm and $\overline{\eta} = 1.44$ mm for angular die with $2\theta = 75°$;
- the dimensional time iteration step is $\overline{\tau} = \overline{t}_{it} = 610\ \mu$s for ECAP die with $2\theta = 75°$;
- the dimensional transition time is $\overline{t}_{tr} = 11.3$ s for ECAP die with $2\theta = 75°$.

Acknowledgements. Authors thank three "anonymous" referees for their valuable notes and suggestions. Authors are thankful to Mechanical Sciences Editors and Copernicus GmbH Team for this great opportunity to publish our original research at your respectful periodical Mechanical Sciences under a Creative Commons License.

References

Boulahia, R., Gloaguen, J.-M., Zaïri, F., Naït-Abdelaziz, M., Seguela, R., Boukharouba, T., Lefebvre, J. M.: Deformation behaviour and mechanical properties of polypropylene processed by equal channel angular extrusion: Effects of back-pressure and extrusion velocity, Polymer, 50, 5508–5517, doi:10.1016/j.polymer.2009.09.050, 2009.

Chijiwa, K., Hatamura, Y., and Hasegawa, N.: Characteristics of plasticine used in the simulation of slab in rolling and continuous casting, T. Iron Steel I. Jpn., 21, 178–186, doi:10.2355/isijinternational1966.21.178, 1981.

Haghighi, R. D., Jahromi, A. J., and Jahromi, B. E.: Simulation of aluminum powder in tube compaction using equal channel angular extrusion, J. Mater. Eng. Perform., 21, 143–152, doi:10.1007/s11665-011-9896-1, 2012.

Han, W. Z., Zhang, Z. F., Wu, S. D., and Li, S. X.: Investigation on the geometrical aspect of deformation during equal-channel angular pressing by in-situ physical modeling experiments, Mat. Sci. Eng. A, 476, 224–229, doi:10.1016/j.msea.2007.04.114, 2008.

Laptev, A. M., Perig, A. V., and Vyal, O. Y.: Analysis of equal channel angular extrusion by upper bound method and rigid blocks model, Mat. Res., São Carlos (Mater. Res.-Ibero-Am. J.), 17, 359–366, doi:10.1590/S1516-14392013005000187, 2014.

Minakowski, P.: Fluid model of crystal plasticity: numerical simulations of 2-turn equal channel angular extrusion, Tech. Mechanik, 34, 213–221, 2014.

Nagasekhar, A. V., Tick-Hon, Y., and Ramakanth, K. S.: Mechanics of single pass equal channel angular extrusion of powder in tubes, Appl. Phys. A, 85, 185–194, doi:10.1007/s00339-006-3677-y, 2006.

Nejadseyfi, O., Shokuhfar, A., Azimi, A., and Shamsborhan, M.: Improving homogeneity of ultrafine-grained/nanostructured materials produced by ECAP using a bevel-edge punch, J. Mater. Sci., 50, 1513–1522, doi:10.1007/s10853-014-8712-3, 2015.

Perig, A. V.: 2D upper bound analysis of ECAE through 2θ-dies for a range of channel angles, Mat. Res., São Carlos (Mater. Res.-Ibero-Am. J.), 17, 1226–1237, doi:10.1590/1516-1439.268114, 2014.

Perig, A. V. and Laptev, A. M.: Study of ECAE mechanics by upper bound rigid block model with two degrees of freedom, J. Braz. Soc. Mech. Sci. Eng., 36, 469–476, doi:10.1007/s40430-013-0121-z, 2014.

Perig, A. V. and Golodenko, N. N.: CFD Simulation of ECAE through a multiple-angle die with a movable inlet wall, Chem. Eng. Commun., 201, 1221–1239, doi:10.1080/00986445.2014.894509, 2014a.

Perig, A. V. and Golodenko, N. N.: CFD 2D simulation of viscous flow during ECAE through a rectangular die with parallel slants, Int. J. Adv. Manuf. Technol., 74, 943–962, doi:10.1007/s00170-014-5827-2, 2014b.

Perig, A. V., Laptev, A. M., Golodenko, N. N., Erfort, Y. A., and Bondarenko, E. A.: Equal channel angular extrusion of soft solids, Mat. Sci. Eng. A, 527, 3769–3776, doi:10.1016/j.msea.2010.03.043, 2010.

Perig, A. V., Zhbankov, I. G., and Palamarchuk, V. A.: Effect of die radii on material waste during equal channel angular extrusion, Mater. Manuf. Process., 28, 910–915, doi:10.1080/10426914.2013.792420, 2013a.

Perig, A. V., Zhbankov, I. G., Matveyev, I. A., and Palamarchuk, V. A.: Shape effect of angular die external wall on strain unevenness during equal channel angular extrusion, Mater. Manuf. Process., 28, 916–922, doi:10.1080/10426914.2013.792417, 2013b.

Perig, A. V., Tarasov, A. F., Zhbankov, I. G., and Romanko, S. N.: Effect of 2θ-punch shape on material waste during ECAE through a 2θ-die, Mater. Manuf. Process., 30, 222–231, doi:10.1080/10426914.2013.832299, 2015.

Rejaeian, M. and Aghaie-Khafri, M.: Study of ECAP based on stream function, Mech. Mater., 76, 27–34, doi:10.1016/j.mechmat.2014.05.004, 2014.

Roache, P. J.: Computational fluid dynamics, Hermosa Publishers, Albuquerque, 1976.

Segal, V. M.: Engineering and commercialization of equal channel angular extrusion (ECAE), Mat. Sci. Eng. A, 386, 269–276, doi:10.1016/j.msea.2004.07.023, 2004.

Sofuoglu, H. and Rasty, J.: Flow behavior of Plasticine used in physical modeling of metal forming processes, Tribol. Int., 33, 523–529, doi:10.1016/S0301-679X(00)00092-X, 2000.

Wu, Y. and Baker, I.: An experimental study of equal channel angular extrusion, Scripta Mater., 37, 437–442, doi:10.1016/S1359-6462(97)00132-2, 1997.

17

A comparison among different Hill-type contraction dynamics formulations for muscle force estimation

F. Romero and F. J. Alonso

Department of Mechanical, Energy and Materials Engineering, University of Extremadura, Avda. de Elvas s/n, 06006 Badajoz, Spain

Correspondence to: F. Romero (fromsan@unex.es)

Abstract. Muscle is a type of tissue able to contract and, thus, shorten, producing a pulling force able to generate movement. The analysis of its activity is essential to understand how the force is generated to perform a movement and how that force can be estimated from direct or indirect measurements. Hill-type muscle model is one of the most used models to describe the mechanism of force production. It is composed by different elements that describe the behaviour of the muscle (contractile, series elastic and parallel elastic element) and tendon. In this work we analyze the differences between different formulations found in the literature for these elements. To evaluate the differences, a flexo-extension movement of the arm was performed, using as input to the different models the surface electromyography signal recorded and the muscle-tendon lengths and contraction velocities obtained by means of inverse dynamic analysis. The results show that the force predicted by the different models is similar and the main differences in muscle force prediction were observed at full-flexion. The results are expected to contribute in the selection of the different formulations of Hill-type muscle model to solve a specific problem.

1 Introduction

The study of mechanical muscle models is one of the major topics in Biomechanics of human movement. The analysis of the forces that produce a given movement (inverse dynamics, ID) or the movement induced by a set of muscle forces or activations (forward dynamics, FD) are typical problems that need the description of muscle mechanical properties. Since the classic work of Hill (1938), the number of models proposed has been grown up, ranging from simple models to the most complex (as e.g. the model developed by Hatze (1977) with up to 50 parameters required to describe the motion of a simple joint). Muscle models are commonly categorized into three groups according to Winters and Stark (1987): (i) simple second-order models: it is a "black box" approximation where the inputs are either the neural signal or the external load and the output corresponds to either the joint position or torque; (ii) Hill-based lumped-parameter model, which is the most widely used and will be described in Sect. 2.1; and (iii) Huxley-based distributed-parameter models that attempts to explain correctly the mechanism of contraction with great accuracy but at a high computational effort. Some recent works, not described nor discussed here, deal with fractional order models of different kinds of muscles, such as the *gastrocnemius* muscle in Sommacal et al. (2007, 2008) or the *hamstring* muscle group in Grahovac and Žigić (2010). This type of muscle attempts to describe the viscoelastic properties of muscle tissue as a whole (HosseinNia et al., 2012).

The mechanical behaviour of muscle tissue can be described by means of passive elements such as springs and damping elements (SE and DE, respectively). These elements, combined properly, allow to understand the response of muscle tissue under compressive and tensile loads. In the literature (see, e.g., Yamaguchi, 2001) it is possible to find different models combining the properties of those mechanical components: the Maxwell model (Fig. 1a) uses both elements attached in series. Contrariwise, in the Voight model (Fig. 1b), those elements are used in parallel. Lastly, the Kelvin model (Fig. 1c), modifies the Voight model to include an additional spring in series with the DE. The dif-

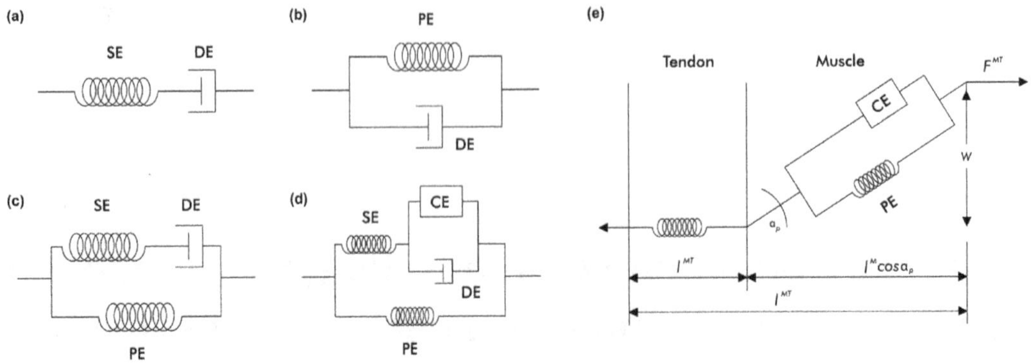

Figure 1. Muscle models (tendon not included in a-d models). (**a**) Maxwell model. (**b**) Voigth model. (**c**) Kelvin model. (**d**) Hill model. (**e**) Nomenclature used in the Hill-type muscle model.

ferent combinations of springs and dashpots are intended to improve the physiological response, however, as these models are composed by passive elements, they are not able to reproduce properly the active muscle contraction. This issue was solved by introducing the contractile element (Hill, 1938). This model (Fig. 1d) combines the passive properties of the Kelvin model with the active properties given by the contractile element (CE). Nowadays, the Hill-type muscle model is the most used in biomechanical studies involving muscular coordination (Zajac, 1989; Van Soest and Bobbert, 1993; Van Den Bogert et al., 1998; Thelen, 2003; Silva, 2003; Ackermann and Schiehlen, 2006; Ackermann, 2007; García-Vallejo, 2010; Alonso et al., 2012).

Figure 1e shows an schematic representation of the muscle-tendon unit using the Hill-type muscle model. In this sense, the elastic properties of the tendon are represented by a spring attached in series with the Hill-type muscle model. Regarding the muscle, the CE is responsible of the active force generated in the muscle, and two non-linear passive springs describe the properties of muscle tissue: on the one hand, the series elastic element (SE) represents the elasticity of the actin-miosyn crossbridges (Yamaguchi, 2001) and, on the other hand, the parallel elastic element (PE) describes the passive elastic properties of the muscle fibers. The SE element can be neglected with little inaccuracy if the study does not involve short-tendon actuators (Silva, 2003; Zajac, 1989).

The contribution of tendon is important for physiological purposes, but is often neglected for simplicity, as the process to distinguish between tendon and muscle length increases the difficulty of the problem. However, as pointed out by Yamaguchi (2001), tendon elasticity is important if the tendon stretches an amount approaching the fiber length of a particular muscle. This fact is relevant in some muscles such as the *soleus* or the *gastrocnemius* at the ankle joint or the *rectus femoris* at the knee. In the other cases, or as a first approximation, it is possible to consider the tendon as a rigid element, and include as the tendon length its slack length (l^{T}_{slack}), that is, the length for which the tendon just begin to resist lengthening.

The models available in the literature, based on the Hill-type description, differ basically in the amount of parameters to define the muscle mechanical behaviour. An extensive review of Hill-based muscle models was performed by Winters (1990b). In the present work, we analyse two parameter-based models derived from the Hill-type description of muscle tissue. Specifically, the model proposed by Van Soest and Bobbert (1993), the one by Thelen (2003), and a mathematical adjustment proposed by Kaplan (2000) and extended by Silva (2003) are presented and discussed here. The objective is to have a reference collecting the most widely used muscle models in biomechanics, highlighting their characteristics and discussing their use for specific problems. To do so, we analyse the differences observed in muscle force production by using the proposed models. Moreover, the variability of each muscle element between models is also studied. The order in which the different elements will be addressed is summarized in Table 1.

2 Methods

2.1 Mathematical description of muscle dynamics

The categories described previously are used depending on the type of problem to solve. From the mechanical point of view, although Huxley's muscle model describes precisely the chemical and mechanical process that take place in the muscle contraction, its use is not recommended in coordination studies that involve several muscle actuators as the complexity of the problem increases considerably. In this type of studies, the mechanical behaviour of muscle tissue can be modelled by means of passive elements such as springs and damping elements (SE and DE, respectively), and therefore, by using Hill-type models. The concepts introduced in this section are deeply described in Winters (1990a) and Zajac (1989).

Table 1. Schematic summary of the different sections addressed.

Formulation of muscle's dynamic equations			
Activation dynamics	Contraction dynamics		
Mechanical expressions of muscle elements			
Tendon	Muscle		
	Parallel element	Contractile element	
		Force–velocity	Force–length

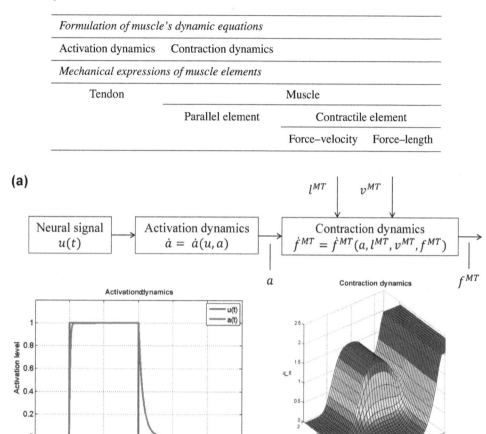

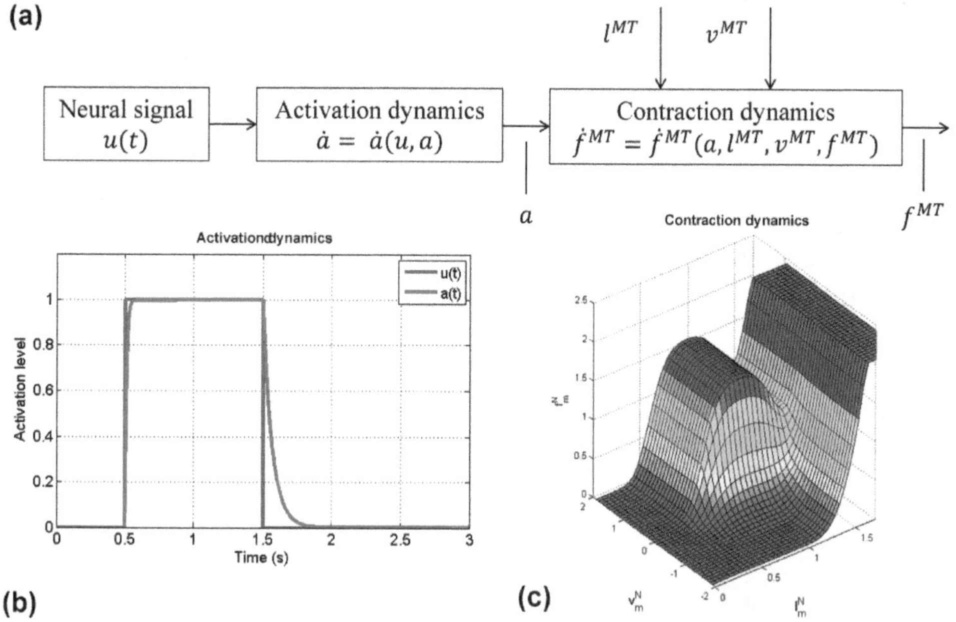

Figure 2. (**a**) Schematic representation of muscle dynamics: inputs and outputs for activation and contraction dynamics. (**b**) Activation dynamics. (**c**) Contraction dynamics including the effects of the parallel elastic element.

In a Hill-type muscle model, the dynamic behaviour of the active element, i.e., the contractile element can be expressed in terms of two cascaded differential equations (2a), the excitation-to-activation dynamics (Eq. 1, Fig. 2b) and the activation-to-force (contraction) dynamics (Eq. 2, Fig. 2c):

$$\dot{a} = f(u, a) \tag{1}$$

$$\dot{f}^{\mathrm{MT}} = g\left(a, l^{\mathrm{MT}}, v^{\mathrm{MT}}, f^{\mathrm{MT}}\right) \tag{2}$$

The expression in Eq. (1) transforms the muscular excitation signal, u, from the central nervous system into an activation signal, a, representing muscle recruitment level. This excitation-to-activation dynamics corresponds to the time lag due to the electrochemical process produced by the action potential that leads to muscle contraction. Equation (2) defines the force exerted by the muscle as a function of the physiological state, that is, dependent of the activation level (a), the muscle length (l^{MT}), the contraction velocity ($v^{\mathrm{MT}} = -\dot{l}^{\mathrm{MT}}$) and the actual force ($f^{\mathrm{MT}}$). The following sections will describe the different dynamics that govern muscle activity and the contribution of the different tissues to the force development.

2.1.1 Excitation-to-activation dynamics

Excitation-to-activation dynamics or, simply, activation dynamics (Eq. 1), represents the muscle fibers recruitment state. The most widely used expression to obtain muscle activation from a set of neural excitation is (Nagano and Gerritsen, 2001):

$$\dot{a}(t) = (u(t) - a(t)) \cdot (t_1 u(t) + t_2), \tag{3}$$

where $t_2 = 1/t_{\mathrm{d}}$ and $t_1 = 1/(t_{\mathrm{a}} - t_2)$ are time constants, $u(t)$ is the neural excitation $0 \leq u \leq 1$, $a(t)$ is the muscle activation, where $0 \leq u \leq 1$, t_{a} is a time activation constant and t_{d} is the deactivation constant. The values for those constants are taken from Umberger et al. (2003).

As depicted on Fig. 2b, activation dynamics represents a delay between input and output, that is, the delay between the reception of the signal (neural excitation) and the transformation of those signals into action potentials to activate the muscle fibers due to electrochemical process that take place in muscle tissue.

The direct acquisition of neural excitations represents a major drawback. Instead, a set of activations can be obtained from indirect measurements such as electromyography (EMG) signals. Buchanan et al. (2004) proposed an EMG-to-activation relationship, as EMG signals can be measured easily. This relationship, namely the A-model, can be written as:

$$a_j(t) = \frac{e^{Au_j(t)} - 1}{e^A - 1}, \tag{4}$$

where $a_j(t)$ is the activation of the jth muscle, $u_j(t)$ the processed sEMG signal of the jth muscle, and A a non-linear shape factor constrained to $-3 < A < 0$ according to Buchanan et al. (2004) or $-5 < A < 0$ as reported by Sartori et al. (2009) (in this work $A = -4$).

2.1.2 Activation-to-force dynamics

The force exerted by a muscle depends on the number of muscle fibers recruited to perform a given movement (activation level), as well as on the actual fiber lengths and the contraction velocity, as expressed in Eq. (2). This mathematical description can be derived from the muscle-tendon components.

According to the distribution of the muscle elements shown in Fig. 1e, and considering tendon and muscle as springs attached in series, it is possible to write:

$$F^{\mathrm{MT}} = F^{\mathrm{T}} = F^{\mathrm{M}}, \tag{5}$$

where superscript MT is referred to muscle-tendon unit, T is related to tendon and M corresponds to muscle. If one expand those equations for tendon and muscle (CE) force, the following expression is obtained:

$$F^{\mathrm{MT}} = F_0^{\mathrm{M}} \cdot f^{\mathrm{T}}\left(l^{\mathrm{T}}\right) = \left(F^{\mathrm{PE}} + F^{\mathrm{M}}\right) \cdot \cos\alpha_p \tag{6}$$

where F_0^{M} is the maximum isometric force, $f^{\mathrm{T}}(l^{\mathrm{T}})$ is the normalized force exerted by the tendon that depends on its length (l^{T}). The terms F^{PE} and F^{M} are, respectively, the forces exerted by the parallel element and the contractile element. The force exerted by this last component can be expressed as (Buchanan et al., 2005):

$$F^{\mathrm{M}} = F_0^{\mathrm{M}} \cdot a \cdot f_{\mathrm{L}}\left(\widetilde{l}^{\mathrm{M}}\right) \cdot f_{\mathrm{V}}\left(\widetilde{v}^{\mathrm{M}}\right), \tag{7}$$

where a is the muscle activation and $f_{\mathrm{L}}(\widetilde{l}^{\mathrm{M}})$ and $f_{\mathrm{V}}(\widetilde{v}^{\mathrm{M}})$ the force–length and force–velocity relationships, respectively. These relationships depend on the normalized muscle length $\widetilde{l}^{\mathrm{M}} = l^{\mathrm{M}}/l_0^{\mathrm{M}}$ and normalized contraction velocity, $\widetilde{v}^{\mathrm{M}} = v^{\mathrm{M}}/v_{\max}$, where $v_{\max} = l_0^{\mathrm{M}}/\tau_c$, being τ_c a time constant (set to 0.1 s in this work).

In the next sections, the mathematical models for each element of the Hill-type muscle model, i.e., tendon, SE, PE and CE will be described, specifically the models proposed by Van Soest and Bobbert (1993), Thelen (2003) and Kaplan (2000)–Silva (2003).

2.2 Tendon

The force exerted by the tendon element is expressed in terms of the tendon strain, ε^{T}, however, other authors use parabolic relationships that depends directly on the tendon length.

The relationship proposed by Thelen (2003) to obtain the force exerted by the tendon as a function of the stiffness is:

$$F^{\mathrm{T}}\left(\varepsilon^{\mathrm{T}}\right) = F^{\mathrm{T}}M_0 \cdot f_{\mathrm{T}}\left(\varepsilon^{\mathrm{T}}\right) = \begin{cases} F_0^{\mathrm{M}} \cdot 0.10377\left(e^{91\varepsilon^{\mathrm{T}}} - 1\right) & \text{if} \quad 0 \le \varepsilon^{\mathrm{T}} < 0.01516, \\ F_0^{\mathrm{M}} \cdot \left(37.526\varepsilon^{\mathrm{T}} - 0.26029\right) & \text{if} \quad 0.01516 \le \varepsilon^{\mathrm{T}} < 0.1 \end{cases} \tag{8}$$

where $\varepsilon^{\mathrm{T}} = \frac{l^{\mathrm{T}} - l_{\mathrm{slack}}}{l_{\mathrm{slack}}}$.

Van Soest and Bobbert (1993) use a simple relationship to obtain the force developed in the tendon. In its work the tendon is presented as a quadratic spring and the expression to obtain the force is defined by:

$$f_{\mathrm{T}}\left(l^{\mathrm{T}}\right) = \begin{cases} 0 & \text{if} \quad l^{\mathrm{T}} < l_{\mathrm{slack}} \\ k^{\mathrm{T}}\left(l^{\mathrm{T}} - l_{\mathrm{slack}}\right)^2 & \text{if} \quad l^{\mathrm{T}} \ge l_{\mathrm{slack}} \end{cases} \tag{9}$$

where $k^{\mathrm{T}} = F_0^{\mathrm{M}} / \left(\varepsilon^{\mathrm{T}} \cdot l_{\mathrm{slack}}\right)^2$.

Kaplan–Silva's description of tendon is not included in this work as they considered in their works that this element could be neglected in slow movements to reduce the dimensionality of the problem. The comparison of the models presented here is shown in Fig. 3b. The relationship given by Thelen is composed by an exponential function followed by a linear expression. On the contrary, the relationship defined by Van Soest and Bobbert is a quadratic curve. Both expressions show slight differences in the region where the muscle works. However, the curves diverge for greater values.

2.3 Muscle

Muscle tissue transforms the muscle fiber recruitment level into muscle contraction. Muscle tissue is defined by the contractile element (CE), that describes the contraction process, and the parallel element (PE), which defines the passive force exerted by the muscle when lengthen over the optimal length. Both elements are described right after.

2.3.1 Parallel element

The parallel element represents the elasticity of the tissue attached in series to the muscle. According to Thelen (2003),

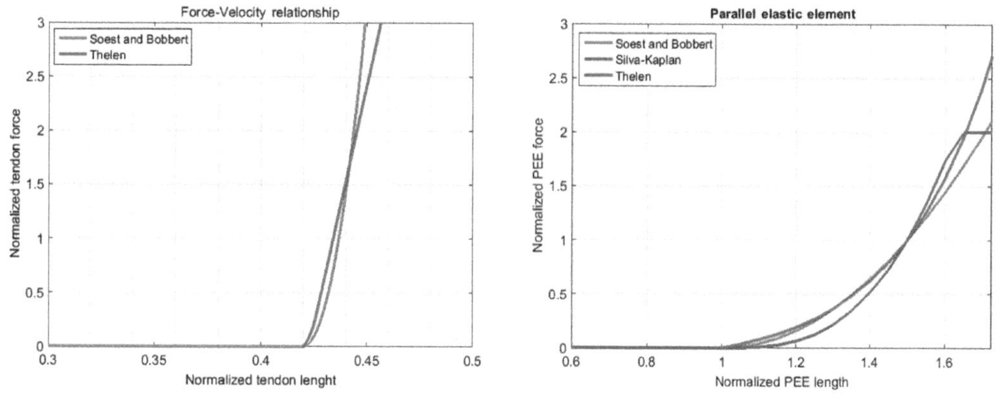

Figure 3. Comparison of the different formulations for **(a)** Tendon and **(b)** parallel elastic element. Green solid line: Van Soest and Bobbert. Blue solid line: Kaplan–Silva. Red solid line: Thelen.

the force–length relationship for the parallel element is defined by:

$$F_{\mathrm{PE}} = F_0^{\mathrm{M}} \cdot f_{\mathrm{PE}}\left(\tilde{l}^{\mathrm{M}}\right) = F_0^{\mathrm{M}} \frac{e^{\left(k_{\mathrm{PE}}\left(\tilde{l}^{\mathrm{M}}\right)/\varepsilon_0^{\mathrm{M}}\right)} - 1}{e^{k_{\mathrm{PE}}} - 1}, \tag{10}$$

where k_{PE} is a shape factor and $\varepsilon_0^{\mathrm{M}}$ is the parallel element strain. In this work, $k_{\mathrm{PE}} = 3.0$ and $\varepsilon_0^{\mathrm{M}} = 0.5$.

Silva (2003), based on the work of Kaplan (2000), proposed an analytical expression for the PE:

$$F_{\mathrm{PE}} = \begin{cases} 0 & \text{for} \quad l^{\mathrm{MT}} \le l_0^{\mathrm{M}} \\ 8\dfrac{F_0^{\mathrm{M}}}{l_0^{M^3}}\left(l^{\mathrm{M}} - l_0^{\mathrm{M}}\right)^3 & \text{for} \quad l_0^{\mathrm{M}} \le l^{\mathrm{M}} \le 1.63 l_0^{\mathrm{M}} \\ 2F_0^{\mathrm{M}} & \text{for} \quad l^{\mathrm{M}} > 1.63 l_0^{\mathrm{M}} \end{cases} \tag{11}$$

Van Soest and Bobbert (1993) does not include an expression for this element, however, a quadratic curve is described in their work. Martins et al. (1998) present a relationship of this type for the PE:

$$F_{\mathrm{PE}} = F_0^{\mathrm{M}} \cdot f_{\mathrm{PE}}(\tilde{l}^{\mathrm{M}}) = \begin{cases} 0 & \text{for} \quad \tilde{l}^{\mathrm{M}} < 1 \\ F_0^{\mathrm{M}} \cdot 4 \cdot \left(\tilde{l}^{\mathrm{M}} - 1\right)^2 & \text{for} \quad \tilde{l}^{\mathrm{M}} \ge 1 \end{cases} \tag{12}$$

The comparison between different formulations is shown on Fig. 3a. The results show a qualitative similarity between curves, specially in the normalized muscle length interval $\tilde{l}^{\mathrm{M}} \in [0, 1.6]$. Values above $\tilde{l}^{\mathrm{M}} = 1.6$ present significant divergences, however values over this limit are rarely reached in normal movements.

2.3.2 Contractile element

The contractile element is responsible of the active force generation in muscle tissue. As expressed by Eq. (7), written below for a better understanding, the force developed by the CE depends on the force–length and force–velocity relationships, that is:

$$F^{\mathrm{M}} = F_0^{\mathrm{M}} \cdot a \cdot f_{\mathrm{L}}\left(\tilde{l}^{\mathrm{M}}\right) \cdot f_{\mathrm{V}}\left(\tilde{v}^{\mathrm{M}}\right) \tag{13}$$

The mathematical expressions for both relationships are detailed below.

Force–length relationship

According to Thelen (2003), the force–length relationship is described by:

$$f_{\mathrm{L}}\left(\tilde{l}^{\mathrm{M}}\right) = \begin{cases} 0 & \text{for} \quad \tilde{l}^{\mathrm{M}} \le 1 \\ e^{-\left(\tilde{l}^{\mathrm{M}}-1\right)^2/\gamma} & \text{for} \quad \tilde{l}^{\mathrm{M}} > 1 \end{cases}, \tag{14}$$

where γ represents the half-width of the curve for $f_{\mathrm{L}} = 1/e$. The value of γ is set to 0.45 in this work.

Silva (2003), again, proposes an analytical expression for the force–length relationship f_{L}:

$$f_{\mathrm{L}}\left(\tilde{l}^{\mathrm{M}}\right) = e^{-\left[\left[-\frac{9}{4}\left(\tilde{l}^{\mathrm{M}} - \frac{19}{20}\right)\right]^4 - \frac{1}{4}\left[-\frac{9}{4}\left(\tilde{l}^{\mathrm{M}} - \frac{19}{20}\right)\right]^2\right]} \tag{15}$$

Contrariwise, Van Soest and Bobbert (1993) use directly a force–length–velocity curve. The expressions for the force–length and force–velocity relationships can be derived from it by setting $\tilde{v}^{\mathrm{M}} = 0$ and $\tilde{l}^{\mathrm{M}} = 1$ respectively (for a detailed explanation see Ackermann, 2007). Both curves are depicted in Fig. 4a and b. The proposed relationship is:

– *Concentric contraction* $\left(v^{\mathrm{M}} < 0\right)$:

$$\frac{f^{\mathrm{M}}\left(\tilde{l}^{\mathrm{M}}, \tilde{v}^{\mathrm{M}}\right)}{F_0^{\mathrm{M}}} = a \cdot \frac{B_{\mathrm{r}}\left(f_{\mathrm{iso}} + A_{\mathrm{r}}\right) - A_{\mathrm{r}}\left(B_{\mathrm{r}} - \frac{\tilde{v}^{\mathrm{M}}}{f_{\mathrm{ac}}}\right)}{B_{\mathrm{r}} - \frac{\tilde{v}^{\mathrm{M}}}{f_{\mathrm{ac}}}}, \tag{16}$$

where a is the muscle activation, $f_{\mathrm{ac}} = \min(1, 3.33 \cdot a)$ and, A_{r} and B_{r} are Hill's constants corrections according to Van Soest and Bobbert (1993). The values for

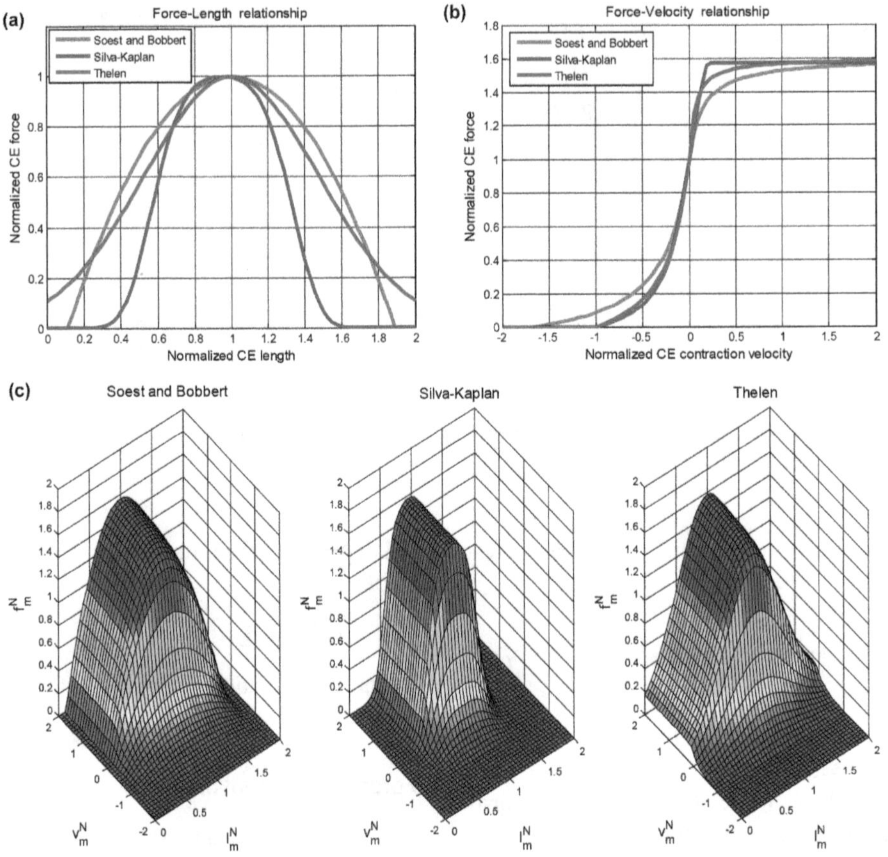

Figure 4. Comparison of the different formulations. **(a)** Force–length relationship. **(b)** Force–velocity relationship. Green solid line: Van Soest and Bobbert. Blue solid line: Kaplan–Silva. Red solid line: Thelen **(c)** Surface plots of the force–length–velocity relationship for the different formulations in the CE (Van Soest and Bobbert, Kaplan–Silva and Thelen, respectively.

those constants are set to $A_\mathrm{r} = 0.41$ and $B_\mathrm{r} = 5.2$. Finally, the isometric force relative to the maximum force, f_iso can be obtained as:

$$f_\mathrm{iso} = \frac{-1}{\mathrm{width}^2} \cdot \tilde{l}^{M^2} + \frac{2}{\mathrm{width}^2} \cdot \tilde{l}^\mathrm{M} - \frac{1}{\mathrm{width}^2} + 1, \quad (17)$$

were width is the maximum range of force production relative to l_0^M.

- *Eccentric contraction* $\left(v^\mathrm{M} > 0\right)$:

$$\frac{f^\mathrm{M}\left(\tilde{l}^\mathrm{M}, \tilde{v}^\mathrm{M}\right)}{F_0^\mathrm{M}} = a \cdot \frac{b_1 - b_2\left(b_3 - \tilde{v}^\mathrm{M}\right)}{b_3 - \tilde{v}^\mathrm{M}}, \quad (18)$$

where shape factors b_1, b_2 and b_3 can be obtained as:

$$\begin{aligned} b_2 &= -f_\mathrm{iso} f_\mathrm{V}^\mathrm{max} \\ b_1 &= \frac{f_\mathrm{ac} B_\mathrm{r}(f_\mathrm{iso} + b_2)^2}{(f_\mathrm{iso} + A_\mathrm{r})\,\mathrm{slope\ factor}} \\ b_3 &= \frac{b_1}{f_\mathrm{iso} + b_2}, \end{aligned} \quad (19)$$

where $f_\mathrm{V}^\mathrm{max}$ is the maximum normalized achievable muscle force when the muscle is lengthening (in this work this value is set to 1.6), f_ac and S_f are shape factors usually set to 1 to reduce computational effort.

The comparison of the different force–length relationship formulations are represented in Fig. 4b.

Force–velocity relationship

According to Thelen (2003), the force–velocity relationship can be written as:

$$f_\mathrm{V}\left(\tilde{v}^\mathrm{M}\right) = \begin{cases} 0 & \text{for} \quad \tilde{v}^\mathrm{M} \leq -1, \\ \dfrac{1 + \tilde{v}^\mathrm{M}}{1 - \tilde{v}^\mathrm{M}/k_\mathrm{CE1}} & \text{for} \quad -1 < \tilde{v}^\mathrm{M} \leq 0, \\ \dfrac{1 + \tilde{v}^\mathrm{M} f_\mathrm{V}^\mathrm{max}/k_\mathrm{CE2}}{1 + \tilde{v}_m/k_\mathrm{CE2}} & \text{for} \quad \tilde{v}^\mathrm{M} > 0 \end{cases}, \quad (20)$$

where k_CE1 and k_CE2 are force–velocity shape factors (0.25 and 0.06 respectively in this work)and $f_\mathrm{V}^\mathrm{max}$ is the maximum normalized achievable muscle force when the muscle is lengthening (in this work this value is set to 1.6).

The analytical expression given by Silva (2003) is defined as:

$$f_V\left(\widetilde{v}^M\right) = \begin{cases} 0 & \text{for} \quad \widetilde{v}^M < -1 \\ -\dfrac{\arctan\left(-5 \cdot \widetilde{v}^M\right)}{\arctan(5)} + 1 & \text{for} \quad -1 \leq \widetilde{v}^M \leq 0.2 \\ \dfrac{\pi}{4\arctan(5)} + 1 & \text{for} \quad \widetilde{v}^M > 0.2 \end{cases} . \quad (21)$$

The expression given by Van Soest and Bobbert (1993) was already addressed on the previous section. The comparison of the different curves for the force–velocity relationship is represented in Fig. 4b. Besides, it is possible to represent in a surface plot the force–length–velocity relationship (Fig. 4c). Results are similar in qualitative terms, and therefore, any of the proposed formulations can be used to obtain muscle forces. However, there are some slight differences that must be mentioned. The force–length relationship described by Silva (2003) is not centred on $\widetilde{l}^M = 1$ and the bell shape is slenderer than the other approaches. However, the force–velocity relationship is quite similar to the model proposed by Thelen (2003). In this case, the model proposed by Van Soest and Bobbert (1993) presents significant differences in the interval $\widetilde{v}^M \in [-0.5, 0.5]$, that may lead to variations of 20 % in F^{CE} compared to the other ones in the same region.

Lastly, it is possible to represent the combined actuation of the PE and CE. As it can be shown in Fig. 2c, if the muscle tissue is stretched beyond its optimal length (l_0^M) the passive force generated by the PE becomes significant.

3 Experimental setup

In order to test the different muscle models analysed in this work, the following experiment was carried out. The idea is to obtain a set of muscle forces to study the variability of the results by using different muscle models in which all the inputs are known. To do so, a simple movement, an arm flexo–extension movement under load was performed. The recordings consist of the acquisition of the kinematics of reflective markers attached to anatomical landmarks and muscle activity (EMG signal) of the long head of the *biceps brachii*. On the one hand, the EMG signal is used as input in the A-model to obtain a set of activations. On the other hand, the kinematics of the flexo–extension movement under load was used to obtain the muscle lengths and contraction velocities to be used with the activations to obtain the muscle forces. This process was performed in OpenSim (Delp et al., 2007) using the upper extremity model from Holzbaur et al. (2005). The model was scaled to subject's dimensions by using the recorded positions of the reflective markers. Parameters such as l_0^M or l_{slack} were obtained after the scale. Others, as F_0^M, were taken directly from literature (Holzbaur et al., 2005).

The experimental test was carried out by a voluntary subject (26 years, healthy with no muscular nor neurological disorders). The test was performed under subject's consent and approved by the local Ethics Committee at University of Extremadura. The acquired movement consists of a dumbbell weightlifting with the right arm, from full extension at anatomical position to full flexion and return to the initial position. A load of 2.5 kg was used. A total of seven recordings were carried out for the experiment. The voluntary was trained previously to maintain constant the velocity of the cycle in order to obtain an adequate repeatability. Seven reflective markers were placed according to Nigg and Herzog protocol (Nigg and Herzog, 1999) in the right arm. The motion was recorded with 12 infra-red light cameras OptiTrack V100:R2 at 100 Hz. The motion capture system was fully synchronized with the Trigno™ Wireless System from Delsys®. The muscle activity was recorded on the superficial long head of the *biceps brachii*. The sEMG electrode was placed following the recommended standard of the Surface EMG for a Non-Invasive Assessment of Muscles (SENIAM) project (Hermens et al., 2000). The skin was abraded and cleaned with isopropyl alcohol. Then, a thin layer of conductive gel was extended along the point of application of the electrode, that is, in the middle zone of the muscle, far from innervated an tendinosus zones (Criswell, 2010). The sEMG signals were filtered by means of singular spectrum analysis (see Romero et al., 2015 for further details) and the first component was retrieved as the trend of the signal, i.e., the input to the A-model. The values were normalized to the maximum voluntary contraction (MVC) following the recommendation given in Konrad (2006). The processing of kinematic and sEMG signals was performed in MATLAB®, running on an Intel® Core™ i5 CPU at 3.20 GHz.

In order to assess the differences in muscle force production between models, a validation metric for curve comparison proposed by Geers (Geers, 1984; Lund et al., 2011) is used. This metric allows to quantify independently differences in magnitude (M) and phase (P). A combined error (C) based on the previous differences is also used. The expressions can be written as:

$$\begin{aligned} M &= \sqrt{\frac{\upsilon_{cc}}{\upsilon_{mm}}} - 1, \\ P &= \frac{1}{\varpi}\cos^{-1}\frac{\upsilon_{mc}}{\sqrt{\upsilon_{mm}\upsilon_{cc}}}, \\ C &= \sqrt{M^2 + P^2}, \end{aligned} \quad (22)$$

where the values υ_{mm}, υ_{cc} and υ_{mc} can be defined as:

$$\begin{aligned} \upsilon_{mm} &= \frac{1}{t_2 - t_1}\int_{t_1}^{t_2} m(t)^2 \mathrm{d}t, \\ \upsilon_{cc} &= \frac{1}{t_2 - t_1}\int_{t_1}^{t_2} c(t)^2 \mathrm{d}t, \end{aligned} \quad (23)$$

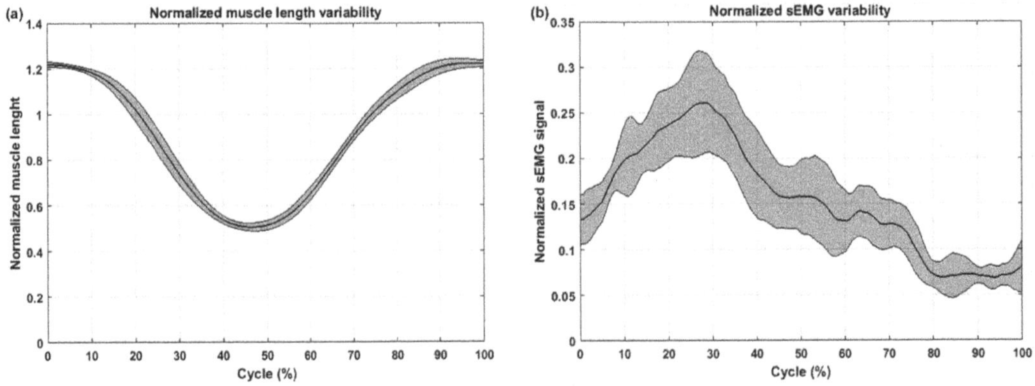

Figure 5. **(a)** Variability in muscle length during tests. **(b)** Variability in the recorded sEMG during tests. Mean ± standard deviation. The normalized muscle lengths were obtained as the ratio $l^{\mathrm{M}}/l_0^{\mathrm{M}}$. The sEMG signals were normalized to the maximum voluntary contraction (MVC).

$$v_{\mathrm{mc}} = \frac{1}{t_2 - t_1} \int_{t_1}^{t_2} m(t)c(t)\mathrm{d}t,$$

where $m(t)$ and $c(t)$ are the measured and the computed signals and $[t_1, t_2]$ is the time interval in which the analysis is performed. In this work we perform a pairwise comparison (see Table 2), therefore $m(t)$ can be defined as the interest signal, which is compared with the signal $c(t)$. Major similarities between signals are related with indices near to zero.

4 Results and discussion

The differences between models have already been highlighted along the methods section. The variability of the acquired signals during the different contractions is represented in Fig. 5. As shown in the figure, the standard deviation is qualitatively low for the reconstructed normalized muscle length (represented by the limits of the shaded region). In the same sense, the measurements of sEMG signals present certain variability, mainly due to differences in muscle fibers recruitment by the central nervous system to prevent fatigue or damage.

The results related to the different elements of the muscle will be discussed in the order they were described in Sect. 2. First, regarding to tendon (see Fig. 3a), the presented models show similarities for typical values of muscle lengths in normal activities, however for greater values the differences are important. However, due to the high slope of the tendon force curve it is usual to consider this component as a stiff element. This fact reduces the dimensionality of the problem, however, as commented before, tendon elasticity is important if the tendon stretches an amount approaching the fiber length of a particular muscle, and therefore it must be considered in these cases (Zatsiorsky and Prilutsky, 2012).

Regarding to muscle elements (see Figs. 3b, 4), the three proposed models show qualitative similarities. For the parallel element, a notable difference between Silva–Kaplan's model and the rest for values below $\tilde{l}^{\mathrm{M}} = 1.5$ (see Fig. 3b). As the range of normalized muscle length of the recorded movement corresponds to $\tilde{l}^{\mathrm{M}} \in [0.5, 1.2]$ (Fig. 5a) significant differences will appear in the normalized PE force, as shown in Fig. 6a, and therefore in the total normalized muscle force (Fig. 6c). In the case of the contractile element, it was pointed out previously that the main variability was found on the force–length relationship, and therefore, differences in muscle force production will be extended to the total muscle force. These differences in the normalized CE force are depicted in Fig. 6b. Considering the range of normalized muscle force and the shape of the force–length relationship (Fig. 4a), the differences in the range 35–60 % are the ones expected. In fact, similar values are observed in the neighbouring of the optimal length, where all the force–length curves present similar values. The major differences are observed at 50 % of the cycle, corresponding to full flexion. These variations correspond to the differences in the bell-shaped curve of the force–length relationship, as for full-flexion values ($\tilde{l}^{\mathrm{M}} \approx 0.5$) there are considerable differences in the force–length relationship. Silva–Kaplan's model presents major variation in this point. Moreover, when computing the total force (PE + CE) the differences are increased in the limits of the cycle as a consequence of the contribution of the PE.

Regarding to the results shown in Table 2 the main differences between models can be observed in magnitude rather than phase. According to the results, Thelen and Silva–Kaplan's models present major similarities between them compared to Thelen and Soest and Bobbert's. These differences are more evident during the extension phase with deviations of 40.68 and 28.38 % between Soest and Bobbert's models and those of Thelen or Silva–Kaplan, respectively. Differences between curves are lower during flexion in both

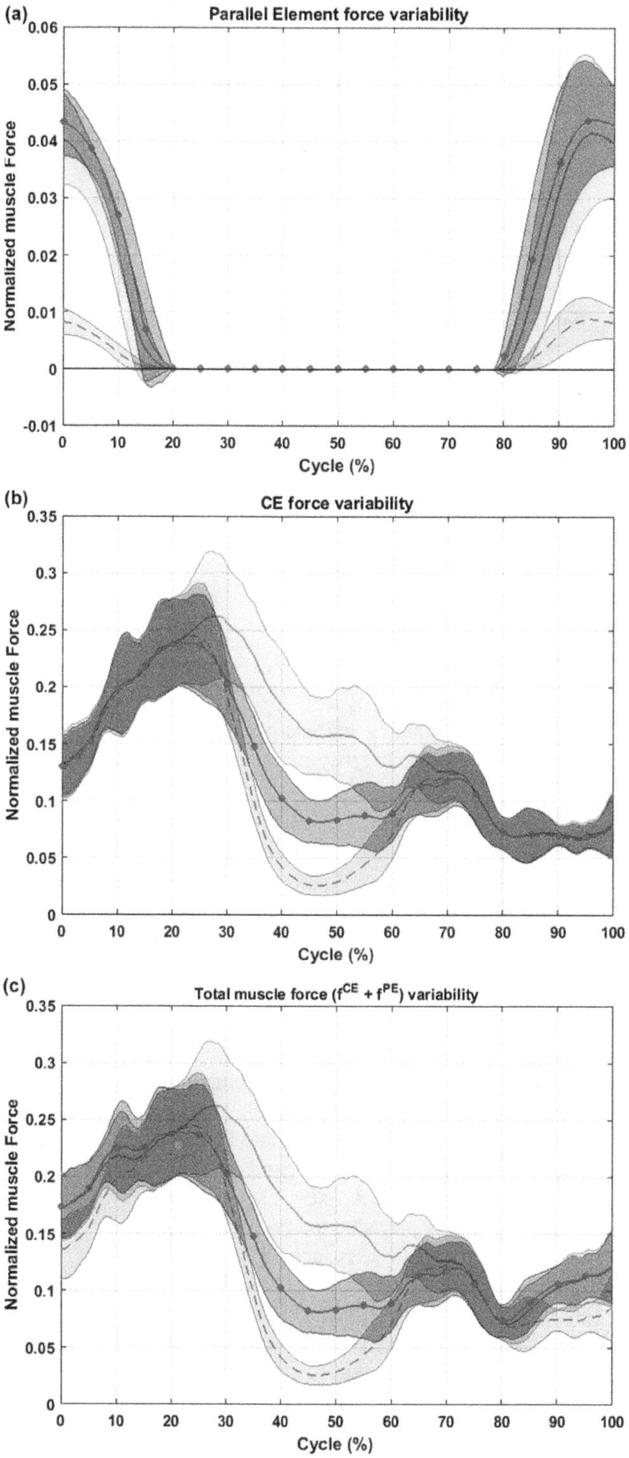

Figure 6. Normalized muscle forces of *biceps brachii* in the flexo–extension movement. Mean ± standard deviation. **(a)** PE force. **(b)** CE force. **(c)** CE + PE force. Green solid line: Van Soest and Bobbert. Red dotted line: Thelen. Blue dashed line: Silva–Kaplan. The normalized muscle forces were obtained as the ratio $F^{\mathrm{PE/CE}}/F_0^{\mathrm{M}}$.

Table 2. Comparison of different formulations by using Geers index.

	(a) Magnitude comparison. Complete cycle.			(b) Phase comparison. Complete cycle.			(c) Combined error. Complete cycle.		
	Soest and Bobbert	Silva–Kaplan	Thelen	Soest and Bobbert	Silva–Kaplan	Thelen	Soest and Bobbert	Silva–Kaplan	Thelen
Soest and Bobbert	0	−0.2035	−0.1545	0	0.1252	0.0669	0	0.2389	0.1683
Silva–Kaplan	0.2555	0	0.0615	0.1252	0	0.0602	0.2846	0	0.0861
Thelen	0.1827	−0.0580	0	0.0669	0.0602	0	0.1946	0.0836	0
	(d) Magnitude comparison. Flexion.			(e) Phase comparison. Flexion.			(f) Combined error. Flexion		
	Soest and Bobbert	Silva–Kaplan	Thelen	Soest and Bobbert	Silva–Kaplan	Thelen	Soest and Bobbert	Silva–Kaplan	Thelen
Soest and Bobbert	0	−0.1764	−0.1330	0	0.1127	0.0620	0	0.2093	0.1467
Silva–Kaplan	0.2142	0	0.0526	0.1127	0	0.0526	0.2420	0	0.0745
Thelen	0.1534	−0.0500	0	0.0620	0.0526	0	0.1655	0.0726	0
	(g) Magnitude comparison. Extension.			(h) Phase comparison. Extension.			(i) Combined error. Extension.		
	Soest and Bobbert	Silva–Kaplan	Thelen	Soest and Bobbert	Silva–Kaplan	Thelen	Soest and Bobbert	Silva–Kaplan	Thelen
Soest and Bobbert	0	−0.2891	−0.2210	0	0.1556	0.0752	0	0.3284	0.2335
Silva–Kaplan	0.4068	0	0.0958	0.1556	0	0.0821	0.4355	0	0.1262
Thelen	0.2838	−0.0874	0	0.0752	0.0821	0	0.2936	0.1199	0

magnitude and phase as expressed by Geer's indices. On the contrary, the major differences observed in the models are related to the full-flexion state, where the models show certain divergences. At full-flexion, Van Soest and Bobebrt's model overestimates the force whereas Kaplan–Silva does the opposite (see Fig. 6c and Table 2). Moreover this last model underestimates also the force at full-extension compared to the other models. The reason of these deviations arises from the muscle description itself. Kaplan–Silva's model is based in the adjustment of Hill's model for multibody formulation purposes and the mathematical description depends only on the normalized muscle length and contraction velocity. On the contrary, Thelen and Van Soest and Bobbert's models contains enough muscle parameters to adjust the mechanical properties of muscle tissue adequately.

On the other hand, if one attends to the computational effort, Kaplan–Silva (0.021811 s) outperforms Thelen (0.023208 s) and Van Soest and Bobbert's (0.026339 s) muscle descriptions. The differences may not be significant in this simple off-line experiment but can make the difference in real time applications. Only the comparison of these results with in vivo measurements will provide the better solution for each case, as a trade-off between accuracy and execution time is required. Finally, this study presents some limitations that must be mentioned. The muscle parameters considered in this study have been taken from literature. Moreover, the methodology proposed to analyze the differences between the presented models is based in a simple exercise for a single muscle. Complex exercises involving several muscles may report different results as the EMG signal used as input to the A-model may contain noise or crosstalk if not filtered properly. Nevertheless, the results presented here can be used as a reference to muscle force production during arm flexo-extension movement and to evaluate differences between models prior their selection for biomechanical studies.

5 Conclusions

In this paper, the mechanical expressions to represent muscle tissue have been presented. To compute muscle efforts, the most significant models available in the literature have been presented. As it has been shown, there are slight differences between the obtained muscle efforts. However, there are some relevant aspects that should be mentioned. The model proposed by Soest and Bobbert gives a detailed description of muscle behaviour, however, this model involves the measurement of too many parameters and most of them are not available in practice. By contrast, Silva–Kaplan's model is based in analytical expressions, and from the computational point of view, their use is more efficient as the expressions are simpler than the other cases, however, for some specific applications it may be interesting to include physiological information which is not provided with this formulation. Thelen's model contains enough parameters to describe muscle force production and it has been implemented in a widely used and validated open source software: OpenSim. The use of each model depends mainly on the computational effort and the control of the parameters involved in the experiments. In this way, Silva–Kaplan's model can be used in experiments in which computing time and computational effort are critical but not the use of physiological parameters whereas Thelen's model provides enough control of physiological parameters to perform simulations in a reasonable time. Therefore, the selection of the most appropriate model depends on the specific conditions of the problem.

Acknowledgements. This work was supported by the Spanish Ministry of Economy and Competitiveness under project DPI2012-38331-C03, co-financed by the European Union through EFRD funds.

References

Ackermann, M.: Dynamics and energetics of walking with prostheses, PhD thesis, University of Stuttgart, Stuttgart, Germany, 2007.

Ackermann, M. and Schiehlen, W.: Dynamic analysis of human gait disorder and metabolical cost estimation, Arch. Appl. Mech., 75, 569–594, 2006.

Alonso, J., Romero, F., Pàmies-Vilà, R., Lugrís, U., and Font-Llagunes, J.: A simple approach to estimate muscle forces and orthosis actuation in powered assisted walking of spinal cord-injured subjects, Multibody Syst. Dyn., 28, 109–124, 2012.

Buchanan, T., Lloyd, D., Manal, K., and Besier, T.: Neuromusculoskeletal modeling: estimation of muscle forces and joint moments and movements from measurements of neural command, J. Appl. Biomech., 20, 367–395, 2004.

Buchanan, T., Lloyd, D., Manal, K., and Besier, T.: Estimation of muscle forces and joint moments using a forward-inverse dynamics model, Med. Sci. Sport. Exer., 37, 1911–1916, 2005.

Criswell, E.: Cram's introduction to surface electromyography, Jones & Bartlett Publishers, Sudbury, USA, 2010.

Delp, S., Anderson, F., Arnold, A., Loan, P., Habib, A., John, C., Guendelman, E., and Thelen, D.: OpenSim: open-source software to create and analyze dynamic simulations of movement, IEEE T. Bio.-Med. Eng., 54, 1940–1950, 2007.

García-Vallejo, D.: Simulación de la marcha humana mediante optimización paramétrica, XVIII Congreso Nacional de Ingeniería Mecánica, Ciudad Real, 3–5 November 2010.

Geers, T. L.: An Objective Error Measure for the Comparison of Calculated and Measured Transient Response Histories, The Shock and Vibration Bulletin, The Shock and Vibration Information Center, Naval Research Laboratory, Washington, D.C., Bulletin 54, Part 2, 99–107, June 1984.

Grahovac, N. and Žigić, M.: Modelling of the hamstring muscle group by use of fractional derivatives, Comput. Math. Appl., 59, 1695–1700, 2010.

Hatze, H.: A myocybernetic control model of skeletal muscle, Biol. Cybern., 25, 103–119, 1977.

Hermens, H., Freriks, B., Disselhorst-Klug, C., and Rau, G.: Development of recommendations for sEMG sensors and sensor placement procedures, J. Electromyogr. Kines., 10, 361–374, 2000.

Hill, A.: The Heat of Shortening and the Dynamic Constants of Muscle, P. Roy. Soc. B-Biol. Sci., 126, 136–195, 1938.

Holzbaur, K., Murray, W., and Delp, S.: A model of the upper extremity for simulating musculoskeletal surgery and analyzing neuromuscular control, Ann. Biomed. Eng., 33, 829–840, 2005.

HosseinNia, S., Romero, F., Tejado, I., Vinagre, B., and Alonso, J.: Effects of Introducing Fractional Dynamics in Hill's Model for Muscle Contraction, in: System Identification, International Federation of Automatic Control, 16, 1743–1748, 2012.

Kaplan, M.: Efficient optimal contro of large-scale biomechanical systems, PhD thesis, Stanford University, USA, 2000.

Konrad, P.: The ABC of EMG, in: A Practical Introduction to Kinesiological Electromyography, Version 1.4, Noraxon INC, USA, 26–43, 2006.

Lund, M. E., de Zee, M., and Rasmussen, J.: Comparing calculated and measured curves in validation of musculoskeletal models, in: XIII International Symposium on Computer Simulation in Biomechanics, Leuven, Belgium, 30 June–2 July 2011.

Martins, J., Pires, E., Salvado, R., and Dinis, P.: A numerical model of passive and active behavior of skeletal muscles, Comput. Method. Appl. M., 151, 419–433, 1998.

Nagano, A. and Gerritsen, K.: Effects of neuromuscular strength training on vertical jumping performance – A computer simulation study, J. Appl. Biomech., 17, 113–128, 2001.

Nigg, B. and Herzog, W.: Biomechanics of the musculo-skeletal system, vol. 192, Wiley New York, 1999.

Romero, F., Alonso, F., Cubero, J., and Galán-Marín, G.: An automatic SSA-based de-noising and smoothing technique for surface electromyography signals, Biomed. Signal Proces., 18, 317–324, 2015.

Sartori, M., Reggiani, M., Mezzato, C., and Pagello, E.: A lower limb EMG-driven biomechanical model for applications in rehabilitation robotics, in: IEEE International Conference on Advanced Robotics, ICAR 2009, Munich, Germany, 22–26 June 2009, 1–7, 2009.

Silva, M.: Human motion analysis using multibody dynamics and optimiation tools, PhD thesis, Instituto Superior Técnico, Universidade Técnica de Lisboa, Lisbon, Portugal, 2003.

Sommacal, L., Melchior, P., Cabelguen, J.-M., Oustaloup, A., and Ijspeert, A.: Fractional multimodels of the gastrocnemius muscle for tetanus pattern, in: Advances in fractional calculus, Springer, 271–285, 2007.

Sommacal, L., Melchior, P., Oustaloup, A., Cabelguen, J.-M., and Ijspeert, A. J.: Fractional multi-models of the frog gastrocnemius muscle, J. Vib. Control, 14, 1415–1430, 2008.

Thelen, D.: Adjustment of muscle mechanics model parameters to simulate dynamic contractions in older adults, J. Biomech. Eng.-T. ASME, 125, 70–77, 2003.

Umberger, B., Gerritsen, K., and Martin, P.: A model of human muscle energy expenditure, Comput. Method. Biomec., 6, 99–111, 2003.

Van Den Bogert, A., Gerritsen, K., and Cole, G.: Human muscle modelling from a users perspective, J. Electromyogr. Kines., 8, 119–124, 1998.

Van Soest, A. and Bobbert, M.: The contribution of muscle properties in the control of explosive movements, Biol. Cybern., 69, 195–204, 1993.

Winters, J. M.: Hill-based muscle models: a systems engineering perspective, Springer-Verlag, 69–93, 1990a.

Winters, J. M.: Hill-based muscle models: a systems engineering perspective, in: Multiple muscle systems, Springer, 69–93, 1990b.

Winters, J. M. and Stark, L.: Muscle models: what is gained and what is lost by varying model complexity, Biol. Cybern., 55, 403–420, 1987.

Yamaguchi, G.: Dynamic modeling of musculoskeletal motion: a vectorized approach for biomechanical analysis in three dimensions, Kluwer Academic Publishers Norwell, MA, 2001.

Zajac, F.: Muscle and tendon: Properties, models, scaling and applications to biomechanics and motor control, Crit. Rev. Biomed. Eng., 17, 359–411, 1989.

Zatsiorsky, V. and Prilutsky, B.: Biomechanics of skeletal muscles, Human Kinetics, Human Kinetics Publisher, Champaign, USA, 2012.

18

A computationally efficient model to capture the inertia of the piezoelectric stack in impact drive mechanism in the case of the in-pipe inspection application

Jin Li[1], Chang Jun Liu[1], Xin Wen Xiong[1], Yi Fan Liu[2], and Wen Jun Zhang[3]

[1]The Complex and Intelligent System Research Center, School of Mechanical and Power Engineering, East China University of Science and Technology, Meilong Road 130, Shanghai, 200237, China

[2]Robotic Systems Laboratory, Ecole Polytechnique Fédérale de Lausanne (EPFL), C/o Nicolas Cantale Avenue de prefaully 56, 1020 Lausanne, Switzerland

[3]Department of Mechanical Engineering, University of Saskatchewan, Saskatoon, S7N5A9, Canada

Correspondence to: Chang Jun Liu (cjliu@ecust.edu.cn)

Abstract. This paper presents a new model for the piezoelectric actuator (PA) in the context of in the impact drive mechanism (IDM) for the in-pipe inspection application. The feature of the model is capturing the inertia of PA stack in a distributed manner as opposed to the lumped manner in literature. The benefit arising from this feature is a balanced trade-off between computational efficiency and model accuracy. The study presented in this paper included both theoretical development (i.e. the model of the piezoelectric actuator and the model of the entire IDM which includes the actuator) and experimental verification of the model. The study has shown that (1) the inertia of the PA in such a robot will significantly affect the accuracy of the entire model of IDM and (2) the simulation of the dynamic behavior with the proposed model is sufficiently accurate by comparing with the experiment. It is thus recommended that the inertia of the PA be considered in the entire model of the IDM robot. The model is an analytical type, which has a high potential to be used for the model-based control of the IDM robot and optimization of its design for a much improved performance of the IDM system.

1 Introduction

A tremendous amount of progresses in micro-robotics have been made in recent years, including micromanipulation for micro or nano-objects handling and nano-positioning. These systems have many applications ranging from in vivo biomedical therapeutic procedures to military reconnaissance (Fatikow and Rembold, 2009; Ouyang et al., 2008). In-pipe robots driven by piezoelectric actuators (PA) have advantages of low energy consumption, small size, and high precision down to the scale of micrometers and even nanometers. In-pipe robots can perform specific functions (scanning, detecting and repairing) in small pipes, such as condenser pipe, blood vessel and intestine (Dario et al., 2003; Yukawa et al., 2006; Xia et al., 2008).

One such type of actuators is based on the stick-slip (S-S) actuation principle (Zhang et al., 2012). In order to realize the autonomous motion, one type of the S-S actuator called impact drive mechanism (IDM) (Dario et al., 2003; Tenzer and Mrad, 2004; Fukui et al., 2001) was proven to be very useful. The schematic diagram of IDM is shown in Fig. 1. The system consists of a main body, an attached block, and a PA (e.g. PZT actuator). Take the motion towards right as a forward motion. The forward motion has three stages: (1) neutral state at which the PA is given the zero voltage; (2) slow extension at which the voltage increases slowly and thus the PA extends slowly to push the block with a small acceleration (note that the main body holds its position by the static friction force from the in-wall of the pipe); (3) quick contraction at which the driving voltage suddenly drops down to zero, and the PA contracts quickly and pulls the main body moving forward. By repeating the foregoing actions, the main body is moved forward continuously (step by step; like a step motor). There is possibility that PA is replaced by other direct

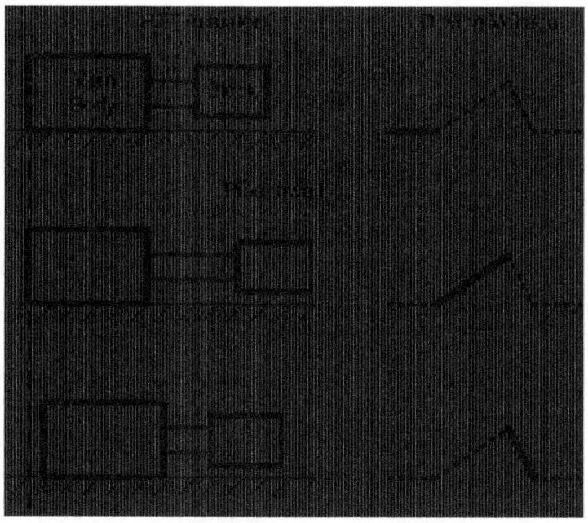

Figure 1. The working principle of the IDM robot.

driven actuators such as shape memory alloy (Hattori et al., 2014). Further, the special requirement on the input voltage (i.e. with one direction of the stroke fast and the opposite direction of the stroke slow) can be realized by the resonant principle (Yokozawa and Morita, 2015).

A model that describes the dynamics of the IDM system is useful to the control of the system and to optimization of the structure of the system, including both the mechanism and controller (Zhang et al., 1999; Li et al., 2001; Liu et al., 2015). Liu et al. (2009) studied the IDM robot actuated by a piezoelectric bimorph (which is of a different configuration of piezoelectric materials to generate motion or deformation) based on a model without consideration of the inertia of the PA. Makkar et al. (2007) and Li et al. (2009) studied the modeling for friction without consideration of the inertia of PA. Ha et al. (2005) optimized the mass ratio of the block and main body and the driving waveform (Fig. 1); however they have not included the inertia of PA in their analysis. Hunstig et al. (2013) investigated the performance and the velocity limitation of the piezoelectric inertia drives under ideal excitations. Sabzehmeidani et al. (2010) studied IDM intelligent control based on mass-spring-damping model with 2 DOFs. Their model has not explicitly included the dynamics of the PA.

In the previous studies where the inertia of the PA is considered, the PA was usually treated as an ideal spring and damping with zero mass (Higuchi et al., 1993; Chang and Li, 1999). However, the inertia of the PAs consumes the driving power and finally influences the dynamic behaviors of the robot. Their effects may be comparable with those out of other components. This means that consideration of the inertia of PA especially in the form of a continuous media may be necessary especially for micro-systems (Liu et al., 2015). Further, aiming at the control of such a system, an analytical model with reasonable accuracy is highly useful (Yokose et al., 2014).

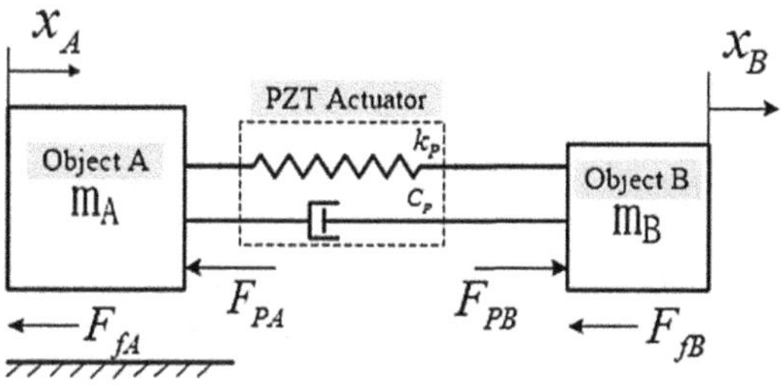

Figure 2. The free-body diagram of the IDM robot.

This paper presents a study on developing an analytical model for the IDM robot in the context of in-pipe inspection. The inertia of PA and that of the other components in this case are comparable. The experiment was carried out to verify the effectiveness of the model. The remainder of the paper is organized in the following. Section 2 presents the model. Section 3 discusses the experimental verification. Section 4 presents some model-predicted results behind this IDM in-pipe inspection robot, which are useful to optimizing the design of the IDM in-pipe robot. Finally, there is a conclusion in Sect. 5.

2 Model development

2.1 Governing equation

The free body diagram of the IDM in-pipe robot is shown in Fig. 2. The entire system moves along the x direction, where $m_{\mathrm{A}}$and m_{B} represent the mass of the main body (Object A) and the attached block (Object B), and x_{A} and x_{B} are the corresponding displacements. From Fig. 2 and by applying the Newtonian second law, one can obtain the following two equations on Objects A and B, respectively

$$m_{\mathrm{A}}\ddot{x}_{\mathrm{A}} = k_P(x_{\mathrm{B}} - x_{\mathrm{A}}) + c_p(\dot{x}_{\mathrm{B}} - \dot{x}_{\mathrm{A}}) - F_{\mathrm{fA}} - F_{\mathrm{PA}} \quad (1)$$

$$m_{\mathrm{B}}\ddot{x}_{\mathrm{B}} = -k_P(x_{\mathrm{B}} - x_{\mathrm{A}}) - c_p(\dot{x}_{\mathrm{B}} - \dot{x}_{\mathrm{A}}) - F_{\mathrm{fB}} + F_{\mathrm{PB}}. \quad (2)$$

In the above equations, the stiffness and damping of the PA are expressed as k_P and c_p, respectively. F_{fA} and F_{fB} are the friction forces acting on the main body and block, respectively. F_{PA} and F_{PB} are the driving forces generated from the PA, respectively. It is noted that $F_{\mathrm{PA}} \neq F_{\mathrm{PB}}$ (because the dynamics of the PA is considered). Further, the in-pipe inspection robot here has revised a little bit on the IDM, that is, having the inertial mass (i.e. block B) down to the ground.

2.2 Dynamic model of the PA

The PA is in itself a dynamics system, which means that the PA has inertia, damping, and stiffness. Suppose the PA is an elastic body and its inertia is uniformly distributed along the x direction. Under the configuration as shown in Fig. 1, the piezoelectric rod extends such that Object A tends to move along the left direction while Object B tends to move along the right direction. As such, there must be an instantaneous center of velocity at a certain time during contraction or at a certain position along the piezoelectric rod. We cut the whole PA rod into two segments on this instantaneous center. Figure 3 shows two separate force diagram of these two parts of the PA, where F_P is the driving force generated in the PA rod by the input voltage, dl_A and dl_B are the displacements of the two ends of the whole PA rod during the time period of dt, at which the instantaneous center forms, m_{PA} and m_{PB} are the masses of the two parts of the PA rod, respectively.

The works with the two parts of the PA rod during dt are:

$$dW_A = (F_{PA} - F_P)dl_A = (F_{PA} - F_P)\left(v_A dt + \frac{1}{2}a_A dt^2\right) \quad (3)$$

$$dW_B = (F_P - F_{PB})dl_B = (F_P - F_{PB})\left(v_B dt + \frac{1}{2}a_B dt^2\right). \quad (4)$$

The kinetic energy of m_{PA} at a certain time is:

$$E_{PA} = \int_0^l \frac{1}{2}v_x^2 \rho dx = \int_0^l \frac{1}{2}\left(\frac{x}{l}v\right)^2 \rho dx = \frac{1}{6}mv^2. \quad (5)$$

The change in the kinetic energy within dt is:

$$dE_{PA} = \frac{1}{6}m_{PA}\left[(v_A + a_A dt)^2 - v_A^2\right]$$
$$= \frac{1}{6}m_{PA}a_A^2 dt^2 + \frac{1}{3}m_{PA}v_A a_A dt. \quad (6)$$

Note that the change in kinetic energy within dt should be equal to the work done during this time. That is,

$$dE_{PA} = dW_A. \quad (7)$$

From Eqs. (3), (6), (7), we have (for the left part of the piezoelectric rod P1):

$$F_{PA} = \frac{1}{3}m_{PA}a_A + F_P = \frac{1}{3}\left(\frac{\dot{x}_A}{\dot{x}_A - \dot{x}_B}\right)m_P\ddot{x}_A + F_P. \quad (8)$$

Similarly, we have (for the right part of the piezoelectric rod P2):

$$F_{PB} = -\frac{1}{3}m_{PB}a_B + F_P = \frac{1}{3}\left(\frac{\dot{x}_B}{\dot{x}_A - \dot{x}_B}\right)m_P\ddot{x}_B + F_P. \quad (9)$$

2.3 Friction model

Without loss of the generality for the purpose of this study, the Karnopp model (Karnopp, 1985) was employed to describe the frictional force in this work, i.e.:

$$F_f = \{F_C\lambda(v) + F_S[1 - \lambda(v)]\}\,\mathrm{sign}(v) \quad (10)$$

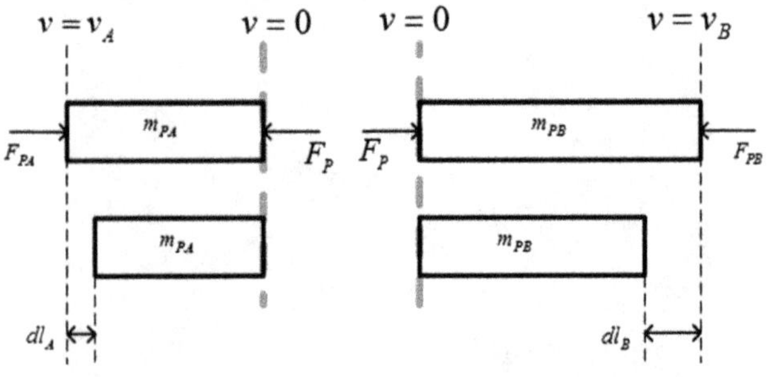

Figure 3. The free-body diagram of the PZT actuator.

where

$$\lambda(v) = \begin{cases} 1 & |v| > \alpha \\ 0 & |v| \le \alpha \end{cases} \quad \alpha > 0.$$

$F_C = \mu_K F_n$ is the Coulomb friction, and $F_S = \mu_S F_n$ is the static friction. F_n denotes the normal force, μ_K and μ_S are the frictional coefficients, respectively. It is noted that there is a friction force on Object A and Object B, respectively.

2.4 Model integration

The ultimate goal of the analysis is to find the motion of the main body (Object A) and the inertial mass or block (Object B), as shown in Figs. 1 and 2. For the convenience of the reader, we put together all the equations as derived before in the following.

$$\begin{cases} m_A\ddot{x}_A = k_P(x_B - x_A) + c_P(\dot{x}_B - \dot{x}_A) - F_{fA} - F_{PA} \\ m_B\ddot{x}_B = -k_P(x_B - x_A) - c_P(\dot{x}_B - \dot{x}_A) - F_{fB} + F_{PB} \\ F_{PA} = \frac{1}{3}m_{PA}a_A + F_P = \frac{1}{3}\left(\frac{\dot{x}_A}{\dot{x}_A - \dot{x}_B}\right)m_P\ddot{x}_A + F_P \\ F_{PB} = -\frac{1}{3}m_{PB}a_B + F_P = \frac{1}{3}\left(\frac{\dot{x}_B}{\dot{x}_A - \dot{x}_B}\right)m_P\ddot{x}_B + F_P \\ F_{fA} = \{F_{CA}\lambda(\dot{x}_A) + F_{SA}[1 - \lambda(\dot{x}_A)]\}\,\mathrm{sign}(\dot{x}_A) \\ F_{fB} = \{F_{CB}\lambda(\dot{x}_B) + F_{SB}[1 - \lambda(\dot{x}_B)]\}\,\mathrm{sign}(\dot{x}_B) \end{cases} \quad (11)$$

Further, in the above equations, the PA driving force is $F_P = k_P d_e V$, where k_P denotes the mechanical stiffness and d_e denotes piezoelectric coefficient (Fung et al., 2008; Low and Guo, 1995). In our case, $k_P = 5 \times 10^7\,\mathrm{N\,m^{-1}}$, $d_e = 3 \times 10^{-7}\,\mathrm{m\,V^{-1}}$, $m_P = 120\,\mathrm{g}$. In the test system, there is a support force from the pipe acting on the block against gravity. Consequently, the block gets a friction force $F_{SB} = 0.1\,\mathrm{N}$, which is however much smaller than F_{SA}. So $F_{CB} \approx 0$ while $F_{CA} = 5/6\,F_{SA}$. The waveform of the input voltage has a strong influence on the performance of the IDM robot. A triangular waveform was employed in this study to drive the PA. As shown in Fig. 4, the duty ratio was 75 and 80 %, respectively, in this study. Duty ratio is defined as the period of time from zero to peak (t_1) of the wave over the total period of operational time (t_2), namely t_1/t_2.

Figure 5 shows the displacements when the driving voltage is 80 v, $t_1 = 1.5$ ms, and $t_2 = 2$ ms. The Object A moves backward a little when the PA extends slowly. The actuator pulls Object A forward during its quick contraction.

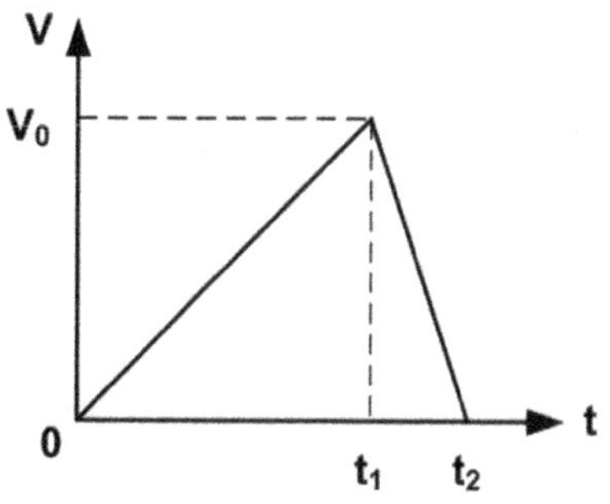

Figure 4. The waveform of the input voltage. The duty ratio is defined as t_1/t_2 and it is 75 and 80 %, respectively, in this study; V_0 is the maximum voltage.

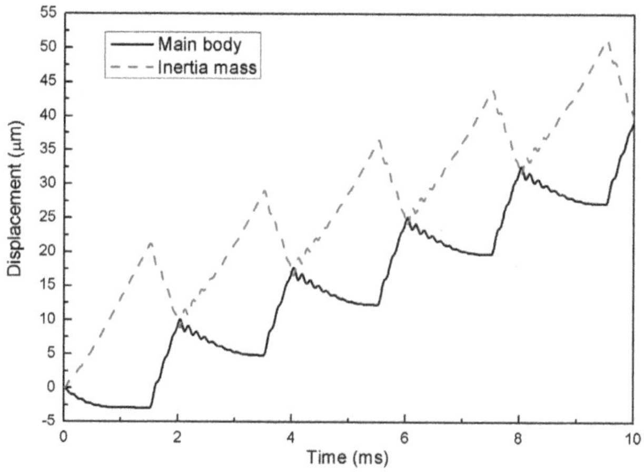

Figure 5. Displacements of Object A and block in 5 cycles.

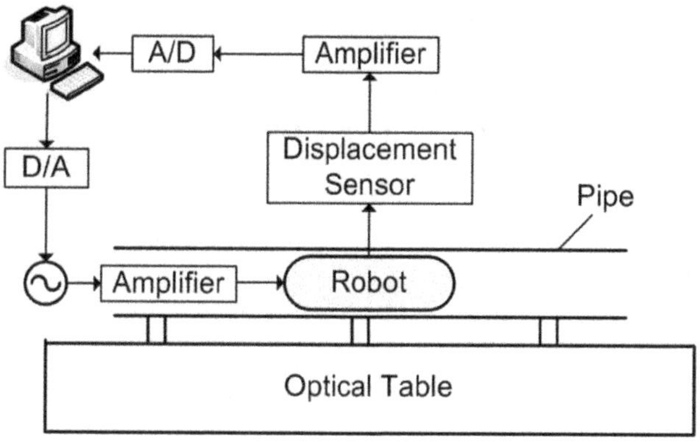

Figure 6. The schematic diagram of the signal flow.

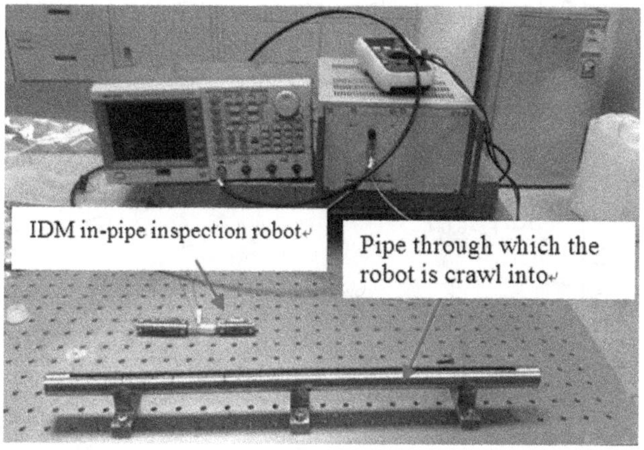

Figure 7. The robot and physical set up for testing.

2.5 Model validation

An experimental validation has been carried out on an optical stage. Figure 6 is a schematic diagram of the test-bed. The robot was driven by a generator and an amplifier (PI-LVPZT). The inner wall of the pipe is made of stainless steel with the radius of 18 mm in Fig. 7. The movement of the robot has been measured by a laser triangulation sensor (Keyence^{TM} LKH008).

Figure 8 shows the velocity of the main body (Object A). It can be seen that the predicted result with the developed model is closer to the measurement than to the model without consideration of the inertia of PA. Further, both models are able to improve the prediction accuracy with the voltage being less than 30 v. This is reasonable, when the input voltage is small, the kinetic energy of the robot is small, and the effect of the inertia is small accordingly. After the voltage is greater than 30 v, the difference between the two models tends to increase with respect to the measured result (i.e. the error of the model without consideration of the inertia of the PA increases) with the increase of the voltage.

Figure 9 shows the result of the velocity with driving frequency. The driving voltage is 80 v, and the duty ratio is

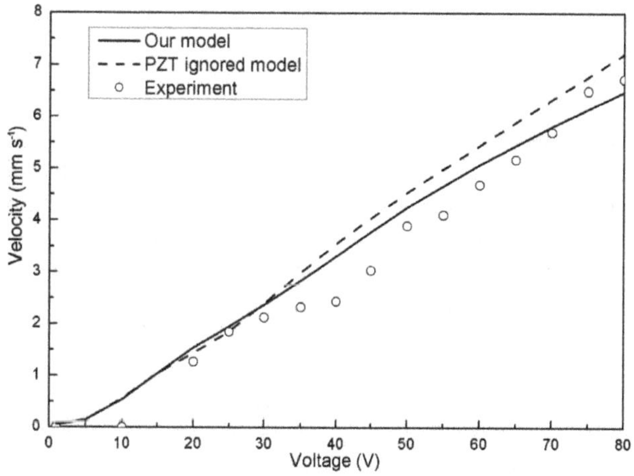

Figure 8. Velocity vs. voltage.

75 %. There are two main peaks of the velocity. It is noted that the frequencies are employed as the excitation signals to examine system behavior in terms of energy efficiency. It can be found that the measured locations of the peak values agree well with the one predicted by the models. At the

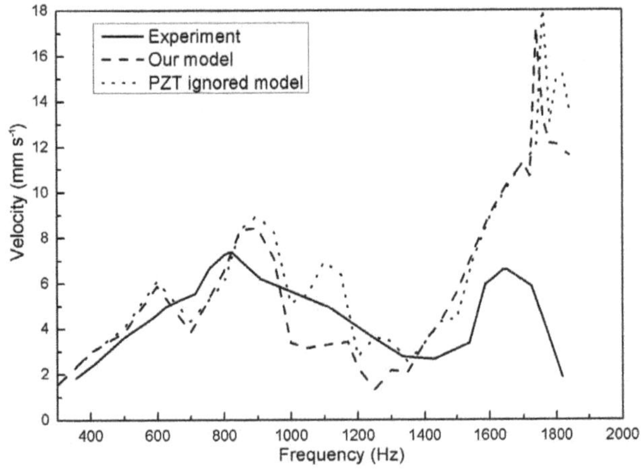

Figure 9. The velocity vs. frequency.

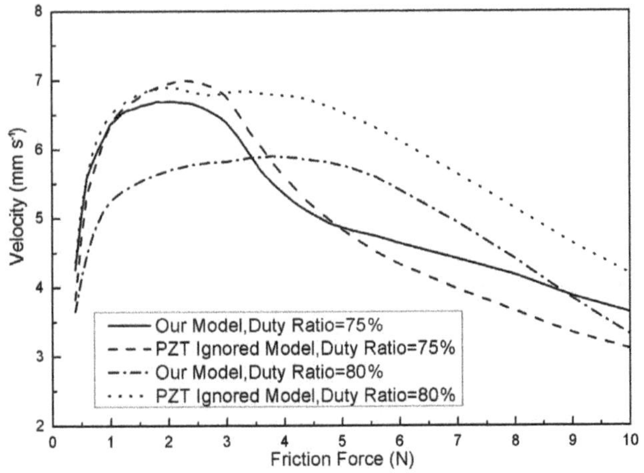

Figure 11. Velocity of the robot versus friction.

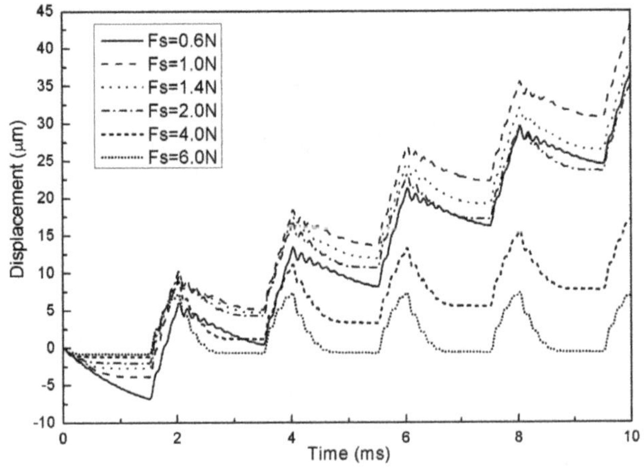

Figure 10. Displacements of the main body versus different friction coefficients.

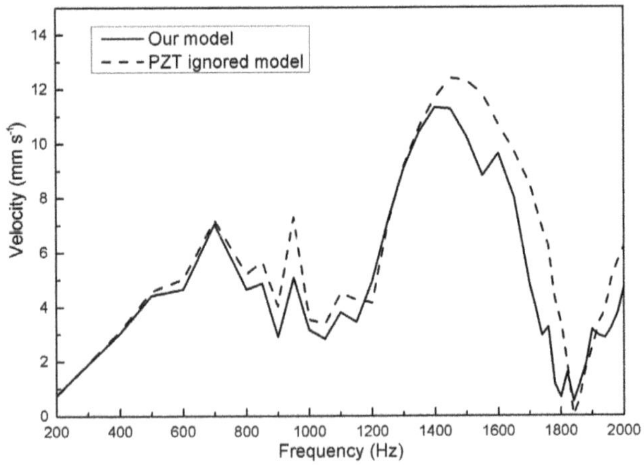

Figure 12. The velocity of the actuator versus the frequency of the input voltage.

second peak value, the one predicted by the models is much larger than the measured one. This is because under a high frequency excitation, the mechanical loss and heat consumption are significant in the real structure, while this feature has not been considered in the model. Non-uniform friction along the pipe is also responsible for the velocity versus frequency relation. Nevertheless, the developed model is better than the model without consideration of the inertia of the PA.

Overall, the error of the model with consideration of the inertia of the PA is 8.65 % while the error of the model without consideration of the inertia of the PA is 14.34 %, with respect to the measured result. In the experiment, the measurement approach has introduced a relatively large error which is up to about 10 % (the displacement measurement is based on the ruler with the resolution of 0.1 mm, which translate to an error of 10 %). As such, the developed model shows an acceptable accuracy.

3 Notes on optimization of the performance of the in-pipe inspection robot

Figure 10 shows the displacement of the main body under different frictions. Too small or too large friction will cause the failure of the robot. The largest step movement happens when the friction force is around 1.0 N. As the friction force reaches 6N, the robot does not move forward in one cycle at all, which implies a failure. Figure 11 shows the velocity of the robot with respect to the friction. It is shown that an optimal velocity can be reached by adjusting the friction force.

Figure 12 shows the velocity with respect to the frequency. It has many velocity peaks from the frequency range from 200 to 2000 Hz. The frequencies at the peaks of velocity are slightly different between two models. When the frequency is high, the velocity predicted with our model is smaller than the model without consideration of the inertia of the PA. This is so because the mass of the PA also consume the driving energy. The dynamic behaviors are quite different when the mass of the PZT actuator is taken into consideration, such as response displacement, resonant frequency. These factors are crucial when we consider the control strategy.

4 Conclusions

This paper presented a work towards the development of a model that can capture the continuous inertia of the PA in IDM in the context of pipe inspection and has a balanced accuracy and computational efficiency. The work also included an experimental study to analyze the effectiveness of the model. It can be concluded from this study that the developed model can predict the system behavior significantly more accurate than the model without consideration of the inertia of the PA. Since the model is also computationally efficient, it has a high potential to be used for feedback control of the system as well as design optimization of the system.

Acknowledgements. This work was supported in part by the National Natural Science Foundation of China (Grant No. 51305138 and 51375166), the Fundamental Research Funds for the Central Universities and Science and Technology Commission of Shanghai Municipality (Grant No. 13ZR1453300).

References

Chang, S. H. and Li, S. S.: A High Resolution Long Travel Friction-drive Micro-positioner with Programmable Step Size, Rev. Sci. Instrum., 70, 2776–2782, 1999.

Dario, P., Hannaford, B., and Menciassi, A.: Smart surgical tools and augmenting devices, IEEE Trans. Robot. Autom., 19, 782–791, 2003.

Fatikow, S. and Rembold, U.: Microsystem Technology and Microrobotics, Springer-Verlag, Berlin, Heidelberg, 303–361, 2009.

Fukui, R., Torii, A., and Ueda, A.: Micro robot actuated by rapid deformation of piezoelectric elements, International Symposium on Micromechatronics and Human Science, 9–12 September 2001, Nagoya, Japan, 117–122, 2001.

Fung, R. F., Han, C. F., and Ha, G. L.: Dynamic Responses of the Impact Drive Mechanism Modeled by the Distributed Parameter System, Appl. Math. Model., 32, 1734–1743, 2008.

Ha, J. L., Fung, R. F., and Han, C. F.: Optimization of an impact drive mechanism based on real-coded genetic algorithm, Sensors Actuat. A, 121, 488–493, 2005.

Hattori, S., Hara, M., Nabae, H., Hwang, D., and Higuchi, T.: Design of an impact drive actuator using a shape memory alloy wire, Sensors Actuat. A, 219, 45–47, 2014.

Higuchi, T., Furutani, K., Yamagata, Y., Kudoh, K., and Ogawa, K.: Improvement of velocity of impact drive mechanism by controlling friction, J. Adv. Automat. Tech., 5, 71–76, 1993.

Hunstig, M., Hemsel, T., and Sextro, W.: Stick-slip and slip-slip operation of piezoelectric inertia drives. Part I: Ideal excitation, Sensors Actuat. A, 200, 90–100, 2013.

Karnopp, D.: Computer Simulation of Stick-slip in Mechanical Dynamic System, J. Dyn. Syst.-ASME, 107, 100–103, 1985.

Li, J. W., Zhang, W. J., Yang, G. S., Tu, S. D., and Chen, X. B.: Thermal-error modeling for complex physical systems the-state-of-arts review, Int. J. Adv. Manufact. Technol., 42, 168–179, 2009.

Li, Q., Zhang, W. J., and Chen, L.: Design for control (DFC): a concurrent engineering approach for mechatronic system design, IEEE-ASME T. Mech., 6, 161–169, 2001.

Liu, P. K., Wen, Z. J., and Sun, L. N.: An in-pipe micro robot actuated by piezoelectric bimorphs, Chinese Sci. Bull., 54, 2134–2142, 2009.

Liu, Y. F., Li, J., Hu, X. H., Zhang, Z. M., Cheng, L., Lin, Y., and Zhang, W. J.: Modeling and control of piezoelectric inertia-friction actuators: review and future research directions, Mech. Sci., 6, 95–107, doi:10.5194/ms-6-95-2015, 2015.

Low, T. S. and Guo, W.: Modeling of Three-layer Piezoelectric Bimorph Beam with hysteresis, IEEE-ASME J. Microelectromech. S., 4, 230–237, 1995.

Makkar, C., Hu, G., Sawyer, W. G., and Dixon, W. E.: Lyapunov-Based Tracking Control in the Presence of Uncertain Nonlinear Parameterizable Friction, IEEE T. Automat. Contr., 52, 1988–1994, 2007.

Ouyang, P. R., Tjiptoprodjo, R. C., Zhang, W. J., and Yang, G. S.: Micro-motion Devices Technology: The State of Arts Review, Int. J. Adv. Manufact. Techol., 38, 463–478, 2008.

Sabzehmeidani, Y., Mailah, M., and Hussein, M.: Intelligent Control and Modelling of a Micro robot for in-pipe application, ECME, Puerto De La Cruz, Spain, p. 30, 2010.

Tenzer, P. E. and Mrad, R. B.: A systematic procedure for the design of piezoelectric inchworm precision positioners, IEEE-ASME T. Mech., 9, 427–435, 2004.

Xia, Q. X., Xie, S. W., and Huo, Y. L.: Numerical simulation and experimental research on the multi-pass neck-spinning of non-axisymmetric offset tube, J. Mater. Process. Tech., 206, 500–508, 2008.

Yokose, T., Hosaka, H., Yoshida, R., and Morita, T.: Resonance frequency ratio control with an additional inductor for a miniaturized resonant-type SIDM actuator, Sensors Actuat. A, 214, 142–148, 2014.

Yokozawa, H. and Morita, T.: Wireguide driving actuator using resonant-type smooth impact drive mechanism, Sensors Actuat. A, 230, 40–44, 2015.

Yukawa, T., Suzuki, M., and Satoh, Y.: Design of magnetic wheels in-pipe inspection robot, IEEE-ASME J. Microelectromech. S., 15, 1289–1298, 2006.

Zhang, W. J., Li, Q., and Guo, S. L.: Integrated Design of Mechanical Structure and Control Algorithm for a Programmable Four-Bar Linkage, IEEE-ASME T. Mech., 4, 354–362, 1999.

Zhang, Z. M., An, Q., Li, J. W., and Zhang, W. J.: Piezoelectric friction-inertia actuator – a critical review and future perspective, Int. J. Adv. Manuf. Tech., 62, 669–685, 2012.

19

A representation of the configurations and evolution of metamorphic mechanisms

W. Zhang[1,2], X. Ding[1], and J. Liu[2]

[1]School of Mechanical Engineering and Automation, Beihang University, Beijing, China
[2]State Key Laboratory of Robotics, Shenyang Institute of Automation, Chinese Academy of Sciences, Shenyang, China

Correspondence to: W. Zhang (zhangwuxiang@buaa.edu.cn)

Abstract. Metamorphic mechanisms are members of the class of mechanisms that are able to change their configurations sequentially to meet different requirements. The paper introduces a comprehensive symbolic matrix representation for characterizing the topology of one of these mechanisms in a single configuration using general information concerning links and joints. Furthermore, a matrix representation of an original metamorphic mechanism that has the ability to evolve is proposed by uniting the matrices representing all of the mechanism's possible configurations. The representation of metamorphic kinematic joints is developed in accordance with the variation laws of these mechanisms. By introducing the joint variation matrices derived from generalized operations on the related symbolic adjacency matrices, evolutionary relationships between mechanisms in adjacent configurations and the original metmaorphic mechanism are made distinctly. Examples are provided to demonstrate the validation of the method.

1 Introduction

In contrast to a traditional mechanism, a metamorphic mechanism is a mechanism with variable topological structures and it is a good approach for resolving the contradiction between economy, adaptation and efficiency. The concept of metamorphic mechanisms was first introduced based on the idea of reconfiguration in 1996 by Jian S. Dai and Rees Jones, which led to a new era of modern mechanism development (Dai and Rees Jones, 1998).

Research on the metamorphic mechanism has been making significant improvements in fundamentals and applications for nearly twenty years. The essence and characteristics of metamorphic mechanisms as well as three metamorphic approaches including variable components, adjacent relations and kinematic joints were introduced by Dai et al. (2005a) and Liu and Yang (2004). In addition, some of the basic constituent elements of these mechanisms, including links and their connectivity relationships, remain unchanged to give the mechanism's adjacent configuration complex coupling features. These two aspects are key factors affecting the study of methods for configuring metamorphic mechanisms (Zhang et al., 2011). Therefore, to create topological variations in the characteristics of mechanisms in different configurations, the appropriate structural representation for a metamorphic mechanism has been researched in recent years.

Mechanism diagrams, topological graphs and conventional adjacency matrices (Tsai, 2001) are simple and intuitive tools for describing the structure of a mechanism in a single configuration. Dai et al. (2005b) and Dai and Rees Jones (2005) were the first to propose an elementary transformation matrix that represented the variation of a mechanism using the adjacency matrix method. Wang and Dai (2007) introduced joint symbols into the adjacency matrix to express the variations of kinematic joints. In the matrix, all of the links were numbered sequentially and placed in principal diagonal positions; the off-diagonal elements were expressed using joint symbols that represented the connectivity relationship. Lan and Du (2008) used -1 as an element indicating a joint frozen into a new adjacency matrix to represent the topological changes of metamorphic mechanisms. Slaboch and Voglewede (2011) and Korves et al. (2012) proposed mechanism state matrices as a novel way to represent the topological characteristics of planar and spatial reconfigurable mechanisms. These matrices can be used as an analysis tool to automatically determine the degrees of freedom of planar mechanisms that only contain one degree of freedom (DOF) joint. Herve (2006) showed how to create translational parallel manipulators using Lie-group algebra, which can give reference to the related research. Yan and Kang (2009) showed how to perform configuration synthesis of mechanisms with variable topologies using graph theory.

However, the axial orientation of a joint and information on link variations were not epitomized in the aforementioned research. Therefore, Yang (2004) introduced the concept of a geometric constraint for expressing the relative positions and orientations of the joint axes and generalized it into six types: parallelism, coincidence, intersection, perpendicularity, coplanarity and randomness. Li et al. (2010) suggested using a constraint graph from computational geometry rather than the traditional topological graph to characterize a metamorphic linkage to simplify the representation of its configuration changes. The adjacency submatrix of the constraint graph provides a convenient description of changes in the topology of links and joints in the operation of the metamorphic linkage. Li et al. (2009) and Li and Dai (2010a, b) developed a topological representation matrix with information on loops, types of links and joints that included orientation information, which has been used in subsequent research. They also introduced a joint-orientation interchanging metamorphic method based on the matrix.

This paper presents a novel method of characterizing the topology of metamorphic mechanisms in all configurations that involves information about links and joints, including their types and axial orientations. Furthermore, a method of constructing an original matrix that represents the original metamorphic mechanism is proposed. Next, the paper proposes two matrix operations that are useful for representing topological changes and evolving features.

2 Configuration characteristics of metamorphic mechanisms

A metamorphic mechanism is a mechanism with variable topological structures that can be transformed from one structure to another continuously. There are variable parts and coupling parts, giving the metamorphic mechanism a variable topological structure and coupling relationship. In particular, variability is the distinguishing feature that separates metamorphic mechanisms from common mechanisms; this is an important area of research. The incorporation of links, the changing relationships of adjacent links and the changing properties of kinematic pairs have been explored to summarize the variable features of metamorphic mechanisms. In essence, a metamorphic kinematic joint is the essential prerequisite for changing the number of and connective relationships among its active links, leading to a transformation of the configuration of the entire mechanism (Zhang et al., 2011).

An example of a planar five-bar metamorphic linkage which has five configurations is shown in Fig. 1. Transformations between them are performed by locking different kinematic joints sequentially. When the mechanism is in configuration 1, as shown in Fig. 1a, slider c is locked at the top end of the slot in link d. In this configuration, the mechanism can be treated as a four-bar mechanism. When the mechanism is in configuration 2, as shown in Fig. 1b, links a and b are fixed together by locking the revolute joint B as well as slider c and link d are unlocked. Therefore the mechanism is transformed into a guide bar mechanism. When revolute joint C is locked, links b and c are fixed to transform mechanism into another guide bar mechanism, as shown in Fig. 1c. When the mechanism is in configuration 4, links a and e are fixed together by locking revolute joint A, as shown in Fig. 1d. Link b becomes the driving link of the mechanism. When link d arrives at the location shown in Fig. 1e, joint D is locked to fix links d and e. This transforms mechanism into a crank slider mechanism. Therefore, the mechanism realizes transformations between different configurations by locking its kinematic joints in particular sequences.

From Fig. 1, we conclude that the structure of the mechanism can be transformed from one to another by locking different kinematic joints accordingly. By applying modes such as the geometric limit, force limit, and variation of the driving kinematic joint, the working conditions of these kinematic joints can be switched between active and locked states. In addition, metamorphic kinematic joints are able to change their types and motion orientations to realize configuration transformations (Yan and Kuo, 2006). In a metamorphic mechanism, there is at least one metamorphic kinematic joint, which can change the number and connectivity relationships of active links. There are some basic constituent elements, including links and their connectivity relationships, which remain unchanged to create complex coupling features among the links in adjacent configurations, as shown in Fig. 1.

Therefore, to understand the configuration characteristics of metamorphic mechanisms, it is necessary to present a configuration representation that can express not only the characteristics of the mechanism in all of its configurations but also the variations during the transformation process intuitively with the help of specific operations.

3 Representing the configurations of metamorphic mechanisms

It is known that an adjacency matrix can be used to represent the topological structures of metamorphic mechanisms. This matrix and an EU-elementary matrix operation were introduced for expressing a configuration transformation (Dai et al., 2005b; Dai and Rees Jones, 2005). Furthermore, a sym-

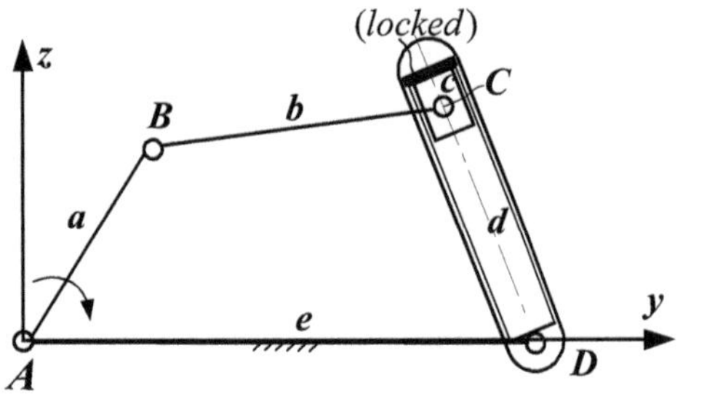

(a) The mechanism in configuration 1

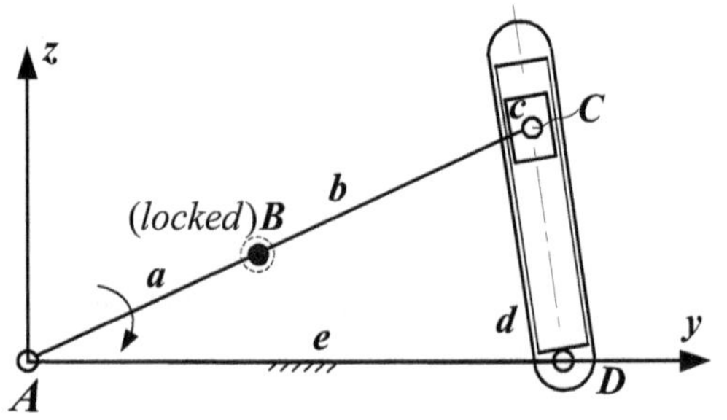

(b) The mechanism in configuration 2

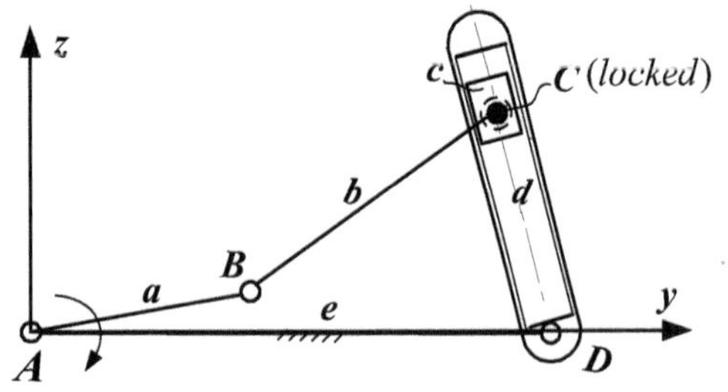

(c) The mechanism in configuration 3

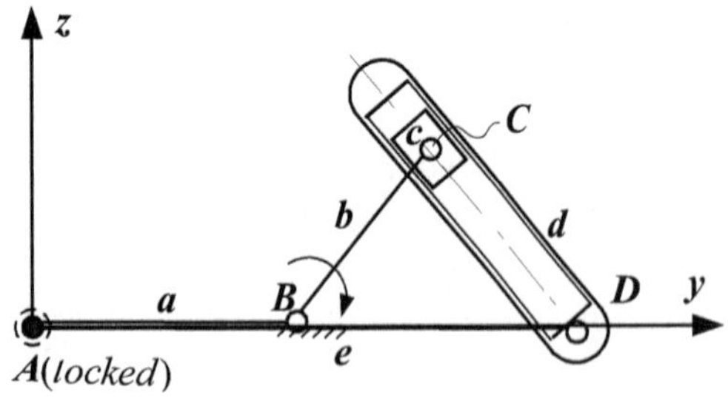

(d) The mechanism in configuration 4

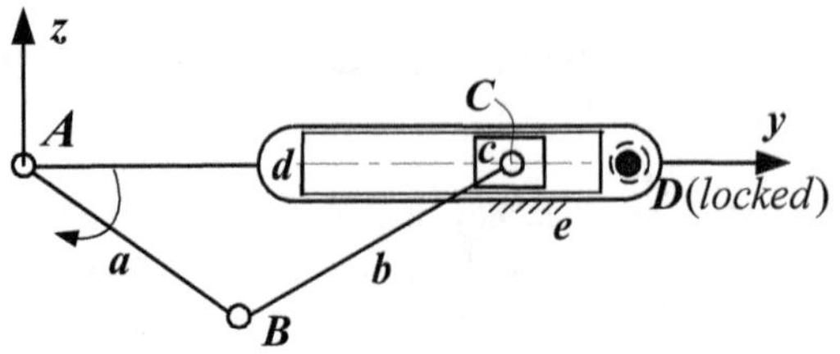

(e) The mechanism in configuration 5

Figure 1. A five-bar planar metamorphic linkage.

bolic adjacency matrix was constructed by introducing information on link variations and joint orientations (Li and Dai, 2010a; Zhang and Ding, 2012). The variations and coupling features of the metamorphic mechanism in adjacent configurations can be determined by applying the generalized difference and intersection operations to the corresponding symbolic matrices.

However, the matrices representing mechanisms in different configurations do not have the same dimension and need to be normalized, increasing the complexity of the representation and operations. Simultaneously, the upper off-diagonal elements in the matrix are the same as the lower off-diagonal elements, which means that the matrix contains information in duplicate. Therefore, to decrease the complexity of expressing the matrix and subsequent operations on it, we improve the symbolic matrix for the mechanism in configuration m and express it as follows:

$$\mathbf{A}^{(m)} = \begin{pmatrix} L_1 & J_{1,2}^{(m)} & \cdots & J_{1,i}^{(m)} & \cdots & J_{1,k-1}^{(m)} & J_{1,k}^{(m)} \\ a_{2,1}^{(m)} & L_2 & \cdots & J_{2,i}^{(m)} & \cdots & J_{2,k-1}^{(m)} & J_{2,k}^{(m)} \\ \vdots & \vdots & \ddots & \vdots & \vdots & \vdots & \vdots \\ a_{i,1}^{(m)} & \cdots & \cdots & L_i & \cdots & \cdots & J_{i,k}^{(m)} \\ \vdots & \vdots & \vdots & \vdots & \ddots & \vdots & \vdots \\ a_{k-1,1}^{(m)} & a_{k-1,2}^{(m)} & \cdots & a_{k-1,i}^{(m)} & \cdots & L_{k-1} & J_{k-1,k}^{(m)} \\ a_{k,1}^{(m)} & a_{k,2}^{(m)} & \cdots & a_{k,i}^{(m)} & \cdots & a_{k,k-1}^{(m)} & L_k \end{pmatrix}, \quad (1)$$

where the principal diagonal element L_i represents the link whose sequence number in the mechanism is i. The numbers of rows and columns are both k, which indicates the number of links in all configurations. Normally, k is greater than or equal to the maximum number of effective links in every configuration. The upper off-diagonal element $J_{i,j}^{(m)}$ denotes the connectivity relationship between links L_i and L_j. It can be represented by a symbol with subscript where the symbol denotes the joint type and the subscript expresses the geometric constraint relationship of the joint axes located at the ends of the link. It is noted that the rule is also applicable for analyzing tertiary links for the essence of the proposed matrix is to record the connectivity relationship between links. A special element -1 is employed here to represent a frozen joint between two links (Lan and Du, 2008) and the element 0 represents the two links that are not connected. Specific expressions for the geometric constraints, including parallelism, intersection, coincidence, perpendicularity and randomicity, are given in Li et al. (2009). The lower off-diagonal element $a_{j,i}^{(m)}$ is the sequence number of the configuration if the state of the corresponding upper off-diagonal element $J_{i,j}^{(m)}$ is changed in configuration m. It should be noted that, for the first configuration matrix $\mathbf{A}^{(1)}$, if there is a joint constraint between links i and j, the value of $a_{j,i}^{(1)}$ is 1. If there is no such constraint, its value is 0. For a matrix $\mathbf{A}^{(m)}$, when only the joint constraint between links i and j is changed in configuration m, the value of $a_{j,i}^{(m)}$ is m. This value is also assigned to the other lower off-diagonal elements to be consistent with the corresponding elements in the previous matrix, $\mathbf{A}^{(m-1)}$. Therefore, dimensional consistency of the matrices for the mechanisms in different configurations is one of the advantages of the proposed symbolic matrix representation. In addition, the symbolic matrix can describe the connectivity relationship of all links synthetically as well as their corresponding variations and provides sufficient information for the subsequent matrix operations.

By applying Eq. (1), we express the five-bar metamorphic linkage, which has the five configurations shown in Fig. 1, as

$$\mathbf{A}^{(1)} = \begin{pmatrix} e & R & 0 & 0 & R_{\parallel R} \\ 1 & a & R_{\parallel R} & 0 & 0 \\ 0 & 1 & b & R_{\parallel R} & 0 \\ 0 & 0 & 1 & c & -1 \\ 1 & 0 & 0 & 1 & d \end{pmatrix},$$
$$\mathbf{A}^{(2)} = \begin{pmatrix} e & R & 0 & 0 & R_{\parallel R} \\ 1 & a & -1 & 0 & 0 \\ 0 & 2 & b & R_{\parallel R} & 0 \\ 0 & 0 & 1 & c & P_{\perp R} \\ 1 & 0 & 0 & 2 & d \end{pmatrix},$$
$$\mathbf{A}^{(3)} = \begin{pmatrix} e & R & 0 & 0 & R_{\parallel R} \\ 1 & a & R_{\parallel R} & 0 & 0 \\ 0 & 3 & b & -1 & 0 \\ 0 & 0 & 3 & c & P_{\perp R} \\ 1 & 0 & 0 & 2 & d \end{pmatrix},$$
$$\mathbf{A}^{(4)} = \begin{pmatrix} e & -1 & 0 & 0 & R_{\parallel R} \\ 4 & a & R & 0 & 0 \\ 0 & 3 & b & R_{\parallel R} & 0 \\ 0 & 0 & 4 & c & P_{\perp R} \\ 1 & 0 & 0 & 2 & d \end{pmatrix},$$
$$\mathbf{A}^{(5)} = \begin{pmatrix} e & R & 0 & 0 & -1 \\ 5 & a & R_{\parallel R} & 0 & 0 \\ 0 & 3 & b & R_{\parallel R} & 0 \\ 0 & 0 & 4 & c & P_{\perp R} \\ 5 & 0 & 0 & 2 & d \end{pmatrix}, \qquad (2)$$

following the configuration transformation sequence. The numbers of rows and columns in all of these matrices are 5, a result that depends on the number of links a, b, c, d, and e occurring in these five configurations. The upper and lower off-diagonal elements record information on the joint constraints and their variations. For example, comparing $A^{(4)}$ and $A^{(5)}$, the elements $J_{1,2}^{(4)}, J_{1,2}^{(5)}$ and $J_{1,5}^{(4)}, J_{1,5}^{(5)}$ differ because they show that joint A and joint D have changed from -1 to R and $R_{\parallel R}$ to -1, respectively. Meanwhile, the corresponding elements $a_{2,1}^{(5)}$ and $a_{5,1}^{(5)}$ have changed from 4 to 5 and 1 to 5 to record the sequence number of the configuration in which joints A and D are in these positions in a working cycle.

4 Matrix operations for metamorphic mechanisms

The proposed symbolic matrix describes the topology of the mechanism in a single configuration. However, exploring the variation laws of these mechanisms in different configurations is very important for developing novel metamorphic mechanisms. Therefore, it is feasible to take advantage of matrix operations for constructing the original metamorphic mechanism and determining the features of its topological variations.

4.1 Constructing the original metamorphic mechanism

The original metamorphic mechanism is able to evolve into any configuration of the mechanism and contains all of the topological elements found in all of configurations in a working cycle. A method for constructing original metamorphic mechanisms from biological modeling and genetic evolution was introduced in Wang and Dai (2007) and Zhang et al. (2008). In this paper, based on Eq. (3), an original matrix $\mathbf{A}^{(0)}$ for representing the original metamorphic mechanism is given by

$$\mathbf{A}^{(0)} = \mathbf{A}^{(1)} \cup \mathbf{A}^{(2)} \cup \cdots \cup \mathbf{A}^{(m)} \cup \cdots \cup \mathbf{A}^{(n)}$$
$$= \begin{pmatrix} L_1 & \cdots & \cdots & \cdots & \cdots & \cdots & \cdots \\ \vdots & \ddots & \vdots & \vdots & \vdots & \vdots & \vdots \\ \cdots & \cdots & L_i & \cdots & \prod_{m=1}^{n-1} J_{i,j}^{(m)} \cup J_{i,j}^{(m+1)} & \cdots & \cdots \\ \vdots & \vdots & \vdots & \ddots & \vdots & \vdots & \vdots \\ \cdots & \cdots & \left\{a_{j,i}^{(1)}, \cdots, a_{j,i}^{(m)}, \cdots, a_{j,i}^{(n)}\right\} & \cdots & L_j & \cdots & \cdots \\ \vdots & \vdots & \vdots & \vdots & \vdots & \ddots & \vdots \\ \cdots & \cdots & \cdots & \cdots & \cdots & \cdots & L_k \end{pmatrix}, \qquad (3)$$

where the operator $\cup$ represents the union of its arguments. The result, $\mathbf{A}^{(0)}$, has the same form as Eq. (1). All of the elements located in the same position in the set of related matrices from $\mathbf{A}^{(1)}$ to $\mathbf{A}^{(n)}$ gradually become united, as shown in Eq. (3). Details of the operative principles are as follows:

1. The principal diagonal elements of $\mathbf{A}^{(0)}$ are the same as those of $\mathbf{A}^{(i)}$ $(i = 1, \ldots, n)$, indicating the links remain unchanged.

2. The operation that unites the lower off-diagonal elements and records the sequence numbers of the configuration is performed by uniting the elements in these matrices as a set of results in $\mathbf{A}^{(0)}$, which can be expressed as

$$\mathbf{A}^{(0)}(j,i) = \left\{a_{j,i}^{(1)}, \ldots, a_{j,i}^{(m)}, \ldots, a_{j,i}^{(n)}\right\} \ (i < j \leq k), \qquad (4)$$

where the number 0 is ignored. If the values of the adjacent elements are same in this set, only one of them should be kept. The information given by this set is very helpful for constructing the matrices for a single configuration of the mechanism.

3. The physical meaning of uniting the upper off-diagonal elements of these matrices is to achieve the most variability in the kinematic joints. The operation starts from the upper off-diagonal elements in the first matrix, $\mathbf{A}^{(1)}$; then, the joint type and orientation are expanded based on the elements of the next adjacency configuration matrix in the sequence. We express the operation as

$$\begin{aligned}\mathbf{A}^{(0)}(i,j) &= \prod_{m=1}^{k-1} \mathbf{A}^{(m)}(i,j) \cup \mathbf{A}^{(m+1)}(i,j) \\ &= \prod_{m=1}^{k-1} J_{i,j}^{(m)} \cup J_{i,j}^{(m+1)} \quad (i < j \leq k). \end{aligned} \tag{5}$$

Basically, the uniting operator is equivalent to an extension of the type and axial orientation of a kinematic joint. If the adjacent elements are same, it represents the corresponding connectivity relationship between the related links keeps unchanged. So these same numbers in the operation result need to be omitted just keeping one.

For example, according to Eqs. (2)–(5), elements $A^{(0)}(4,3)$ and $A^{(0)}(3,4)$ of matrix $A^{(0)}$ can be calculated as follows:

$$A^{(0)}(4,3) = \{1,1,3,4,4\} = \{1,3,4\} \tag{6}$$

$$\begin{aligned}A^{(0)}(3,4) &= \prod_{m=1}^{4} J_{3,4}^{(m)} \cup J_{3,4}^{(m+1)} \\ &= R_{\parallel R} \cup R_{\parallel R} \cup -1 \cup R_{\parallel R} \cup R_{\parallel R} \\ &= R_{\parallel R} \cup -1 \cup R_{\parallel R}. \end{aligned} \tag{7}$$

Therefore, the joint between links b and c changes twice during the configuration transformations from 1 to 3 and from 3 to 4 while its axial orientation remains unchanged during the working cycle.

The construction procedure of the metamorphic kinematic joints can be illustrated in Fig. 2. Firstly, the joints should be listed according to the sequence indicated in the corresponding operation result. Further, geometric limit is used to realize the transformations between these adjacent joints in sequence. Geometric limit is a most common way of making the type of kinematic joints to be changed by releasing or adding appropriate constraints at suitable geometric locations. Such as in Fig. 2a, the kinematic joint between links a and b is a revolute joint whose axis is parallel to the adjacent revolute joint, R_1. In the next configuration, the revolute joint is locked. Therefore, two limiting stoppers are laid on the two links a and b, respectively. When the two stoppers are contacted, the two links are fixed together and the number of DOF of the revolute joint is changed to zero in Fig. 2a. Figure 2b shows that the joint is performing translating motions with arrows denoting the direction of pin's motion and indicating the number of DOFs the joint possesses. When the pin reaches the position shown in the second figure, it stops translating but remains rotating as shown. This is identified as a typical metamorphic kinematic joint that varies from a prismatic joint to a rotating pair. Similarly, Fig. 2c demonstrates a series of varying orientations of a revolute pair undergoing the orientations about different axes, successively.

According to the construction process described above, the matrix of the original metamorphic mechanism for the five-bar metamorphic linkage shown in Fig. 1 is

$$\mathbf{A}_0 = \begin{pmatrix} e & R \cup -1 \cup R & 0 & 0 & R_{\parallel R} \cup -1 \\ \{1,5\} & a & R_{\parallel R} \cup -1 \cup R_{\parallel R} & 0 & 0 \\ 0 & \{1,2,3\} & b & R_{\parallel R} \cup -1 \cup R_{\parallel R} & 0 \\ 0 & 0 & \{1,3,4\} & c & -1 \cup P_{\perp R} \\ \{1,5\} & 0 & 0 & \{1,2\} & d \end{pmatrix}. \tag{8}$$

Therefore, the original metamorphic mechanism can be generated by applying the uniting operator to all of the mechanism's configurations and using link and joint information. In particular, the matrix which includes the information of all links and their connectivity relationships can make us identify all possible combinations between links for creating different mechanisms. So the mechanism is helpful to develop novel metamorphic mechanisms using the representation method.

4.2 The joint variation matrix

The essential method for realizing configuration transformation of metamorphic mechanisms is to change the characteristics of kinematic joints, which lead to variations in the topology of the entire mechanism. Therefore, to determine the joint variation rule for two adjacent configurations of a mechanism, a generalized difference operation for two adjacency matrices is proposed.

Let $\mathbf{A}_{\text{var}}^{(m+1,m)}$ be the joint variation matrix, which can be described as the result of applying the generalized difference operator to the topological representation matrices $\mathbf{A}^{(m+1)}$ and $\mathbf{A}^{(m)}$, that is

$$\mathbf{A}_{\text{var}}^{(m+1,m)} = \mathbf{A}^{(m+1)} - \mathbf{A}^{(m)}, \tag{9}$$

where – represents the generalized difference operator (Lan and Du, 2008; Li et al., 2010). The resulting matrix contains information about the joint variation when the mechanism is transformed from configuration m to configuration $m+1$. If the mechanism is transformed from configuration $m+1$ to configuration m, the joint variation matrix $\mathbf{A}_{\text{var}}^{(m,m+1)}$ can be expressed as

$$\mathbf{A}_{\text{var}}^{(m,m+1)} = \mathbf{A}^{(m)} - \mathbf{A}^{(m+1)}. \tag{10}$$

The purpose of this operation is to record variations in the upper and lower off-diagonal elements. And the joint variation matrix achieved is the key procedure for constructing the matrix represents the original metamorphic mechanism which will be discussed in Sect. 4.3. If the two elements located at the same position in matrices $\mathbf{A}^{(m)}$ and $\mathbf{A}^{(m+1)}$ are

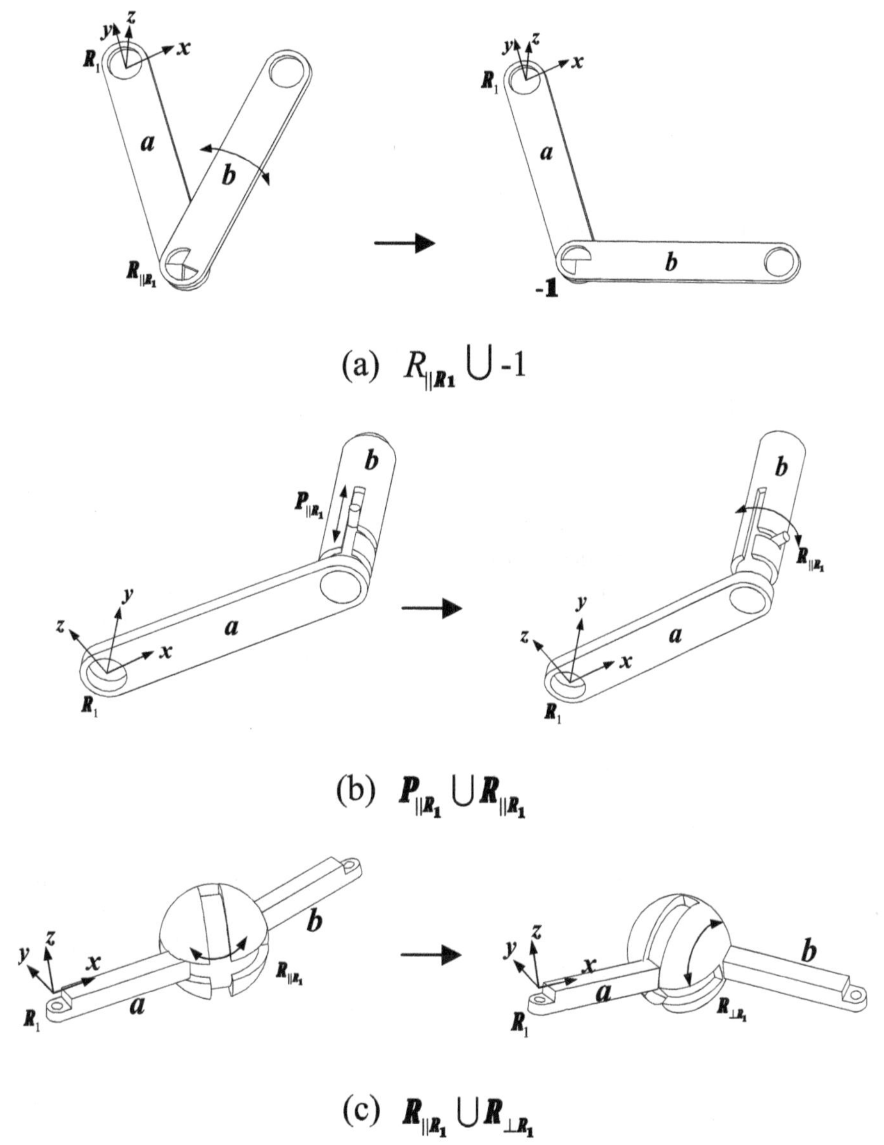

Figure 2. The result of uniting different kinematic joints.

equal, the corresponding element in matrix $\mathbf{A}_{\text{var}}^{(m+1,m)}$ is assigned the number 0. Conversely, the elements in the minuend matrix are reserved directly, and the principal diagonal elements remain unchanged for recognition purposes. The same rule is used for the lower off-diagonal elements. If an upper off-diagonal element is unchanged, the corresponding lower off-diagonal element needs to be assigned the value 0 regardless of its actual value. The joint variation matrix can be constructed directly from the physical meaning of the joint variation rule. In addition, analysing the existing joints with the characteristic of metamorphosis is one of the most important approaches for achieving the principle of constructing the corresponding joint variation matrix.

Therefore, joint variation matrices for configurations 1 to 5 are given as follows:

$$\mathbf{A}_{\text{var}}^{(2,1)} = \begin{pmatrix} e & 0 & 0 & 0 & 0 \\ 0 & a & -1 & 0 & 0 \\ 0 & 2 & b & 0 & 0 \\ 0 & 0 & 0 & c & P_{\perp R} \\ 0 & 0 & 0 & 2 & d \end{pmatrix},$$

$$\mathbf{A}_{\text{var}}^{(3,2)} = \begin{pmatrix} e & 0 & 0 & 0 & 0 \\ 0 & a & R_{\parallel R} & 0 & 0 \\ 0 & 3 & b & -1 & 0 \\ 0 & 0 & 3 & c & 0 \\ 0 & 0 & 0 & 0 & d \end{pmatrix},$$

$$\mathbf{A}_{\text{var}}^{(4,3)} = \begin{pmatrix} e & -1 & 0 & 0 & 0 \\ 4 & a & 0 & 0 & 0 \\ 0 & 0 & b & R_{\parallel R} & 0 \\ 0 & 0 & 4 & c & 0 \\ 0 & 0 & 0 & 0 & d \end{pmatrix},$$

$$\mathbf{A}_{\text{var}}^{(5,4)} = \begin{pmatrix} e & R & 0 & 0 & -1 \\ 5 & a & 0 & 0 & 0 \\ 0 & 0 & b & 0 & 0 \\ 0 & 0 & 0 & c & 0 \\ 5 & 0 & 0 & 0 & d \end{pmatrix}. \tag{11}$$

4.3 The relationship between the original metamorphic mechanism and the mechanism in any configuration

Because an original metamorphic mechanism provides a foundation for a mechanism to transform itself into any configuration and expresses the joint variation characteristics from the symbolic adjacency matrices and the corresponding operations, the relationships between these matrices is as shown in Fig. 3.

1. The relationship between adjacent configurations: the two adjacent matrices shown in Fig. 3 can be transformed into each other using a joint variation matrix. From Eq. (9), the matrix $\mathbf{A}^{(m+1)}$ can be expressed as

$$\mathbf{A}^{(m+1)} = \mathbf{A}^{(m)} + \mathbf{A}_{\text{var}}^{(m+1,m)}, \tag{12}$$

where $+$ represents the generalized addition operator, which changes the elements in matrix $\mathbf{A}^{(m)}$ according to the corresponding elements in the joint variation matrix of $\mathbf{A}_{\text{var}}^{(m+1,m)}$. Comparing the corresponding elements in the two matrices, the lower off-diagonal elements in $\mathbf{A}_{\text{var}}^{(m+1,m)}$ containing the value m are selected, with the corresponding symmetrical upper triangular elements, to replace the corresponding elements in matrix $\mathbf{A}^{(m)}$ while leaving the other elements unchanged. Similarly, matrix $\mathbf{A}^{(m)}$ can be expressed as

$$\mathbf{A}^{(m)} = \mathbf{A}^{(m+1)} + \mathbf{A}_{\text{var}}^{(m,m+1)}. \tag{13}$$

For example, the relationship between matrices $A^{(1)}$ and $A^{(2)}$ is

$$\mathbf{A}^{(2)} = \mathbf{A}^{(1)} + \mathbf{A}_{\text{var}}^{(2,1)} \tag{14}$$

$$\mathbf{A}^{(1)} = \mathbf{A}^{(2)} + \mathbf{A}_{\text{var}}^{(1,2)}. \tag{15}$$

2. The relationships of the original metamorphic mechanism and the mechanism in a single configuration: the original metamorphic mechanism is able to evolve into any configuration. Therefore, the information on the mechanism in configuration m can be extracted from the matrix $\mathbf{A}^{(0)}$ to construct the corresponding matrix $\mathbf{A}^{(m)}$. The process of evolution from $\mathbf{A}^{(0)}$ to $\mathbf{A}^{(m)}$ follows from Eq. (3).

First, the principal diagonal elements denoting the links in $\mathbf{A}^{(0)}$ are placed in their corresponding positions in $\mathbf{A}^{(m)}$ directly. Then, the lower off-diagonal elements containing the value 1 and their corresponding upper off-diagonal elements, which represent constraints on the joints of links in matrix

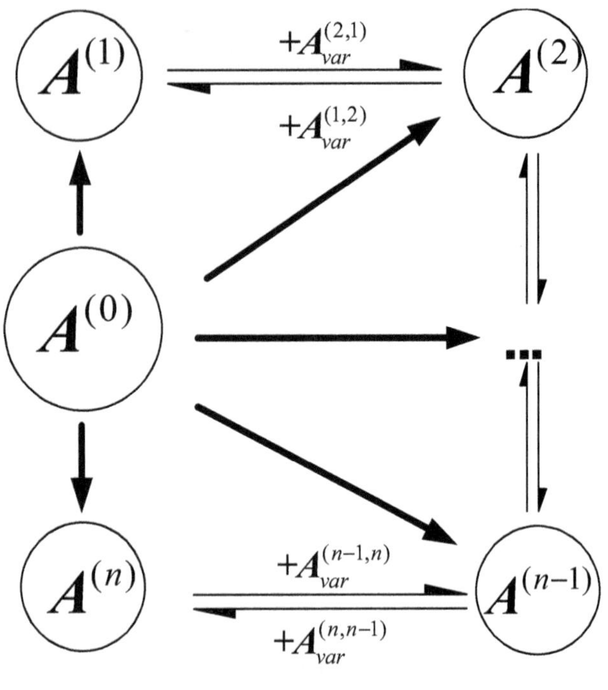

Figure 3. The relationship between the original metamorphic mechanism and the mechanism in any configuration.

$\mathbf{A}^{(0)}$, are similarly mapped to positions in $\mathbf{A}^{(m)}$ as long as the value of the corresponding element is not m. The next important step is to select a number m from the elements comprising sets of numbers and then, to identify its sequence number in the set $\{a_{j,i}^{(1)}, \ldots, a_{j,i}^{(m)}, \ldots, a_{j,i}^{(n)}\}$. The sequence number can be used to determine the corresponding joint constraint conveniently using the element $\prod_{m=1}^{n-1} J_{i,j}^{(m)} \cup J_{i,j}^{(m+1)}$. These elements are then placed into $\mathbf{A}^{(m)}$, the other elements of which are assigned a value of 0.

For example, the elements marked by black triangles ▼ in Eq. (16) are extracted to construct the matrix $\mathbf{A}^{(2)}$, which represents the topology of the mechanism in configuration 2 according to the above procedure.

$$\mathbf{A}_0 = \begin{pmatrix} \underset{\blacktriangledown}{e} & \underset{\blacktriangledown}{R} \cup -1 \cup R & 0 & 0 & \underset{\blacktriangledown}{R_{\parallel R}} \cup -1 \\ \{\underset{\blacktriangledown}{1}, 5\} & \underset{\blacktriangledown}{a} & R_{\parallel R} \cup \underset{\blacktriangledown}{-1} \cup R_{\parallel R} & 0 & 0 \\ 0 & \{1, \underset{\blacktriangledown}{2}, 3\} & \underset{\blacktriangledown}{b} & \underset{\blacktriangledown}{R_{\parallel R}} \cup -1 \cup R_{\parallel R} & 0 \\ 0 & 0 & \{\underset{\blacktriangledown}{1}, 3, 4\} & \underset{\blacktriangledown}{c} & -1 \cup P_{\underset{\blacktriangledown}{\perp} R} \\ \{\underset{\blacktriangledown}{1}, 5\} & 0 & 0 & \{1, \underset{\blacktriangledown}{2}\} & \underset{\blacktriangledown}{d} \end{pmatrix} \tag{16}$$

The diagram in Fig. 3 shows that the evolutionary relationships between the original metamorphic mechanism and all of its configurations can be determined by applying matrix operations to the appropriate matrices.

5 Case study

A spatial four-bar metamorphic mechanism that has two configurations is shown in Fig. 4. When the mechanism is in configuration 1, as shown in Fig. 4a, it can be treated as an

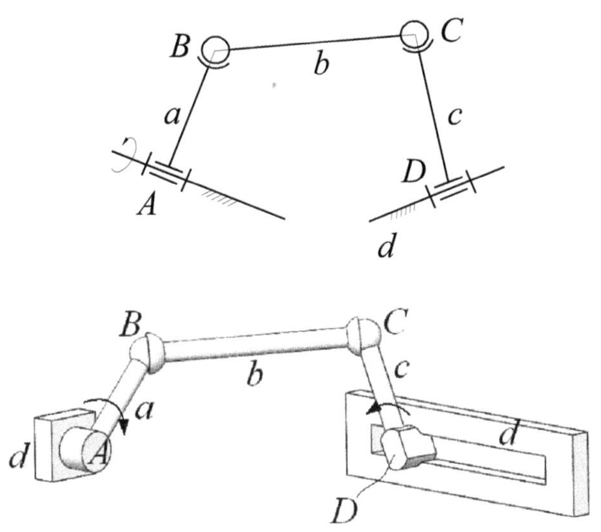

(a) The mechanism in configuration 1

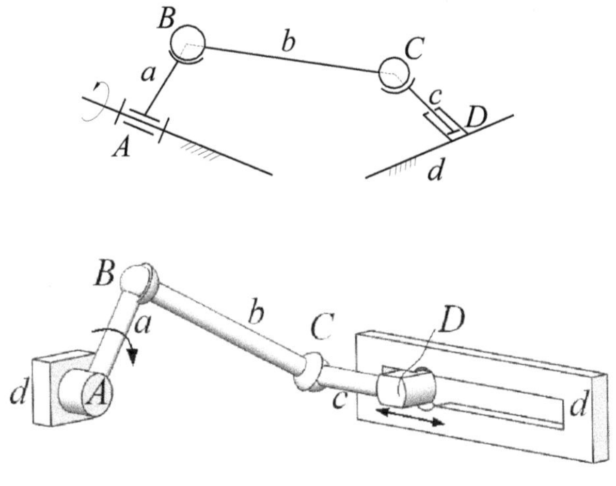

(b) The mechanism in configuration 2

Figure 4. A four-bar spatial metamorphic mechanism.

RSSR mechanism. The axis of joint D between links c and d is perpendicular to the axis of joint A between links a and d. When revolute joint D is transformed into a prismatic joint, the mechanism becomes an RSSP mechanism, as shown in Fig. 4b.

The topological structures of the metamorphic mechanism can be expressed in matrix form as follows:

$$\mathbf{A}^{(1)} = \begin{pmatrix} d & R & 0 & R_{\perp R} \\ 1 & a & S & 0 \\ 0 & 1 & b & S \\ 1 & 0 & 1 & c \end{pmatrix} \tag{17}$$

$$\mathbf{A}^{(2)} = \begin{pmatrix} d & R & 0 & P_{\parallel R} \\ 1 & a & S & 0 \\ 0 & 1 & b & S \\ 2 & 0 & 1 & c \end{pmatrix}. \tag{18}$$

The origin matrix of the original metamorphic mechanism and the joint variation matrix can be expressed as

$$\mathbf{A}_{\mathrm{var}}^{(2,1)} = \mathbf{A}^{(2)} - \mathbf{A}^{(1)} = \begin{pmatrix} d & 0 & 0 & R_{\perp R} \\ 0 & a & 0 & 0 \\ 0 & 0 & b & 0 \\ 2 & 0 & 0 & c \end{pmatrix} \tag{19}$$

$$\mathbf{A}^{(0)} = \mathbf{A}^{(1)} \cup \mathbf{A}^{(2)} = \begin{pmatrix} d & R & 0 & R_{\perp R} \cup P_{\parallel R} \\ 1 & a & S & 0 \\ 0 & 1 & b & S \\ \{1,2\} & 0 & 1 & c \end{pmatrix}. \tag{20}$$

The element $R_{\perp R} \cup P_{\parallel R}$ in matrix $\mathbf{A}^{(0)}$ represents the way in which both the axial orientation and the type of joint D have changed. There, the joint can be considered a metamorphic kinematic joint and be developed according to the variation sequence for the kinematic behaviours of the entire mechanism.

6 Conclusions

The paper proposed a comprehensive symbolic matrix for characterizing the topology of a metamorphic mechanism that involved information on the variations of links and the axial orientations of the kinematic joints. In addition, operations on the matrices of the adjacent configuration mechanisms are defined to construct an origin matrix and joint variation matrices. In particular, the construction and evolution of the matrix representation for an original metamorphic mechanism show how it can be transformed into any configuration matrix. The relationship between the original metamorphic mechanism and all of its possible configurations and methods of moving between them were presented. Examples illustrate the effectiveness of this approach in characterizing metamorphic mechanisms. The configuration representation of metamorphic mechanisms provides a foundation for the analysis and synthesis of novel metamorphic mechanisms.

Acknowledgements. The authors gratefully acknowledge the support of the National Natural Science Foundation of China (Project No. 51575018, No. 51275015 and No. 51175494) and the Foundation of State Key Laboratory of Robotics (No. 2014).

References

Dai, J. S. and Rees Jones, J.: Mobility in metamorphic mechanisms of foldable/erectable kind, in: Proceedings of the 25th ASME Biennial Mechanisms and Robotics Conference, Baltimore, 1998.

Dai, J. S. and Rees Jones, J.: Matrix representation of topological changes in metamorphic mechanisms, ASME Trans. J. Mech. Design, 127, 610–619, 2005.

Dai, J. S., Ding, X. L., and Zou, H. J.: Fundamentals and categorization of metamorphic mechanisms, Chinese J. Mech. Eng., 41, 7–12, 2005a.

Dai, J. S., Ding, X. L., and Wang, D. L.: Topological changes and the corresponding matrix operations of a spatial metamorphic mechanism, Chinese J. Mech. Eng., 41, 30–35, 2005b.

Herve, J. M.: Translational parallel manipulators with douple planar limbs, Mech. Mach. Theory, 41, 433–455, 2006.

Korves, B. A., Slaboch, B. J., and Voglewede, P. A.: Mechanism state matrices for spatial reconfigurable mechanisms, in: Proceedings of the ASME 2012 International Design Engineering Technical Conferences & Computers and Information in Engineering Conference, Chicago, 2012.

Lan, Z. H. and Du, R.: Representation of Topological Changes in Metamorphic Mechanisms with Matrices of the Same Dimension, ASME Trans. J. Mech. Design, 130, 074501-1–074501-4, 2008.

Li, D. L., Zhang, Z. H., and McCarthy, J. M.: A constraint graph representation of metamorphic linkages, Mech. Mach. Theory, 4, 228–238, 2010.

Li, S. J. and Dai, J. S.: Configuration transformation matrix of metamorphic mechanisms and joint-orientation change metamorphic method, China Mech. Eng., 21, 1698–1703, 2010a.

Li, S. J. and Dai, J. S.: Structure of Metamorphic Mechanisms Based on Augmented Assur Groups, China Mech. Eng., 46, 22–41, 2010b.

Li, S. J., Wang, D. L., and Dai, J. S.: Topology of kinematic chains with loops and orientation of joints axes, Chinese J. Mech. Eng., 45, 34–40, 2009.

Liu, C. H. and Yang, T. L.: Essence and characteristics of metamorphic mechanism and their metamorphic ways, in: Proceedings of the 11th World Congress in Mechanism and Machine Science, Tianjin, 2004.

Slaboch, B. and Voglewede, P.: Mechanism state matrices for planar reconfigurable mechanisms, ASME Trans. J. Mech. Robot., 3, 011012-1–011012-7, 2011.

Tsai, L.-W.: Mechanism Design: Enumeration of kinematic structures according to function. CRC Press LLC, Boca Raton, FL, 2001.

Wang, D. L. and Dai, J. S.: Theoretical foundation of metamorphic mechanism and its synthesis, Chinese J. Mech. Eng., 43, 32–42, 2007.

Yan, H. S. and Kuo, C. H.: Topological representations and characteristics of variable kinematics joints, ASME Trans. J. Mech. Design, 128, 384–391, 2006.

Yan, H. S. and Kang, C.-H.: Configuration synthesis of mechanisms with variable topologies, Mech. Mach. Theory, 44, 896–911, 2009.

Yang, T. L.: Topology structure design of robot mechanisms, China Machine Press, 2004.

Zhang, L. P., Wang, D. L., and Dai, J. S.: Biological modeling and evolution based synthesis of metamorphic mechanisms, ASME Trans. J. Mech. Design, 30, 1–11, 2008.

Zhang, W. X. and Ding, X. L.: A method for configuration representation of metamorphic mechanism with information of component variation, in: Advances in Reconfigurable Mechanisms and Robots I, edited by: Dai, J. S., Zoppi, M., and Kong, X., Springer, London, 2012.

Zhang, W. X., Ding, X. L., and Dai, J. S.: Morphological synthesis of metamorphic mechanisms based on Constraint Variation, Proceedings of the Institution of Mechanical Engineers, Part C, J. Mech. Eng. Sci., 225, 2997–3010, 2011.

20

Output decoupling property of planar flexure-based compliant mechanisms with symmetric configuration

Y. S. Du[1,2], T. M. Li[1,2], Y. Jiang[1,2], and J. L. Zhang[1,2]

[1]Manufacturing Engineering Institute, Department of Mechanical Engineering, Tsinghua University, Beijing, 100084, China

[2]Beijing Key Lab of Precision/Ultra-precision Manufacturing Equipment and Control, Tsinghua University, Beijing, 100084, China

Correspondence to: T. M. Li (litm@mail.tsinghua.edu.cn)

Abstract. This paper presents the output decoupling property of planar flexure-based compliant mechanisms with symmetric configuration. Compliance/stiffness modeling methods for flexure serial structures and flexure parallel structures are first derived according to the matrix method. Analytical model of mechanisms with symmetric configuration is then developed to analyze the output decoupling property. The proposed analytical model shows that mechanisms are output decoupled when they are symmetry about two perpendicular axes or when they are composed of either three or an even number of identical fundamental forms distributed evenly around the center. Finally, output compliances of RRR and 4-RRR compliant micro-motion stages are derived from the analytical model and finite element analysis (FEA). The comparisons indicate that the results obtained from the proposed analytical model are in good agreement with those derived from FEA, which validates the proposed analytical model.

1 Introduction

Flexure-based compliant mechanisms, which have advantages of no friction, no backlash, compact and monolithic structure, and ease of fabrication, are usually used as micro-positioning stages (Acer and Sabanovic, 2011; Li et al., 2013; Yong and Lu, 2009). They can provide smooth motions through deflections of flexure hinges (Yong et al., 2008; Handley et al., 2004). Due to these advantages, coupled with nanometer positioning accuracy, flexure-based compliant mechanisms have been widely used in many industrial applications, such as scanning probe microscopy (Schitter et al., 2007; Leang and Fleming, 2009), lithography (Choi and Lee, 2005), nano-manipulation and manufacturing (Lai et al., 2012; Verma et al., 2005), and biological science (Ando et al., 2002; Kim et al., 2012). In addition, the flexure-based compliant positioning mechanisms with ultrahigh precision play more and more important roles in applications where a high resolution motion is desirable, such as the MEMS sensors and actuators, optical fiber alignment, and biological cell manipulation (Wang and Zhang, 2008; Li et al., 2012; Zubir et al., 2009).

Decoupling is the crucial property for parallel flexure-based compliant mechanisms. Generally, a decoupled parallel flexure-based compliant mechanism means that one actuator produces only one directional output motion without affecting the motion of other axes. The decoupled parallel flexure-based compliant mechanism should possess input decoupling and output decoupling properties (Li et al., 2012). The input decoupling can be defined as actuator isolation, which means that each actuator would not suffer extra loads induced by the actuation of other actuators. The output decoupling means that one actuator only drives the output platform in one axial direction. Generally, input coupling would impose undesired loads on the actuator and may even damage it. Output coupling would lead to complex kinematic models, thus making precise control difficult to be implemented.

Numerous studies have been carried out to design decoupled flexure-based compliant mechanisms. Among the investigations of flexure-based compliant parallel mecha-

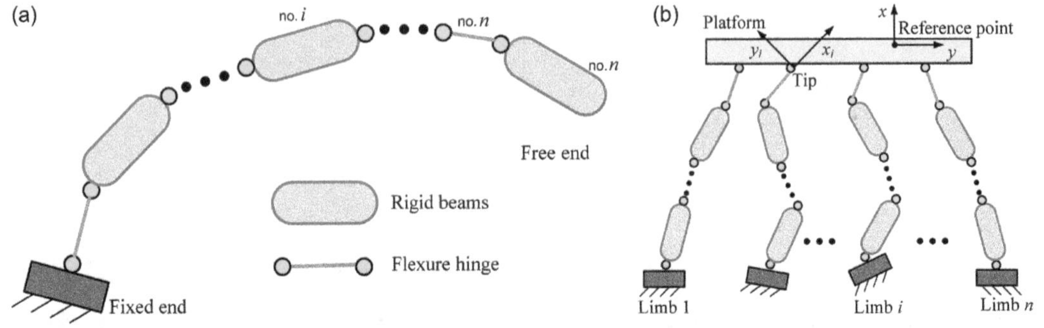

Figure 1. Flexure-based compliant mechanisms: **(a)** flexure serial structure; **(b)** flexure parallel structure.

nisms, the design of a totally decoupled one is first proposed in the literature (Awtar and Slocum, 2007). It presented parallel kinematic XY flexure mechanism designs based on systematic constraint patterns. These constraint arrangements allowed large primary motions and small error motions without running into over constraint problems. Kenton and Leang (2012) built a mechanism by serially stacking or nesting multiple one degree-of-freedom modules to eliminate the cross-axis coupling. Polit and Dong (2011) presented the design, analysis, fabrication, and testing of a high-bandwidth parallel kinematic nano-positioning XY stage. The monolithic stage had two axes, and each axis was composed of a doubly clamped beam and a parallelogram hybrid flexure. The parallelogram hybrid flexures were used to decouple the actuation effect from the other axis. Thus, the mechanism design decoupled the motion in the X and Y directions and restricted parasitic rotations in the XY plane. Li and Xu (2008, 2010) designed a decoupled XY flexure parallel kinematic manipulator. The output decoupling was allowed by the employment of compound parallelogram flexure. By contrast, the input decoupling was implemented by actuation isolation, which was enabled by the double compound parallelogram flexure with large transverse stiffness. Qin et al. (2014) introduced a 2-DOF monolithic mechanism. The statically indeterminate leaf parallelograms were used to provide the decoupling effect, and the displacement of piezoelectric actuator was amplified with a statically indeterminate lever mechanism. The cross-axis coupling ratio was experimentally measured to be below 1 %.

In the past decades, various decoupled compliant mechanisms have been developed. Nevertheless, there is rare analytical model which can prove the output decoupling property of flexure-based compliant mechanisms. Therefore, the present study addresses the output decoupling property of planar flexure-based compliant mechanisms composed of notch flexure hinges. Based on compliance/stiffness modeling methods, analytical model of mechanisms with symmetric configuration is derived. To verify the output decoupling property, the output compliances of the RRR (revolute-revolute-revolute) and 4-RRR compliant micromotion stages are derived from the analytical model and FEA.

The remaining sections of this paper are organized as follows. In Sect. 2, compliance/stiffness modeling methods are derived, including the methods for flexure serial structures and flexure parallel structures. In Sect. 3, the output decoupling properties of the mechanisms, which are symmetric about two perpendicular axes or composed of several fundamental forms, are analyzed. And analytical model utilized for explaining the output decoupling property is derived. In Sect. 4, analytical model and FEA are used to analyze the output decoupling property of the RRR and 4-RRR micromotion stages. Finally, conclusions are drawn in Sect. 5.

2 Compliance/stiffness modeling

Flexure-based compliant mechanisms can be classified into two categories according to the kinematic structure: the flexure serial structure (Fig. 1a) and the flexure parallel structure (Fig. 1b). Serial structures are easier to design and have substantially decoupled degree-of-freedom, whereas the closed-loop kinematic features of flexure parallel structures impart excellent performance in terms of high rigidity, high load carrying capacity, and high accuracy (Li et al., 2013; Pham and Chen, 2005). To analyze the output decoupling property of planar flexure-based compliant mechanism with symmetrical configuration, the compliance/stiffness modeling methods are derived firstly.

2.1 Flexure serial structure

A flexure serial structure comprised of several rigid beams and notch flexure hinges is illustrated in Fig. 1a. The relationship between the elastic deformations of a flexure member and the total deformation at the free end of the serial structure can be given as

$$\boldsymbol{\delta}_i = \mathbf{J}_{di}\boldsymbol{\delta}_i^l, \tag{1}$$

where $\mathbf{J}_{di}$ is the Jacobian matrix transforming a 3×1 vector δ_i^l representing the elastic deformations of flexure members to a 3×1 vector $\boldsymbol{\delta}_i$ at the free end of the flexure serial structure.

According to the zero virtual work principle, the relationship between the external force at the free end and the reaction force at a flexure member can be given as

$$\boldsymbol{F}_i^l = \mathbf{J}_{fi}\boldsymbol{F}, \tag{2}$$

where $\mathbf{J}_{fi}$ is the Jacobian matrix transforming a 3×1 vector $\boldsymbol{F}$ of the external force at the free end to a vector of reaction force $\boldsymbol{F}_i^l$ at a flexure member.

The accumulation of both rotational and translational deformations at the free end of the flexure serial structure is

$$\boldsymbol{\delta} = \sum_{i=1}^{n} \mathbf{J}_{di}\boldsymbol{\delta}_i^l = \sum_{i=1}^{n} \mathbf{J}_{di}\mathbf{C}_i^l\boldsymbol{F}_i^l = \left(\sum_{i=1}^{n} \mathbf{J}_{di}\mathbf{C}_i^l\mathbf{J}_{fi}\right)\boldsymbol{F}, \tag{3}$$

where $\mathbf{C}_i^l$ is the local compliance matrix established in the local frame attached to a flexure member.

Therefore, the compliance of the flexure serial structure is

$$\mathbf{C} = \frac{\partial \boldsymbol{\delta}}{\partial \boldsymbol{F}} = \sum_{i=1}^{n} \mathbf{J}_{di}\mathbf{C}_i^l\mathbf{J}_{fi} = \mathbf{J}_d\mathbf{C}^*\mathbf{J}_f, \tag{4}$$

where

$$\mathbf{J}_d = \left[\mathbf{J}_{d_1}\ \mathbf{J}_{d_2}\ \mathbf{J}_{d_3} \ldots \mathbf{J}_{d_n}\right],$$

and

$$\mathbf{C}^* = \mathrm{diag}\left[\mathbf{C}_1\ \mathbf{C}_2\ \mathbf{C}_3 \ldots \mathbf{C}_n\right].$$

For flexure serial structures, the compliance can be derived through the matrix method. The compliance shown in Eq. (4) can also be obtained through the screw theory based approach (Yu et al., 2011). It indicates that the proposed analysis of compliance modeling has close relations to the syntheses of compliant mechanism based on screw theory.

2.2 Flexure parallel structure

A flexure parallel structure can be considered as an infinitely rigid platform supported by n limbs, as illustrated in Fig. 1b. We denote $\mathbf{K}_i$ as the stiffness matrix of limb i. The platform and the tips of the limbs have equivalent angular displacements. However, the linear displacements of the reference point of the platform are determined based on the linear and angular displacements of the tips of the limbs; hence,

$$\boldsymbol{\delta} = \mathbf{J}_{Pi}\boldsymbol{\delta}_i, \tag{5}$$

where $\boldsymbol{\delta}$ and $\boldsymbol{\delta}_i$ are, respectively, the 3×1 displacement vectors of the platform at the reference point and at the tip of the ith limb, and $\mathbf{J}_{Pi}$ is the amplification matrix of the displacement.

The 3×1 vector of the force and moment F applied to the reference point of the platform is distributed to the tip of each limb as a 3×1 vector of force and moment F_i, hence,

$$F = \sum_{i=1}^{n} \mathbf{J}_{fi}F_i = \sum_{i=1}^{n} \mathbf{J}_{fi}\mathbf{K}_i\boldsymbol{\delta}_i = \left(\sum_{i=1}^{n} \mathbf{J}_{fi}\mathbf{K}_i\mathbf{J}_{Pi}^{-1}\right)\boldsymbol{\delta}, \tag{6}$$

where $\mathbf{J}_{fi}$ is the transformation matrix when F_i is moved from the tip of the ith limb to the reference point.

Hence, the stiffness of the flexure parallel structure can be given as

$$\mathbf{K} = \frac{\partial F}{\partial \boldsymbol{\delta}} = \sum_{i=1}^{n} \mathbf{J}_{fi}\mathbf{K}_i\mathbf{J}_{Pi}^{-1} = \mathbf{J}_F^*\mathbf{K}^*\mathbf{J}_P^*, \tag{7}$$

where

$$\mathbf{J}_F^* = \left[\mathbf{J}_{F_1}\ \mathbf{J}_{F_2}\ \mathbf{J}_{F_3} \ldots \mathbf{J}_{F_n}\right],$$

$$\mathbf{J}_p^* = \left[\left(\mathbf{J}_{P1}^{-1}\right)^T \left(\mathbf{J}_{P2}^{-1}\right)^T \left(\mathbf{J}_{P3}^{-1}\right)^T \ldots \left(\mathbf{J}_{Pn}^{-1}\right)^T\right]^T,$$

and

$$\mathbf{K}^* = \mathrm{diag}(\mathbf{K}_1\ \mathbf{K}_2\ \mathbf{K}_3 \ldots \mathbf{K}_n).$$

Obviously, it is more convenient to use the compliance matrices for the flexure serial structure, while use the stiffness matrices for the flexure parallel structure. To achieve motion, the stiffness terms along the motion should be small, while the other stiffness terms should be large. Compliance/stiffness modeling can represent the relationship between the compliance/stiffness and geometrical parameters, and thus it can be applied to optimize the geometrical parameters. Therefore, Compliance/stiffness modeling methods are crucial to design planar flexure-based compliant mechanisms with finite deformations.

3 Output decoupling property

To analyze the output decoupling property, the influence of symmetrical configuration could be investigated. For a notch flexure hinge, there are strong couplings between the rotational and translational motions. Furthermore, the rotation center of a flexure hinge drifts whenever the hinge works, resulting in motion errors. To obtain decoupling properties, multiple flexure hinges can be combined to form certain structures (Qin et al., 2013). In practice, the parallelogram joints are usually used as prismatic joints. The parallelogram joints can keep the orientation of the platform invariant when the mechanism is actuated by the force F_x along one axis. However, undesired cross-coupling error e_y along the other axis is generated at the same time, as depicted in Fig. 2a, which makes it difficult for the kinematic analysis and control of such a mechanism. Compared with one parallelogram joint, two parallelogram joints shown in Fig. 2b can eliminate the coupling error completely, and thus the symmetric configuration could be regarded as two ideal prismatic joints to form certain flexure-based compliant structures to obtain output decoupling property (Yong and Lu, 2008). Obviously, the motion illustrated in Fig. 2a is a pure translation. By contrast, the motion in Fig. 2b is a pure translation without undesired cross-coupling errors. Therefore, flexure-based compliant mechanisms are usually designed with symmetric configuration to obtain the output decoupling property.

3.1 Planar flexure-based compliant mechanisms with symmetric configuration

3.1.1 Symmetric configuration about x axis

As shown in Fig. 3, the planar flexure-based compliant mechanism is symmetric about the x axis and all ends are fixed. We can see that the mechanism is composed of one platform and four limbs, and each limb is composed of several notch flexure hinges and rigid beams. The output point is located at the center of the platform.

The mechanism can be divided into two identical parts: the upper part and the lower part. $\mathbf{J}_{Fs}$ and $\mathbf{J}_{Ps}$ denote the Jacobian matrices transforming forces/moments and deformations of the upper part, $\mathbf{K}_1$ denotes the stiffness matrix at the point o of half part ($K_{1,m-n}$ represents the stiffness in the direction of m caused by the force/moment n), while $\mathbf{J}_{Fx}$ and $\mathbf{J}_{Px}$ are defined as the Jacobian matrices transforming forces/moments and deformations of the lower part. Then, the stiffness matrices of the upper part and lower part can be given as

$$\mathbf{K}_{ts} = \mathbf{J}_{Fs} \cdot \mathbf{K}_1 \cdot \mathbf{J}_{Ps} = \mathbf{K}_1, \tag{8}$$

$$\mathbf{K}_{tx} = \mathbf{J}_{Fx} \cdot \mathbf{K}_1 \cdot \mathbf{J}_{Px}, \tag{9}$$

where

$$\mathbf{J}_{Fs} = \mathbf{J}_{Ps}^T = \mathbf{E} \text{ (identity matrix)},$$

$$\mathbf{K}_1 = \begin{bmatrix} K_{1,x-Fx} & K_{1,x-Fy} & K_{1,x-Mz} \\ K_{1,y-Fx} & K_{1,y-Fy} & K_{1,y-Mz} \\ K_{1,\alpha-Fx} & K_{1,\alpha-Fy} & K_{1,\alpha-Mz} \end{bmatrix},$$

and

$$\mathbf{J}_{Fx} = \mathbf{J}_{Px}^T = \begin{bmatrix} 1 & 0 & 0 \\ 0 & -1 & 0 \\ 0 & 0 & -1 \end{bmatrix}.$$

Thus, the stiffness matrix of the flexure-based compliant mechanism is given as

$$\mathbf{K} = \mathbf{K}_{ts} + \mathbf{K}_{tx} = \begin{bmatrix} 2K_{1,x-Fx} & 0 & 0 \\ 0 & 2K_{1,y-Fy} & 2K_{1,y-Mz} \\ 0 & 2K_{1,\alpha-Fy} & 2K_{1,\alpha-Mz} \end{bmatrix}. \tag{10}$$

We can see from Eq. (10) that the terms $K_{1,y-Fx}$ and $K_{1,\alpha-Fx}$ are both zero. It indicates that the flexure-based compliant mechanisms with symmetric configuration about the x axis are output decoupled along the x axis.

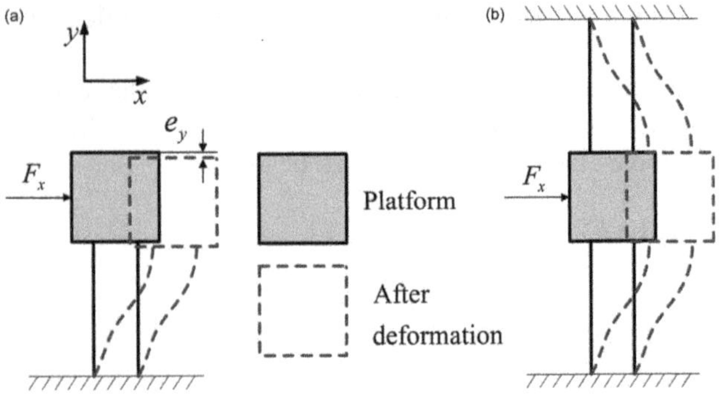

Figure 2. Parallelogram joints for planar parallel mechanisms: **(a)** one parallelogram joint; **(b)** two parallelogram joints with symmetric configuration.

3.1.2 Symmetric configuration about y axis

Figure 4 illustrates the flexure-based compliant mechanism which is symmetric about the y axis and all ends are fixed. The mechanism consists of one platform and four limbs, and the output point is located at the center of the platform.

The mechanism can be divided into two identical parts: the left part and the right part. $\mathbf{J}_{Fl}$ and $\mathbf{J}_{Pl}$ denote the Jacobian matrices transforming forces/moments and deformations of the left part, $\mathbf{K}_2$ denotes the stiffness matrix at the point o of half part ($K_{2,m-n}$ represents the stiffness in the direction of m caused by the force/moment n), whereas $\mathbf{J}_{Fr}$ and $\mathbf{J}_{Pr}$ are defined as the Jacobian matrices transforming forces/moments and deformations of the right part. Then, the stiffness matrices of the left part and right part can be given as

$$\mathbf{K}_{tl} = \mathbf{J}_{Fl} \cdot \mathbf{K}_2 \cdot \mathbf{J}_{Pl}, \tag{11}$$

$$\mathbf{K}_{tr} = \mathbf{J}_{Fr} \cdot \mathbf{K}_2 \cdot \mathbf{J}_{Pr} = \mathbf{K}_2, \tag{12}$$

where

$$\mathbf{J}_{Fl} = \mathbf{J}_{Pl}^T = \begin{bmatrix} -1 & 0 & 0 \\ 0 & 1 & 0 \\ 0 & 0 & -1 \end{bmatrix},$$

$$\mathbf{K}_2 = \begin{bmatrix} K_{2,x-Fx} & K_{2,x-Fy} & K_{2,x-Mz} \\ K_{2,y-Fx} & K_{2,y-Fy} & K_{2,y-Mz} \\ K_{2,\alpha-Fx} & K_{2,\alpha-Fy} & K_{2,\alpha-Mz} \end{bmatrix},$$

and

$$\mathbf{J}_{Fr} = \mathbf{J}_{Pr}^T = \mathbf{E} \text{ (identity matrix)}.$$

Thus, the stiffness matrix of the mechanism can be expressed as

$$\mathbf{K} = \mathbf{K}_{tl} + \mathbf{K}_{tr} = \begin{bmatrix} 2K_{2,x-Fx} & 0 & 2K_{2,x-Mz} \\ 0 & 2K_{2,y-Fy} & 0 \\ 2K_{2,\alpha-Fx} & 0 & 2K_{2,\alpha-Mz} \end{bmatrix}. \tag{13}$$

We can see from Eq. (13) that the terms $K_{2,x-Fy}$ and $K_{2,\alpha-Fy}$ are both zero. It indicates that the flexure-based compliant mechanisms with symmetric configuration about the y axis are output decoupled along the y axis.

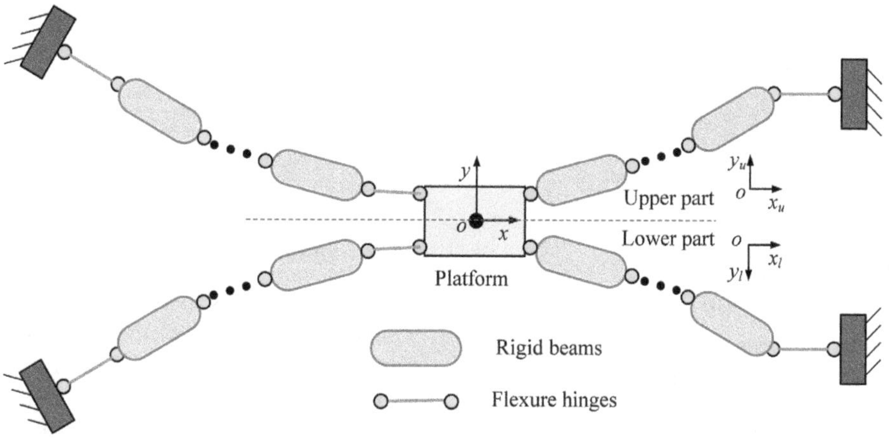

Figure 3. Mechanism with symmetric configuration about the x axis.

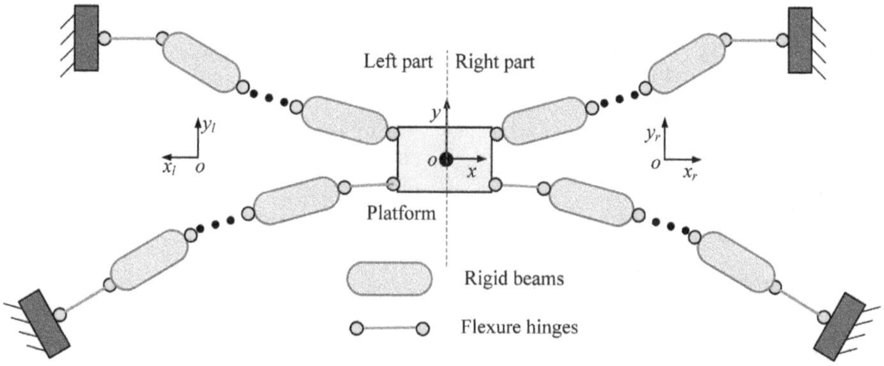

Figure 4. Mechanism with symmetric configuration about the y axis.

3.1.3 Symmetric configuration about x axis and y axis

The flexure-based compliant mechanism is symmetric about two perpendicular axes (the x axis and y axis), and the output point is located at the center of the platform. The flexure-based compliant mechanism can be divided into two identical parts: the left part and the right part. Denote $\mathbf{K}_2$ as the stiffness matrix at the point o of half part ($K_{2,m-n}$ is the stiffness in the direction of m caused by the force/moment n), the stiffness matrix of the flexure-based compliant mechanism can be expressed as

$$\mathbf{K} = \mathbf{J}_{Fl} \cdot \mathbf{K}_2 \cdot \mathbf{J}_{Pl} + \mathbf{J}_{Fr} \cdot \mathbf{K}_2 \cdot \mathbf{J}_{Pr} = \begin{bmatrix} 2K_{2,x-Fx} & 0 & 0 \\ 0 & 2K_{2,y-Fy} & 0 \\ 0 & 0 & 2K_{2,\alpha-Mz} \end{bmatrix}. \quad (14)$$

We can see from Eq. (14) that the stiffness matrix consists of only main diagonal components. Therefore, the analyses above indicate that the planar flexure-based compliant mechanisms are output decoupled when they are symmetric about two perpendicular axes (the x axis and y axis).

3.2 Variations of the fundamental form

3.2.1 The fundamental form

The fundamental form is composed of several flexure hinges, and it can be regarded as the limb of flexure parallel structures. The fundamental form can be extended according to its designated function, such as the parallelogram joints and prismatic joints mentioned above. In general, the mechanisms may be composed of several fundamental forms. In order to analyze the performances of flexure-based compliant mechanisms, the fundamental form, which is actually a flexure serial structure, should be first investigated. In practice, the notch flexure hinges of the flexure serial structure are all along one straight line, as shown in Fig. 5.

According to compliance/stiffness modeling methods discussed above, the compliance of the practical flexure serial structure can be obtained from Eq. (4), and $\mathbf{J}_{di}$, $\mathbf{J}_{fi}$, and $\mathbf{C}_i^l$ can be given as

$$\mathbf{J}_{di} = \mathbf{J}_{fi}^T = \begin{bmatrix} 1 & 0 & r_y \\ 0 & 1 & -r_x \\ 0 & 0 & 1 \end{bmatrix} \begin{bmatrix} -1 & 0 & 0 \\ 0 & 1 & 0 \\ 0 & 0 & -1 \end{bmatrix}, \quad (15)$$

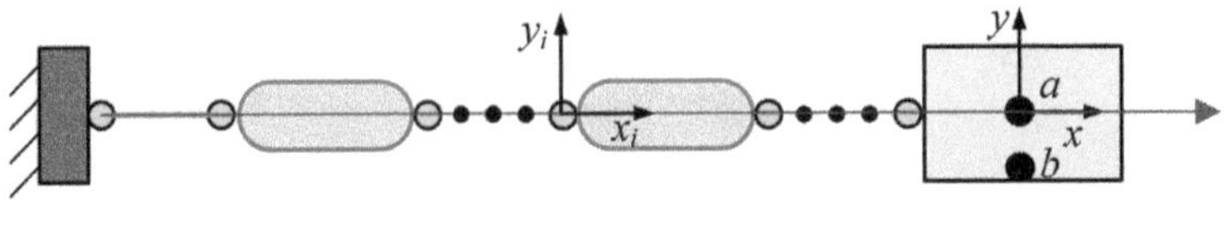

Figure 5. The practical flexure serial structure.

$$\mathbf{C}_i^l = \begin{bmatrix} C_{i,x-Fx} & 0 & 0 \\ 0 & C_{i,y-Fy} & C_{i,y-Mz} \\ 0 & C_{i,\alpha-Fy} & C_{i,\alpha-Mz} \end{bmatrix}, \tag{16}$$

where r_x and r_y are the distances from the platform at the reference point to the ith flexure hinge along the x axis and y axis, respectively, $C_{i,m-n}$ is the compliance of ith hinge in the direction of m caused by the force/moment n.

Then, when the output point of the structure is located at the reference point b, the compliance of the flexure serial structure can be given as

$$\mathbf{C}_b = \sum_{i=1}^{n} \mathbf{J}_{di}\mathbf{C}_i^l\mathbf{J}_{fi} = \begin{bmatrix} C_{b,x-Fx} & C_{b,x-Fy} & C_{b,x-Mz} \\ C_{b,y-Fx} & C_{b,y-Fy} & C_{b,y-Mz} \\ C_{b,\alpha-Fx} & C_{b,\alpha-Fy} & C_{b,\alpha-Mz} \end{bmatrix}, \tag{17}$$

where

$$C_{b,x-Fx} = \sum_{i=1}^{n}\left(C_{i,x-Fx} + r_y^2 C_{i,\alpha-Mz}\right),$$

$$C_{b,y-Fy} = \sum_{i=1}^{n}\left(C_{i,y-Fy} + 2r_x C_{i,y-Mz} + r_x^2 C_{i,\alpha-Mz}\right),$$

$$C_{b,y-Mz} = \sum_{i=1}^{n}\left(-C_{i,y-Mz} + r_x C_{i,\alpha-Mz}\right),$$

and

$$C_{b,\alpha-Mz} = \sum_{i=1}^{n}\left(C_{i,\alpha-Mz}\right).$$

When the output point of the structure is located at the reference point a, the distance r_y is zero, and the compliance of the flexure serial structure can be given as

$$\mathbf{C}_a = \sum_{i=1}^{n} \mathbf{J}_{di}\mathbf{C}_i^l\mathbf{J}_{fi} = \begin{bmatrix} C_{a,x-Fx} & 0 & 0 \\ 0 & C_{a,y-Fy} & C_{a,y-Mz} \\ 0 & C_{a,\alpha-Fy} & C_{a,\alpha-Mz} \end{bmatrix}, \tag{18}$$

where

$$C_{a,x-Fx} = \sum_{i=1}^{n} C_{i,x-Fx},$$

$$C_{a,y-Fy} = C_{b,y-Fy},$$

$$C_{a,y-Mz} = C_{b,y-Mz},$$

and

$$C_{a,\alpha-Mz} = C_{b,\alpha-Mz}.$$

Therefore, the analyses show that the compliance/stiffness matrix can be expressed as Eq. (18) when the flexure hinges and output reference point are all along one straight line, and it can be expressed as Eq. (17) when the flexure hinges and output reference point are not along one straight line.

3.2.2 Variations

Besides the symmetric configuration about x axis and y axis, without loss of generality, the flexure-based compliant mechanisms may be composed of several identical fundamental forms which are distributed evenly around the center of the platform, as shown in Fig. 6 (Yong and Lu, 2008). When the output point is located at the center of the platform and the number of the fundamental forms is $2n$ or 3, the output decoupling of them could be discussed as follows.

As the analyses mentioned above, the mechanism can be divided into $2n$ ($n > 1$) identical fundamental forms. Denote $\mathbf{K}_m$ as the stiffness matrix at the point o of one fundamental form ($K_{m,w-u}$ represents the stiffness in the direction of w caused by the force/moment u). When the flexure hinges and output reference point of one fundamental form are along one straight line, the stiffness matrix of the ith fundamental form can be given as

$$\mathbf{K}_i = \mathbf{J}_{Fi} \cdot \mathbf{K}_m \cdot \mathbf{J}_{Pi}, \tag{19}$$

where

$$\mathbf{K}_m = \begin{bmatrix} K_{m,x-Fx} & 0 & 0 \\ 0 & K_{m,y-Fy} & K_{m,y-Mz} \\ 0 & K_{m,\alpha-Fy} & K_{m,\alpha-Mz} \end{bmatrix},$$

$$\mathbf{J}_{Fi} = \begin{bmatrix} \cos[(i-1)\pi/n] & \sin[(i-1)\pi/n] & 0 \\ -\sin[(i-1)\pi/n] & \cos[(i-1)\pi/n] & 0 \\ 0 & 0 & 1 \end{bmatrix},$$

and

$$\mathbf{J}_{Pi} = \begin{bmatrix} \cos[(i-1)\pi/n] & -\sin[(i-1)\pi/n] & 0 \\ \sin[(i-1)\pi/n] & \cos[(i-1)\pi/n] & 0 \\ 0 & 0 & 1 \end{bmatrix}.$$

Hence, the stiffness matrix of the mechanism at center of the platform is given as

$$\mathbf{K}_{2n} = \sum_{1}^{2n} \mathbf{K}_i = \begin{bmatrix} nK_{m,x-Fx} + nK_{m,y-Fy} & 0 & 0 \\ 0 & nK_{m,y-Fy} + nK_{m,x-Fx} & 0 \\ 0 & 0 & 2nK_{m,\alpha-Mz} \end{bmatrix}. \tag{20}$$

We can see from Eq. (20) that the stiffness matrix is a diagonal matrix. It indicates that the flexure-based compliant mechanisms composed of $2n$ ($n > 1$) fundamental forms are output decoupled. The stiffness matrix of the mechanism can be obtained according to Eq. (20) when the number of the

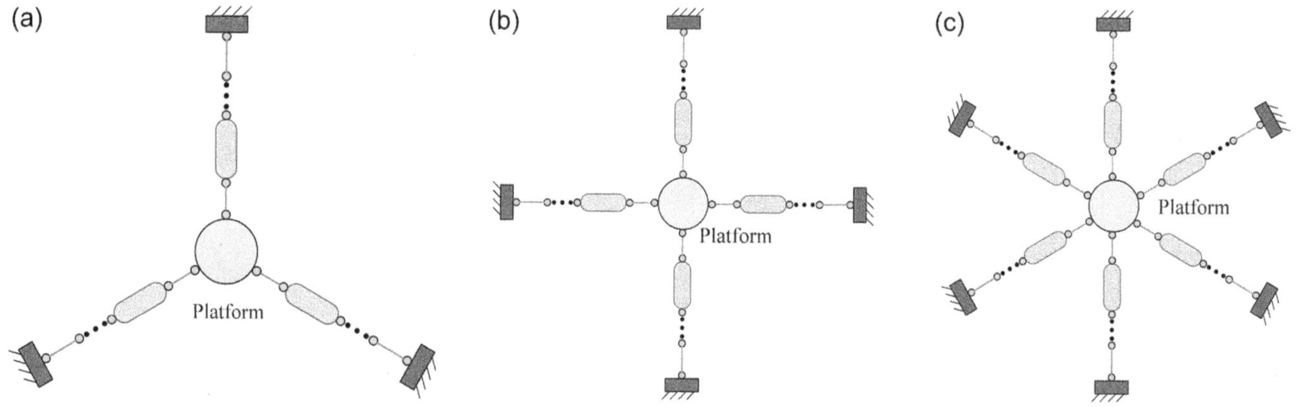

Figure 6. Fundamental forms distributed evenly around the center of the platform: **(a)** $n=3$; **(b)** $n=4$; **(c)** $n=6$.

fundamental forms is 4 and 6, as shown in Fig. 7b and c. In addition, there are some special conditions should be taken into account. When the number of the fundamental forms is 3, as shown in Fig. 7a, the stiffness matrix of the mechanism can be obtained from Eq. (19), and it can be expressed as

$$\mathbf{K}_3 = \sum_1^3 \mathbf{K}_i = \begin{bmatrix} \frac{3}{2}K_{m,x-Fx} + \frac{3}{2}K_{m,y-Fy} & 0 & 0 \\ 0 & \frac{3}{2}K_{m,y-Fy} + \frac{3}{2}K_{m,x-Fx} & 0 \\ 0 & 0 & 3K_{m,\alpha-Mz} \end{bmatrix}. \quad (21)$$

Similarly, we can see from Eq. (21) that the stiffness matrix is a diagonal matrix. It indicates that the flexure-based compliant mechanisms composed of 3 fundamental forms are output decoupled. In addition, when the mechanism is symmetric about y axis and the number of the fundamental forms is 2, the stiffness matrix of the mechanism can be obtained from Eq. (14), and it can be expressed as

$$\mathbf{K}_2 = \sum_1^2 \mathbf{K}_i = \begin{bmatrix} 2K_{m,x-Fx} & 0 & 0 \\ 0 & 2K_{m,y-Fy} & 0 \\ 0 & 0 & 2K_{m,\alpha-Mz} \end{bmatrix}. \quad (22)$$

Therefore, the flexure-based compliant mechanisms composed of $2n$ or 3 identical fundamental forms which are distributed evenly around the center of the platform are output decoupled. In addition, the stiffness of mechanisms composed of $2n$ ($n>1$) or 3 fundamental forms can be calculated from Eqs. (20) and (21), while the stiffness of mechanisms composed of 2 fundamental forms which are symmetric about y axis can be calculated from Eq. (22).

4 Application

To validate the analytical model derived above, simple closed-form equations are derived to determine the output compliances of the RRR and 4-RRR stages (Yong and Lu, 2008). The RRR flexure-based compliant structure is composed of three constant rectangular cross-section flexure hinges, and the 4-RRR structure is composed of four RRR limbs connected together in parallel, as shown in Fig. 7. The parallel configuration of the 4-RRR structure is advantageous over a serial configuration structure.

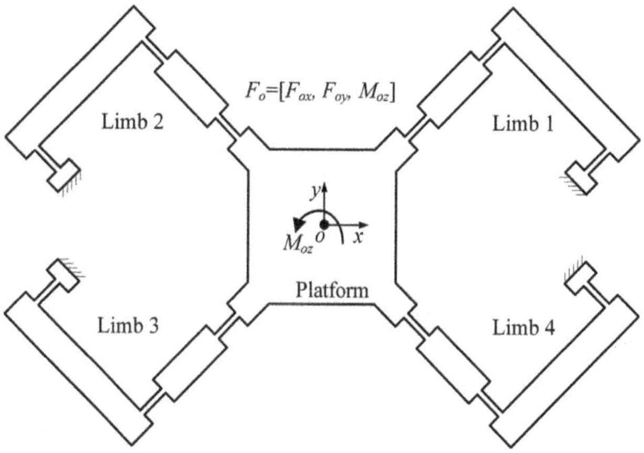

Figure 7. 4-RRR flexure-based compliant micro-motion stage.

4.1 Output compliance of the RRR micro-motion stage

Notch flexure hinges are produced by drilling or milling two closely spaced holes, forming a circular or constant rectangular cross-section cutout (Tseytlin, 2002). The geometrical parameters of the constant rectangular cross-section flexure hinge are illustrated in Fig. 8. The geometrical dimensions include the hinge thickness t, the hinge length L, the side height h, the rigid beam width W, the total length L_a, the total height H, and the depth D.

The in-plane compliance matrix of the constant rectangular cross-section flexure hinge is shown in Eq. (16). The exponential relationship between geometrical parameters and deformations can be derived through FEA, and the proposed compliance matrix can be obtained based on the exponential model. The compliance terms can be expressed as follows.

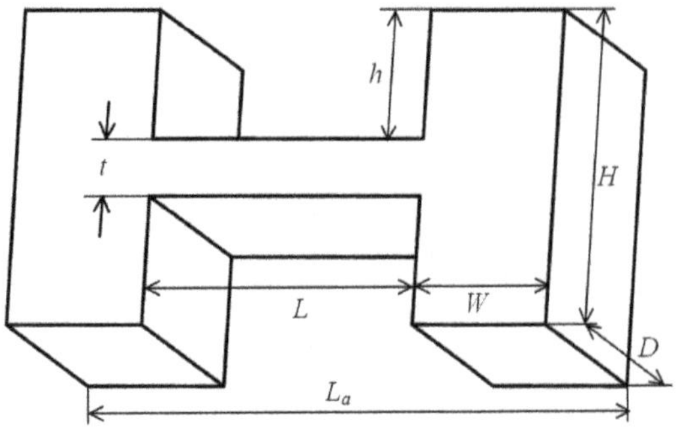

Figure 8. Geometrical dimensions of the constant rectangular cross-section flexure hinge.

$$C_{i,x-Fx} = \frac{4}{EDH \cdot t^{0.82}} \left[h^{1.07} \cdot L^{0.75} + L_a \right] \tag{23}$$

$$C_{i,\alpha-Mz} = \frac{16.5 \cdot L^{0.88}}{ED \cdot t^{2.88}} \tag{24}$$

$$C_{i,y-Mz} = C_{i,\alpha-Fy} = \frac{8.3 \cdot L^{1.88}}{ED \cdot t^{2.88}} \tag{25}$$

$$C_{i,y-Fy} = \frac{5.3 \cdot h^{0.01} \cdot L^{2.9}}{ED \cdot t^{2.91}} + \frac{\alpha \cdot E \cdot C_{i,x-Fx}}{G} \tag{26}$$

where E is the Yong's modulus and α is the shear coefficient.

Note that all flexure hinges are modeled to have 3-DOF (i.e. bending compliance about the z axis, and axial and shear compliances along the x and y axes, respectively). Flexure hinges are modeled as having 3-DOF rather than 6-DOF because both the RRR and 4-RRR compliant stages considered here are planar. The RRR flexure-based compliant stage is shown in Fig. 9, together with its dimensions, displacements, local coordinates of flexure hinges, and the applied forces/moments.

The compliances at point o_1 (Fig. 9a) contributed by each flexure hinge in the structure are firstly calculated. The overall compliances of the RRR compliant stage at point o (Fig. 9b) are then obtained by summing all the contributions to the compliances in the corresponding directions of each individual flexure hinge. The compliance matrix of each flexure hinge is shown in Eq. (16), and the geometrical parameters are illustrated in Fig. 8. Considering that b_o and h_o are the key distances of the RRR stage, the output compliance matrix at o_1 can be expressed as

$$\mathbf{C}_{O1} = \sum_{i=1}^{3} \mathbf{J}_{di} \mathbf{C}_i^l \mathbf{J}_{fi}. \tag{27}$$

When the output forces are applied to point o rather than o_1, and the displacements at this point are desired, the transformation matrix $\mathbf{J}_{fo}$ can be used to transfer the output forces/moments from o_1 to o, while $\mathbf{J}_{do}$ can be used to transfer the displacements from o_1 to o, where b_1 and h_1 are the components of the vector $\boldsymbol{r}$ pointing from o_1 to o along x axis and y axis, respectively. Thus, the output compliance at point o can be expressed as

$$\mathbf{C}_{\mathrm{RRR},1O} = \mathbf{J}_{do} \mathbf{C}_{O1} \mathbf{J}_{fo}, \tag{28}$$

where

$$\mathbf{J}_{do} = \mathbf{J}_{fo}^T = \begin{bmatrix} 1 & 0 & -h_1 \\ 0 & 1 & b_1 \\ 0 & 0 & 1 \end{bmatrix}.$$

Table 1. Geometrical parameters and material properties of the compliant micro-motion stage.

Parameters/material properties	
Young's modulus, E (GPa)	210
Poisson's ratio, ν	0.30
h_o (mm)	15
b_o (mm)	50
h_1 (mm)	25
b_1 (mm)	25
L (mm)	10
t (mm)	2
h (mm)	4
W (mm)	5
D (mm)	10

4.2 Output compliance of the 4-RRR micro-motion stage

The 4-RRR compliant micro-motion stage is generated by arranging the four RRR compliant stages 90° apart, and it is symmetric about x axis and y axis respectively, as shown in Fig. 7. According to the symmetric configuration, the output compliance matrix at point o of limbs 2, 3 and 4, respectively, are

$$\mathbf{C}_{\mathrm{RRR},2O} = \mathbf{J}_2 \mathbf{C}_{\mathrm{RRR},1O} \mathbf{J}_2^T, \tag{29}$$

$$\mathbf{C}_{\mathrm{RRR},3O} = \mathbf{J}_\pi \mathbf{C}_{\mathrm{RRR},1O} \mathbf{J}_\pi^T, \tag{30}$$

$$\mathbf{C}_{\mathrm{RRR},4O} = \mathbf{J}_\pi \mathbf{C}_{\mathrm{RRR},2O} \mathbf{J}_\pi^T, \tag{31}$$

where

$$\mathbf{J}_2 = \begin{bmatrix} -1 & 0 & 0 \\ 0 & 1 & 0 \\ 0 & 0 & -1 \end{bmatrix}$$

and

$$\mathbf{J}_\pi = \begin{bmatrix} \cos(\pi) & -\sin(\pi) & 0 \\ \sin(\pi) & \cos(\pi) & 0 \\ 0 & 0 & 1 \end{bmatrix}.$$

Then, the compliance matrix of the 4-RRR flexure-based compliant micro-motion stage can be given as

$$\mathbf{C}_{4\mathrm{RRR}} = \left(\mathbf{C}_{\mathrm{RRR},1O}^{-1} + \mathbf{C}_{\mathrm{RRR},2O}^{-1} + \mathbf{C}_{\mathrm{RRR},3O}^{-1} + \mathbf{C}_{\mathrm{RRR},4O}^{-1} \right)^{-1}. \tag{32}$$

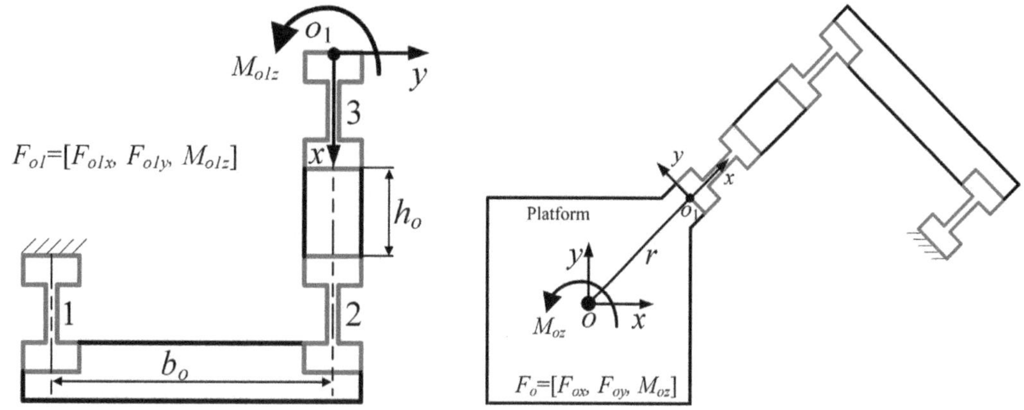

Figure 9. RRR flexure-based compliant mechanism: **(a)** applied forces/moments at output point o_1; **(b)** applied forces/moments at output point o.

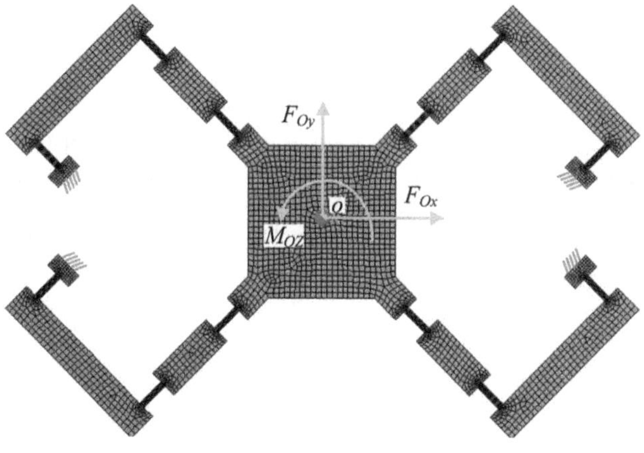

Figure 10. FEA model of a 4-RRR flexure-based compliant micro-motion stage.

4.3 Validation

Considering the accuracy of FEA, it is used as a benchmark. The output compliances of the RRR and 4-RRR micro-motion stages are calculated to study the output decoupling property. The results calculated by the analytical model are compared with those derived from FEA. Table 1 lists the geometrical parameters and material properties of the RRR and 4-RRR compliant micro-motion stages employed, where all flexure hinges have equivalent geometrical parameters and material properties. A FEA model was generated using ANSYS. Meshes, constraints, and the applied forces/moments of the FEA model are shown in Fig. 10.

The output compliance of the RRR and 4-RRR micro-motion stages determined using FEA can be expressed as, respectively

$$\mathbf{C}_{\mathrm{RRR,FEA}} = \begin{bmatrix} 3.820\times10^{-5} & -5.057\times10^{-5} & 8.889\times10^{-4} \\ -5.057\times10^{-5} & 1.029\times10^{-4} & -1.457\times10^{-3} \\ 8.889\times10^{-4} & -1.457\times10^{-3} & 2.410\times10^{-2} \end{bmatrix}. \quad (33)$$

$$\mathbf{C}_{\mathrm{4RRR,FEA}} = \begin{bmatrix} 1.184\times10^{-6} & 0 & 0 \\ 0 & 3.239\times10^{-6} & 0 \\ 0 & 0 & 3.076\times10^{-4} \end{bmatrix}. \quad (34)$$

Since the off-diagonal compliances terms in Eq. (34) are very small and insignificant compared to the diagonal terms, they are assumed to be zero. It indicates that deformations occur only along the direction of the applied force/moments.

Figure 11 shows the FEA results of the force-displacement relationship of the 4RRR micro-motion stage. It can be seen that the displacement of the platform along x axis increases linearly with the increase of the x axis actuating force. Furthermore, the displacement of x axis is constant (approaching zero) when the actuation force of y axis increases, while the displacement of y axis is constant (approaching zero) when the actuation force of x axis increases. It indicates that the 4-RRR micro-motion stage has output decoupling property, which agrees with the analyses above.

Compared with Eqs. (33) and (34), the output compliance of the RRR and 4-RRR micro-motion stages determined using analytical model can be expressed as, respectively

$$\mathbf{C}_{\mathrm{RRR}} = \begin{bmatrix} 3.938\times10^{-5} & -5.200\times10^{-5} & 9.161\times10^{-4} \\ -5.200\times10^{-5} & 1.058\times10^{-4} & -1.499\times10^{-3} \\ 9.161\times10^{-4} & -1.499\times10^{-3} & 2.482\times10^{-2} \end{bmatrix}, \quad (35)$$

$$\mathbf{C}_{\mathrm{4RRR}} = \begin{bmatrix} 1.205\times10^{-6} & 0 & 0 \\ 0 & 3.273\times10^{-6} & 0 \\ 0 & 0 & 3.096\times10^{-4} \end{bmatrix}. \quad (36)$$

We can see from Eq. (36) that the compliance matrix is a diagonal matrix, and it indicates that the 4-RRR compliant micro-motion stage is output decoupled. Analytical output compliances were compared with those derived from FEA. The differences between the analytical and FEA results are shown in Table 2.

It is found that values obtained from the proposed analytical model are almost similar to the FEA values of the RRR and 4-RRR compliant micro-motion stages, which validates the proposed analytical model.

Table 2. The differences between the analytical and FEA results.

Output compliance matrix			% difference of RRR			% difference of 4-RRR		
C_{x-Fx}	C_{y-Fx}	$C_{\alpha-Fx}$	3.08	2.83	3.05	1.77	–	–
C_{x-Fy}	C_{y-Fy}	$C_{\alpha-Fy}$	2.83	2.83	2.88	–	1.05	–
C_{x-Mz}	C_{y-Mz}	$C_{\alpha-Mz}$	3.05	2.88	2.98	–	–	0.65

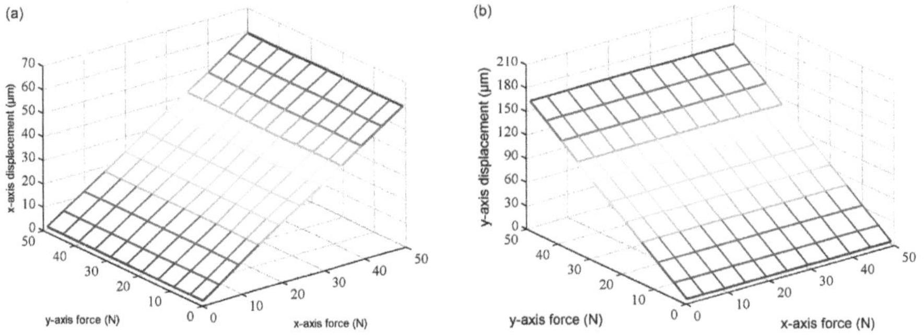

Figure 11. FEA results of force-displacement relationship: **(a)** displacement of x axis; **(b)** displacement of y axis.

5 Conclusions

This paper presented the output decoupling property of planar flexure-based compliant mechanisms. The compliance/stiffness modeling methods were first derived. Then, analytical model utilized for analyzing the output decoupling property of flexure-based compliant mechanisms was obtained in terms of the influences of symmetric configuration. It involved the mechanisms which were symmetric about two perpendicular axes or composed of several fundamental forms. To validate the output decoupling property, the output compliances of the RRR and 4-RRR compliant micro-motion stages were analyzed with the proposed analytical model and FEA. Based upon the results obtained, we can draw the following conclusions:

1. flexure-based compliant mechanisms are output decoupled when they are symmetric about two perpendicular axes;

2. flexure-based compliant mechanisms composed of either three or an even number of identical fundamental forms distributed evenly around the center are output decoupled;

3. comparisons with the results obtained by FEA demonstrate that the proposed analytical model of the 4-RRR micro-motion stage with symmetric configuration is valid.

Acknowledgements. This work was supported by the National Basic Research Program of China (Grant No. 2011CB302400), the National Science Foundation of China (Grant No. 51275260), and the National Science and Technology Major Project of China (Grant No. 2015ZX04001002).

References

Acer, M. and Sabanovic, A.: Comparison of circular flexure hinge compliance modeling methods, in: Proceedings of the 2011 IEEE International Conference on Mechatronics, 13–15 April 2011, Istanbul, Turkey, 271–276, 2011.

Ando, T., Kodera, N., Maruyama, D., Takai, E., Saito, K., and Toda, A.: A high-speed atomic force microscope for studying biological macromolecules in action, Jpn. J. Appl. Phys/, 41, 4851–4856, doi:10.1143/JJAP.41.4851, 2002.

Awtar, S. and Slocum, A. H.: Constraint-based design of parallel kinematic XY flexure mechanisms, J. Mech. Des., 129, 816–830, doi:10.1115/1.2735342, 2007.

Choi, K.-B. and Lee, J. J.: Passive compliant wafer stage for single-step nano-imprint lithography, Rev. Sci. Instrum., 76, 075106, doi:10.1063/1.1948401, 2005.

Handley, D. C., Lu, T.-F., Yong, Y. K., and Hales, C.: Workspace investigation of a 3 DOF compliant micro-motion stage, in: Proceedings of the International Conference on Control, Automation, Robotics and Vision Conference, 6–9 December 2004, Kunming, China, 1279–1284, 2004.

Kenton, B. J. and Leang, K. K.: Design and control of a three-axis serial-kinematic high-bandwidth nanopositioner, IEEE/ASME T. Mechatron., 17, 356–369, doi:10.1109/TMECH.2011.2105499, 2012.

Kim, H.-Y., Ahn, D.-H., and Gweon, D.-G.: Development of a novel 3 degrees of freedom flexure based positioning system, Rev. Sci. Instrum, 83, 055114, doi:10.1063/1.4720410, 2012.

Lai, L.-J., Gu, G.-Y., and Zhu, L.-M.: Design and control of a decoupled two degree of freedom translational parallel micro-positioning stage, Rev. Sci. Instrum., 83, 045105, doi:10.1063/1.3700182, 2012.

Leang, K. K. and Fleming, A. J.: High-speed serial-kinematic SPM scanner: design and drive considerations, Asian. J. Control, 11, 144–153, doi:10.1002/asjc.090, 2009.

Li, C.-X., Gu, G.-Y., Yang, M.-J., and Zhu, L.-M.: Design, analysis and testing of a parallel-kinematic high-bandwidth XY nanopositioning stage, Rev. Sci. Instrum., 84, 125111, doi:10.1063/1.4848876, 2013.

Li, Y. and Xu, Q.: Design of a new decoupled XY flexure parallel kinematic manipulator with actuator isolation, in: Proceedings of the International Conference on Intelligent Robots and Systems, 22–26 September 2008, Nice, France, 470–475, 2008.

Li, Y. and Xu, Q.: Development and assessment of a novel decoupled XY parallel micropositioning platform, IEEE/ASME T. Mechatron., 15, 125–135, doi:10.1109/TMECH.2009.2019956, 2010.

Li, Y., Huang, J., and Tang, H.: A compliant parallel XY micromotion stage with complete kinematic decoupling, IEEE T. Automat. Sci. Eng., 9, 538–553, doi:10.1109/TASE.2012.2198466, 2012.

Pham, H.-H. and Chen, I.-M.: Stiffness modeling of flexure parallel mechanism, Precis. Eng., 29, 467–478, doi:10.1016/j.precisioneng.2004.12.006, 2005.

Polit, S. and Dong, J.: Development of a high-bandwidth XY nanopositioning stage for high-rate micro-/nanomanufacturing, IEEE/ASME T. Mechatron., 16, 724–733, doi:10.1109/TMECH.2010.2052107, 2011.

Qin, Y., Shirinzadeh, B., Zhang, D., and Tian, Y.: Compliance modeling and analysis of statically indeterminate symmetric flexure structures, Precis. Eng., 37, 415–424, doi:10.1016/j.precisioneng.2012.11.004, 2013.

Qin, Y., Shirinzadeh, B., Tian, Y., Zhang, D., and Bhagat, U.: Design and computational optimization of a decoupled 2-DOF monolithic mechanism, IEEE/ASME T. Mechatron., 19, 872–881, doi:10.1109/TMECH.2013.2262801, 2014.

Schitter, G., Åström, K. J., DeMartini, B. E., Thurner, P. J., Turner, K. L., and Hansma, P. K.: Design and modeling of a high-speed AFM-scanner, IEEE T. Control Syst. Technol., 15, 906–915, doi:10.1109/TCST.2007.902953, 2007.

Tseytlin, Y. M.: Notch flexure hinges: an effective theory, Rev. Sci. Instrum., 73, 3363–3368, doi:10.1063/1.1499761, 2002.

Verma, S., Kim, W.-J., and Shakir, H.: Multi-axis maglev nanopositioner for precision manufacturing and manipulation applications, IEEE T. Indust. Appl., 41, 1159–1167, doi:10.1109/TIA.2005.853374, 2005.

Wang, H. and Zhang, X.: Input coupling analysis and optimal design of a 3-DOF compliant micro-positioning stage, Mech. Mach. Theory, 43, 400–410, doi:10.1016/j.mechmachtheory.2007.04.009, 2008.

Yong, Y. K. and Lu, T.-F.: The effect of the accuracies of flexure hinge equations on the output compliances of planar micro-motion stages, Mech. Mach. Theory, 43, 347–363, doi:10.1016/j.mechmachtheory.2007.03.007, 2008.

Yong, Y. K., and Lu, T.-F.: Comparison of circular flexure hinge design equations and the derivation of empirical stiffness formulations, in: Proceedings of the 2009 IEEE International Conference on Advanced Intelligent Mechatronics, 14–17 July 2009, Singapore, 510–515, 2009.

Yong, Y. K., Lu, T.-F., and Handley, D. C.: Review of circular flexure hinge design equations and derivation of empirical formulations, Precis. Eng., 32, 63–70, doi:10.1016/j.precisioneng.2007.05.002, 2008.

Yu, J., Li, S., Su, H., and Culpepper, M. L.: Screw theory based methodology for the deterministic type synthesis of flexure mechanisms, J. Mech. Robot., 3, 031008, doi:10.1115/1.4004123, 2011.

Zubir, M. N. M., Shirinzadeh, B., and Tian, Y.: Development of a novel flexure-based compliant microgripper for high precision micro-object manipulation, Sens. Actuators A, 150, 257–266, doi:10.1016/j.sna.2009.01.016, 2009.

21

A continuum anisotropic damage model with unilateral effect

A. Alliche

Sorbonne Universités, UPMC Univ Paris 06, CNRS, UMR 7190, Institut Jean Le Rond d'Alembert, 75005 Paris, France

Correspondence to: A. Alliche (abdenour.alliche@upmc.fr)

Abstract. A continuum damage mechanics model has been derived within the framework of irreversible thermodynamics with internal variables in order to describe the behaviour of quasi-brittle materials under various loading paths. The anisotropic character induced by the progressive material degradation is explicitly taken into account, and the Helmholtz free energy is a scalar function of the basic invariants of the second order strain and damage tensors. The elastic response varies depending on the closed or open configuration of defects. The constitutive laws derived within the framework of irreversible thermodynamics theory display a dissymmetry as well as unilateral effects under tensile and compressive loading conditions. This approach verifies continuity and uniqueness of the potential energy. An application to uniaxial tension-compression loading shows a good adequacy with experimental results when available, and realistic evolutions for computed stresses and strains otherwise.

1 Introduction

Most geomaterials and concrete are regarded as isotropic and heterogeneous materials before any mechanical loading at mesoscopic scale. The application of a mechanical loading causes occurence of defects whose directions of propagation depends on the local stress field, Microcracks propagate in a direction normal to tension, but tend to close in the case of compression with possible frictional slip on the lips of discontinuities. Complete crack closure causes a recovery of the material stiffness in the direction of compressive stress, this phenomenon is called unilateral effect (Ramtani et al., 1992; Yazdani and Schreyer, 1988; Torrenti and Djebri, 1990; Krajcinovic, 1989). In addition, under a simple mechanical test one observes that cancelation of loading leads to more or less important irreversible strains (Ortiz, 1985). These effects are caused by frictions at crack closure.

Damage mechanics offer a convenient theoretical tool for describing the complex mechanisms associated with damage and failure processes observed under mechanical loading. Many works are reported concerning damage in concrete and geomaterials (Yazdani and Schreyer, 1988; Ortiz, 1985; Alliche and Dumontet, 2011). The production of numerous macroscopic models is mainly imposed by the complexity and the variety of the observed behaviors (Pigeon, 1969; Chaboche, 1993; Rabier, 1989; Marigo, 1985). Most approaches favor the simplicity of the formulation by using a single scalar damage parameter to describe density of defects. More realistic damage model requires a tensorial formulation for a system of defects strongly influenced by the local field of stress. Some models take into account the unilateral effect (Badel et al., 2007; Halm and Dragon, 1998; Desmorat et al., 2007; Challamel et al., 2005; Alliche and Dumontet, 2011) and the dissymmetry between tension and compression.

Therefore, the formulation of any continuum damage model must account for the principal characteristics described previously, which are summarized below:

- degradation of material properties induced by the creation or propagation of defects,
- anisotropic behavior as a consequence of damage,
- dissymmetry in behavior between tension and compression,

- occurrence of irreversible strains after complete unloading.

Furthermore the potential of free energy must verify the property of continuity. This implies conditions on the expression of the jump of the elastic stiffness tensor in the open and closed configuration of the discontinuities.

The present model takes in account the various characteristics quoted above. The potential of free energy is considered as a scalar function of the strain and damage tensors. The constitutive equations are applied to describe the anisotropic elastic damage behavior of concrete. A confrontation is made with existing experimental measurements (Murakami and Kamiya, 1997).

The model has been implemented in the finite element software Plaxis, in order to investigate the model response under more general stress state conditions, as well as to assess the applicability of the model to geotechnical case studies such as deep underground excavations in rocks.

2 Model formulation

2.1 Tensorial damage variable

The presence of microdiscontinuities and their evolution in the material structure leads to an alteration of its mechanical properties. These microcracks exist in different forms and at different scales, in particular in heterogeneous brittle media which geomaterials constitute one of the representative classes.

The need for predicting the mechanical response of geomaterials under various loading paths leads to the representation of such defects by a damage variable, which can take a scalar or tensorial character depending on the degree of characterization of defects. For this class of materials, damage can be represented by a distribution of defects depending on the local field of stress and strain.

Anisotropic damage results from a distribution system of n defects. In the case of a configuration of parallel defects, the associated damage parameter can be written as follows:

$$\mathbf{D} = D_i \boldsymbol{n}_i \otimes \boldsymbol{n}_i, \tag{1}$$

where D_i is the defects density for the ith family of microcracks of orientation n_i. Consequently, for all the defects contained within the microstructure, we obtain by accounting for all the possible orientations:

$$\mathbf{D} = \sum_i D_i \boldsymbol{n}_i \otimes \boldsymbol{n}_i. \tag{2}$$

The second order damage tensor in Eq. (2) is symmetric and therefore has three eigenvalues D_k associated with three eigenvectors E_k $(k = 1, 2, 3)$. Consequently, any system of discontinuities can be represented by p parallel families of microcracks, which can be reduced to three mutually orthogonal defect densities D_k $(k = 1, 3)$:

$$\mathbf{D} = \sum_{k=1}^{3} D_k \boldsymbol{n}_k \otimes \boldsymbol{n}_k. \tag{3}$$

2.2 Thermodynamic potential and state laws

In the case of an elastic damaged material, the thermodynamic potential of free energy w is a scalar function of two state variables, namely the strain tensor $\boldsymbol{\varepsilon}$ and the damage tensor $\mathbf{D}$:

$$w = w(\boldsymbol{\varepsilon}, \mathbf{D}). \tag{4}$$

The scalar function in Eq. (4) is taken to be linear with respect to $\mathbf{D}$ in the case of non-interacting defects and quadratic with respect to $\boldsymbol{\varepsilon}$ to express the linear nature of the behavior law for this class of materials. A possible form of the potential of free energy is written below:

$$w(\boldsymbol{\varepsilon}, \mathbf{D}) = \frac{1}{2} \boldsymbol{\varepsilon} : \mathbb{C}(\mathbf{D}) : \boldsymbol{\varepsilon}, \tag{5}$$

where $\mathbb{C}(\mathbf{D})$ is the fourth order stiffness tensor of an isotropic material for a given damage state. In particular, for the virgin material prior to damage onset, we have:

$$\mathbb{C}(\mathbf{D} = \mathbf{0}) = \mathbb{C}_0. \tag{6}$$

Damage can be viewed as a perturbation of the material structure, resulting in a decrease in the potential of free energy. Therefore the potential of the damaged material is considered to be equal to the potential of the undamaged material $w_0(\boldsymbol{\varepsilon}, \mathbf{D} = \mathbf{0})$ reduced by the energy associated with damage $w_D(\boldsymbol{\varepsilon}, \mathbf{D})$:

$$w(\boldsymbol{\varepsilon}, \mathbf{D}) = w_0(\boldsymbol{\varepsilon}, \mathbf{D} = \mathbf{0}) - w_D(\boldsymbol{\varepsilon}, \mathbf{D}). \tag{7}$$

2.3 Identification and explicit formulation of an anisotropic damage model

2.3.1 Linear elasticity and continuity conditions

Macroscopic modeling of damage unilateral effects constitutes an open research field. Several formulations have been proposed to solve the problem of the damage activation-deactivation process, also called unilateral effect (Chaboche, 1992). In a critical paper review, Cormery and Welemane (2002) have examined several existing formulations and displayed some inconsistencies in existing models. In particular, their theoretical investigation has demonstrated that the formulations proposed by Chaboche (1993) and Halm and Dragon (1996), which are based on a spectral decomposition of the potential to represent the damage activation-deactivation process, lead to an unacceptable thermodynamic potential.

A damage brittle material such as geomaterials exhibit different stiffness in compressive and tensile loading with a kink at the origine (Ramtani et al., 1992; Curnier et al., 1995). For modelling such an unilateral response, we consider an elastic potential energy which is continuously differentiable, and piecewise twice continuously differentiable. Furthermore we assume that for a given damage state $\mathbf{D}$, the strain space E can be partitioned into two half-spaces E^- (for compression) and E^+ (for tension) by means of a hypersurface $\mathcal{J}$ characterized by a scalar-valued function $\Gamma(\boldsymbol{\varepsilon})$:

$$\begin{aligned} \mathcal{J} &= \{\boldsymbol{\varepsilon} / \Gamma(\boldsymbol{\varepsilon}) = 0\} \\ E^- &= \{\boldsymbol{\varepsilon} \in E / \Gamma(\boldsymbol{\varepsilon}) < 0\} \\ E^+ &= \{\boldsymbol{\varepsilon} \in E / \Gamma(\boldsymbol{\varepsilon}) \geq 0\} \end{aligned} \tag{8}$$

Let $w(\boldsymbol{\varepsilon}, \mathbf{D})$ designate the continuously differentiable energy function defined over the strain space E.

$w^+(\boldsymbol{\varepsilon}, \mathbf{D})$ and $w^-(\boldsymbol{\varepsilon}, \mathbf{D})$ are twice continuously differentiable energy functions, such that:

$$w(\boldsymbol{\varepsilon}, \mathbf{D}) = \begin{cases} w^+(\boldsymbol{\varepsilon}, \mathbf{D}) & \text{if } \Gamma(\boldsymbol{\varepsilon}) \geq 0 \\ w^-(\boldsymbol{\varepsilon}, \mathbf{D}) & \text{if } \Gamma(\boldsymbol{\varepsilon}) < 0 \end{cases}$$

$\boldsymbol{\sigma}^+(\boldsymbol{\varepsilon}, \mathbf{D})$ and $\boldsymbol{\sigma}^-(\boldsymbol{\varepsilon}, \mathbf{D})$ represent the first order derivatives of the energy function over their respective subdomains E^+ and E^-:

$$\sigma(\boldsymbol{\varepsilon}, \mathbf{D}) = \begin{cases} \sigma^+(\boldsymbol{\varepsilon}, \mathbf{D}) = \dfrac{\partial w^+(\boldsymbol{\varepsilon}, \mathbf{D})}{\partial \boldsymbol{\varepsilon}} & \text{if } \Gamma(\boldsymbol{\varepsilon}) \geq 0 \\ \sigma^-(\boldsymbol{\varepsilon}, \mathbf{D}) = \dfrac{\partial w^-(\boldsymbol{\varepsilon}, \mathbf{D})}{\partial \boldsymbol{\varepsilon}} & \text{if } \Gamma(\boldsymbol{\varepsilon}) < 0 \end{cases}$$

In the same way, $\mathbb{C}^+(\mathbf{D})$ and $\mathbb{C}^-(\mathbf{D})$ represent the second order partial derivatives with respect to the strain tensor, defined respectively over E^+ and E^- :

$$\mathbb{C}(\mathbf{D}) = \begin{cases} \mathbb{C}^+(\mathbf{D}) = \dfrac{\partial^2 w^+(\boldsymbol{\varepsilon}, \mathbf{D})}{\partial \boldsymbol{\varepsilon}^2} & \text{if } \Gamma(\boldsymbol{\varepsilon}) \geq 0 \\ \mathbb{C}^-(\mathbf{D}) = \dfrac{\partial^2 w^-(\boldsymbol{\varepsilon}, \mathbf{D})}{\partial \boldsymbol{\varepsilon}^2} & \text{if } \Gamma(\boldsymbol{\varepsilon}) < 0 \end{cases}$$

The continuity of the stress-strain response at the transition between the two states of damage (tension and compression) requires the thermodynamic potential $w(\boldsymbol{\varepsilon}, \mathbf{D})$ to be continuously differentiable. Curnier et al. (1995) and Welemane (2002) have demonstrated that the necessary and sufficient condition for $w(\boldsymbol{\varepsilon}, \mathbf{D})$ to be C^1 – continuous can be expressed as follows:

$$[[\mathbb{C}(\mathbf{D})]] = \mathbb{C}^+(\mathbf{D}) - \mathbb{C}^-(\mathbf{D}) = \left[\left[\frac{\partial^2 w}{\partial \boldsymbol{\varepsilon}^2}\right]\right] = s \cdot (\mathbf{D}) \frac{\partial \Gamma(\boldsymbol{\varepsilon})}{\partial \boldsymbol{\varepsilon}} \otimes \frac{\partial \Gamma(\boldsymbol{\varepsilon})}{\partial \boldsymbol{\varepsilon}}, \forall \boldsymbol{\varepsilon} / \Gamma(\boldsymbol{\varepsilon}) = 0 \tag{9}$$

where $[[\mathbb{C}(\mathbf{D})]]$ represents the jump of the mechanical stiffness tensor through the hyperplane $\Gamma(\boldsymbol{\varepsilon}) = 0$.

s is a positive scalar quantity. The previous condition (9) expresses the fact that the jump in the elasticity tensor across the interface is normal to the interface.

2.3.2 Formulation of the elastic damage model

The most general form of the thermodynamic potential energy $w(\boldsymbol{\varepsilon}, \mathbf{D})$ can be expressed by a combination the invariants of the strain and damage tensors $\boldsymbol{\varepsilon}$ and $\mathbf{D}$:

$$\begin{aligned} w(\boldsymbol{\varepsilon}, \mathbf{D}) = {} & w_0(\boldsymbol{\varepsilon}, \mathbf{D} = \mathbf{0}) + \alpha \mathrm{tr}(\mathbf{D})(\mathrm{tr}(\boldsymbol{\varepsilon}))^2 \\ & + \beta \mathrm{tr}(\mathbf{D}) \mathrm{tr}\left((\boldsymbol{\varepsilon})^2\right) + \gamma \mathrm{tr}(\boldsymbol{\varepsilon} \cdot \mathbf{D}) \mathrm{tr}(\boldsymbol{\varepsilon}) \\ & + \delta \mathrm{tr}\left((\boldsymbol{\varepsilon})^2 \cdot \mathbf{D}\right), \end{aligned} \tag{10}$$

where α, β, γ and δ are material constant parameters.

According to Eq. (10), the potential of free energy can be expressed in the configuration where $\boldsymbol{\epsilon}$ belongs to the tension subdomain E^+ ($\Gamma(\boldsymbol{\epsilon}) > 0$):

$$\begin{aligned} w^+ = {} & w_0^+ + \alpha^+ \mathrm{tr}(\mathbf{D})(\mathrm{tr}(\boldsymbol{\epsilon}))^2 + \beta^+ \mathrm{tr}(\mathbf{D}) \mathrm{tr}\left((\boldsymbol{\epsilon})^2\right) \\ & + \gamma^+ \mathrm{tr}(\boldsymbol{\epsilon} \cdot \mathbf{D}) \mathrm{tr}(\boldsymbol{\epsilon}) + \delta^+ \mathrm{tr}\left(\boldsymbol{\epsilon}^2 \cdot \mathbf{D}\right). \end{aligned} \tag{11}$$

When $\boldsymbol{\epsilon}$ belongs to the compression subdomain E^- ($\Gamma(\boldsymbol{\epsilon}) < 0$):

$$\begin{aligned} w^- = {} & w_0^- + \alpha^- \mathrm{tr}(\mathbf{D})(\mathrm{tr}(\boldsymbol{\epsilon}))^2 + \beta^- \mathrm{tr}(\mathbf{D}) \mathrm{tr}\left((\boldsymbol{\epsilon})^2\right) \\ & + \gamma^- \mathrm{tr}(\boldsymbol{\epsilon} \cdot \mathbf{D}) \mathrm{tr}(\boldsymbol{\epsilon}) + \delta^- \mathrm{tr}\left(\boldsymbol{\epsilon}^2 \cdot \mathbf{D}\right). \end{aligned} \tag{12}$$

From Eqs. (11) and (12) we obtain the following expression for the stiffness tensor discontinuity:

$$\begin{aligned} [[\mathbb{C}(\mathbf{D})]] = {} & 2\left(\alpha^+ - \alpha^-\right) \mathrm{tr}(\mathbf{D}) \mathbf{1} \otimes \mathbf{1} + 2\left(\beta^+ - \beta^-\right) \mathbf{1} \overline{\underline{\otimes}} \mathbf{1} \\ & + \left(\gamma^+ - \gamma^-\right)(\mathbf{1} \otimes \mathbf{D} + \mathbf{D} \otimes \mathbf{1}) \\ & + \left(\delta^+ - \delta^-\right)\left(\mathbf{1} \overline{\underline{\otimes}} \mathbf{D} + \mathbf{D} \overline{\underline{\otimes}} \mathbf{1}\right). \end{aligned} \tag{13}$$

By comparing Eqs. (9) and (10), the function $\Gamma(\boldsymbol{\epsilon}) = 0$ can be identified as:

$$\Gamma(\boldsymbol{\epsilon}) = \mathrm{tr}(\boldsymbol{\epsilon}) \tag{14}$$

The function associated with the hyperplane $\mathcal{J}$ is therefore defined by Eq. (14). This implies that the criterion associated with the transition between the tension hyperspace E^+ and the compression hyperspace E^-, which is expressed for a given state of the damage variable $\mathbf{D}$, depends solely on the sign of $\mathrm{tr}(\boldsymbol{\epsilon})$. This may be illustrated through the analysis of an uniaxial damage loading test, Fig. 1. The solid lines represent typical virgin loading curves under an uniaxial state of tension or compression alone. The dash lines represent the qualitative response of the model during unloading in the tension domain (after reaching a certain amount of damage) and subsequent compression. During crossing of the hyperplane $\Gamma(\boldsymbol{\epsilon}) = \mathrm{tr}(\boldsymbol{\epsilon}) = 0$, the tensorial damage variable $\mathbf{D}$ remains constant, resulting in a null damage rate.

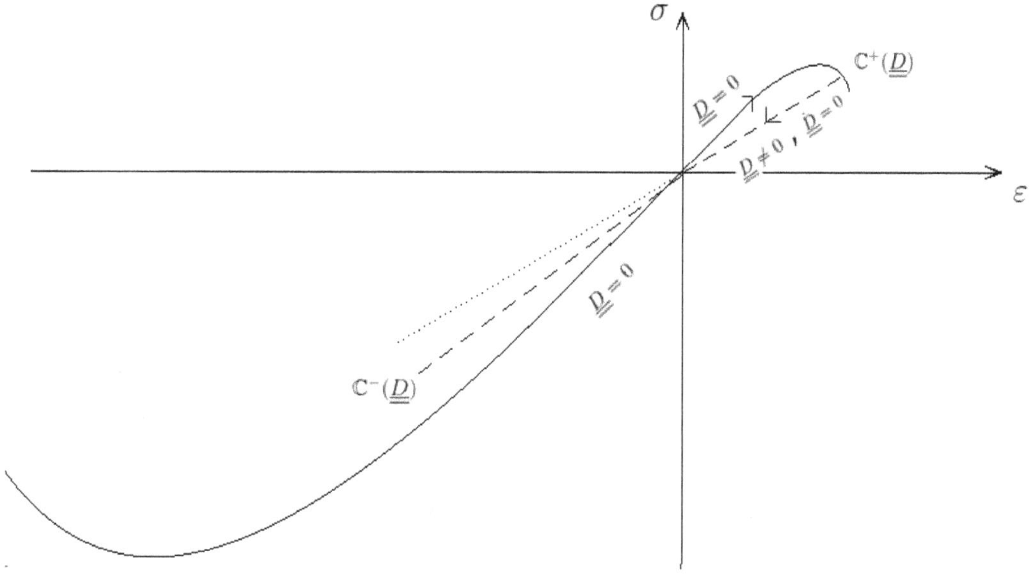

Figure 1. Schematic representation of the transition state between tension and compression loading.

The continuity condition (9) and Eq. (13) imply the following set of relationships, that must be verified by parameters $\alpha^+, \alpha^-, \beta^+, \beta^-, \gamma^+, \gamma^-$ and δ^+, δ^-:

$$\alpha^+ \neq \alpha^-$$
$$\beta^+ = \beta^- = \beta$$
$$\gamma^+ = \gamma^- = \gamma$$
$$\delta^+ = \delta^- = \delta$$

Equations (11) and (12) can be summarized in the following expression, which expresses the potential of free energy $w(\boldsymbol{\epsilon}, \mathbf{D})$ in a unified way:

$$w(\boldsymbol{\epsilon}, \mathbf{D}) = G_0 \mathrm{tr}\left(\boldsymbol{\epsilon}^D \cdot \boldsymbol{\epsilon}^D\right) + \frac{K_0}{2}(\mathrm{tr}(\boldsymbol{\epsilon}))^2 + \alpha^+\left[(\mathrm{tr}(\boldsymbol{\epsilon}))_+\right]^2 \mathrm{tr}(\mathbf{D}) + \alpha^-\left[(-\mathrm{tr}(\boldsymbol{\epsilon}))_+\right]^2 \mathrm{tr}(\mathbf{D}) + \beta \mathrm{tr}(\mathbf{D}) \mathrm{tr}\left((\boldsymbol{\epsilon})^2\right) + \gamma \mathrm{tr}(\boldsymbol{\epsilon} \cdot \mathbf{D}) + \delta \mathrm{tr}\left(\boldsymbol{\epsilon}^2 \cdot \mathbf{D}\right), \quad (15)$$

where $\boldsymbol{\epsilon}^D = \boldsymbol{\epsilon} - \frac{1}{3}\mathrm{tr}(\boldsymbol{\epsilon})\mathbf{1}$ is the deviatoric strain tensor, and $K_0 = \frac{3\lambda_0 + 2\mu_0}{3}$ the compressibility modulus.

The terms in α^+ and α^- in Eq. (15) reflect the unilateral effect due to possible partial deactivation of defects.

3 State laws and damage rate evolution

3.1 State laws : stress tensor and thermodynamic force

The macroscopic tensor $\boldsymbol{\sigma}$ is obtained by partial derivation of $w(\boldsymbol{\epsilon}, \mathbf{D})$ with respect to the strain tensor $\boldsymbol{\epsilon}$:

$$\boldsymbol{\sigma} = \frac{\partial w}{\partial \boldsymbol{\epsilon}} = 2G_0 \boldsymbol{\epsilon}^D + K_0 \mathrm{tr}(\boldsymbol{\epsilon})\mathbf{1} + 2\alpha^+ (\mathrm{tr}(\boldsymbol{\epsilon}))_+ \mathrm{tr}(\mathbf{D})\mathbf{1} - 2\alpha^-(-\mathrm{tr}(\boldsymbol{\epsilon}))_+ \mathrm{tr}(\mathbf{D})\mathbf{1} + 2\beta \mathrm{tr}(\mathbf{D})\boldsymbol{\epsilon} + \gamma\left[\mathrm{tr}(\boldsymbol{\epsilon} \cdot \mathbf{D})\mathbf{1} + \mathrm{tr}(\boldsymbol{\epsilon})\mathbf{D}\right] + \delta(\boldsymbol{\epsilon} \cdot \mathbf{D} + \mathbf{D} \cdot \boldsymbol{\epsilon}). \quad (16)$$

The second state law allows to introduce the thermodynamic force associated to the second order damage tensor $\mathbf{D}$:

$$\mathbf{Y} = -\frac{\partial w}{\partial \mathbf{D}} = -\alpha^+ (\mathrm{tr}(\boldsymbol{\epsilon}))_+^2 \mathbf{1} - \alpha^-(-\mathrm{tr}(\boldsymbol{\epsilon}))_+^2 \mathbf{1} - \beta \mathrm{tr}(\boldsymbol{\epsilon} \cdot \boldsymbol{\epsilon})\mathbf{1} - \gamma \mathrm{tr}(\boldsymbol{\epsilon})\boldsymbol{\epsilon} - \delta \boldsymbol{\epsilon} \cdot \boldsymbol{\epsilon} \quad (17)$$

3.2 Damage criterion

Experimental tests performed on geomaterials show the existence of an area in the strain space inside which damage is negligible. Initiation or evolution of defects may appear only if the state of strain reaches the limit of this area. We assume the existence of a damage criterion in the space of thermodynamic forces written in the following form:

$$F(\mathbf{Y}, \mathbf{D}) = \|\mathbf{Y}\| - \chi(\mathbf{D}), \quad (18)$$

$\chi(\mathbf{D})$ is a scalar function, taken to be linear with respect to $\mathbf{D}$:

$$\chi(\mathbf{D}) = a_1 \mathrm{tr}(\mathbf{D}) + a_0, \quad (19)$$

where a_0 and a_1 are material constants: a_0 characterizes the initial damage threshold, while a_1 describes the manner in which the surface evolves with damage.

The choice of the criterion (18) indicates that the model is associated, and damage evolution is assumed to follow the normality rule:

$$\dot{D} = \begin{cases} 0, & \text{if } F < 0, \text{ or } F = 0 \text{ and } \dot{F} < 0; \\ \dot{\lambda}\dfrac{\partial F(\mathbf{Y},\mathbf{D})}{\partial \mathbf{Y}}, & \text{if } F = 0 \text{ and } \dot{F} = 0. \end{cases} \tag{20}$$

The damage multiplier $\dot{\lambda}$ is determined by the consistency equations $F(\mathbf{Y},\mathbf{D}) = 0$ and $\dot{F}(\mathbf{Y},\mathbf{D}) = 0$, leading to:

$$\dot{D} = \frac{\mathrm{tr}(\mathbf{Y}\cdot\dot{\mathbf{Y}})}{a_1 \mathrm{tr}(\dot{\mathbf{Y}})\,\|\mathbf{Y}\|}\mathbf{Y}. \tag{21}$$

From Eq. (21), the positivity of the dissipation $\mathcal{D}$ is immediately verified:

$$\mathcal{D} = \mathbf{Y}\cdot\dot{\boldsymbol{D}}. \tag{22}$$

4 Application to uniaxial tension

The proposed model can be developed analytically and explicitly in the case of uniaxial monotonic tension. Let $\boldsymbol{e}_1$ be the direction of tension loading. The stress, strain and damage tensors are given below:

$$\boldsymbol{\sigma} = \begin{pmatrix} \sigma_1 & 0 & 0 \\ 0 & 0 & 0 \\ 0 & 0 & 0 \end{pmatrix}, \boldsymbol{\epsilon} = \begin{pmatrix} \epsilon_1 & 0 & 0 \\ 0 & \epsilon_2 & 0 \\ 0 & 0 & \epsilon_3 \end{pmatrix},$$

$$\mathbf{D} = \begin{pmatrix} D_1 & 0 & 0 \\ 0 & D_2 & 0 \\ 0 & 0 & D_3 \end{pmatrix}.$$

We have: $\sigma_1 \geq 0, \epsilon_1 \geq 0, \epsilon_2 = \epsilon_3 \leq 0$, which implies $D_2 = D_3$, and we consider that $\mathrm{tr}(\boldsymbol{\epsilon}) \geq 0$. Using the previous expressions, constitutive relations (16) simplify in the following way:

$$\sigma_1 = \frac{4G_0}{3}(\epsilon_1 - \epsilon_2) + K_0 \mathrm{tr}(\boldsymbol{\epsilon}) + 2\left[\alpha^+ \mathrm{tr}(\boldsymbol{\epsilon}) + \beta\epsilon_1\right]\mathrm{tr}(\mathbf{D}) + \gamma\left[\mathrm{tr}(\boldsymbol{\epsilon}\cdot\mathbf{D}) + \mathrm{tr}(\boldsymbol{\epsilon})D_1\right] \tag{23}$$

$$\epsilon_2 = \frac{\frac{2G_0}{3} - K_0 - 2\alpha^+(D_1 + 2D_2) - \gamma(D_1 + D_2)}{\frac{2G_0}{3} + 2K_0 + 2(2\alpha^+ + \beta)(D_1 + 2D_2) + 2(2\gamma + \delta)D_2}\epsilon_1. \tag{24}$$

The components of the thermodynamic damage force are obtained from Eq. (17):

$$Y_1 = -\alpha^+ \mathrm{tr}((\boldsymbol{\epsilon})_+)^2 - \beta \mathrm{tr}\left(\boldsymbol{\epsilon}^2\right) - \gamma \mathrm{tr}(\boldsymbol{\epsilon})\epsilon_1 - \delta(\epsilon_1)^2, \tag{25}$$

$$Y_2 = Y_3 = -\alpha^+ \mathrm{tr}((\boldsymbol{\epsilon})_+)^2 - \beta \mathrm{tr}\left(\boldsymbol{\epsilon}^2\right) - \gamma \mathrm{tr}(\boldsymbol{\epsilon})\epsilon_2 - \delta(\epsilon_2)^2. \tag{26}$$

The damage criterion is then expressed as:

$$F(\mathbf{Y},\mathbf{D}) = \left(Y_1^2 + 2Y_2^2\right)^{\frac{1}{2}} - a_{1t}\mathrm{tr}(\mathbf{D}) + a_{0t} \tag{27}$$

Constant a_{0t} is identified at the onset of damage by writing:

$$F(\mathbf{Y},\mathbf{D}=\mathbf{0}) = \|\mathbf{Y}\| - a_{0t} = 0$$

and thus,

$$a_{0t} = \left(\chi_{1t}^2 + 2\chi_{2t}^2\right)(\epsilon_1^2)_t. \tag{28}$$

In the previous relation,

$$\chi_{1t} = \alpha^+(1-2\nu_0)^2 + \beta\left(1+2\nu_0^2\right) + \gamma(1-2\nu_0) + \delta,$$

$$\chi_{2t} = \alpha^+(1-2\nu_0)^2 + \beta\left(1+2\nu_0^2\right) + \gamma(1-2\nu_0)\nu + \delta\nu_0^2,$$

where ν_0 is Poisson's ratio for the undamaged material, and $(\epsilon_1)_t$ corresponds to the threshold tensile strain at the onset of damage.

Equation (21) gives the components of the damage rate tensor:

$$\dot{D}_1 = \frac{Y_1\dot{Y}_1 + 2Y_2\dot{Y}_2}{a_1 \mathrm{tr}(\mathbf{Y})\,\|\mathbf{Y}\|}Y_1, \tag{29}$$

$$\dot{D}_2 = \dot{D}_3 = \frac{Y_1\dot{Y}_1 + 2Y_2\dot{Y}_2}{a_1 \mathrm{tr}(\mathbf{Y})\,\|\mathbf{Y}\|}Y_2. \tag{30}$$

We have identified the five parameters of the model in the case of the experimental compression and numerical tension curves of Murakami and Kamiya (1997). The identification of the model parameters has been performed with the numerical optimization software BIANCA (Biological Analysis of Composite Assemblages), which is based on genetical algorithms (Vincenti et al., 2010). The first step consists in identifying the limit elastic strains in tension and compression directly on the experimental curves. The main objective function is the difference between the experimental axial stress values and their numerical counterparts. The five independent damage model parameters are computed by running a large series of simulation, and minimizing the difference between the axial strain-axial stress curve and the corresponding numerical curve in Murakami and Kamiya (1997) in a least square sense.

The elastic material parameters are taken from Murakami and Kamiya (1997): $E = 21.4$ GPa, $\nu = 0.2$. The values of the five independent damage parameters are the following ones: $\alpha^+ = -8.015$ GPa, $\beta = -3.001$ GPa, $\gamma = 11.587$ GPa, $\delta = -5.425$ GPa, $a_{1t} = 0.0195$ MPa.

Figure 2 illustrates the evolution of the longitudinal stress computed from Eq. (23), the computed stress σ_1 is very close to the evolution obtained by Murakami and Kamiya (1997) using their model, which shows the consistency of the identification procedure.

In Fig. 3, we have reported the evolutions of the longitudinal stiffness modulus E and Poisson's ratio ν, computed respectively from Eqs. (23) and (24). As expected, Fig. 3 shows a continuous decrease in Young's modulus and Poisson's ratio starting from the damage threshold strain ($\epsilon_1 = 4\times10^{-4}$).

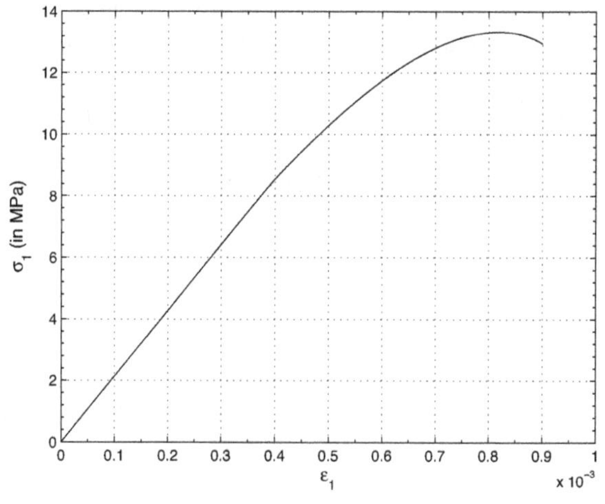

Figure 2. Uniaxial tension. Uniaxial stress-strain curve.

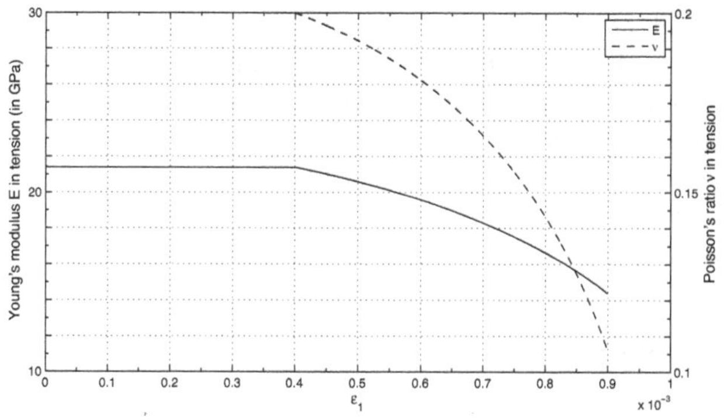

Figure 3. Uniaxial tension. Evolutions of Young's modulus and Poisson's ratio.

5 Application to uniaxial compression

Similar to tensile loading, the equations of the proposed model can be developed analytically in the case of uniaxial compression. With $\boldsymbol{e}_1$ the direction of the compressive load, the expressions of the stress, strain and damage tensors are identical to the uniaxial tension case, the only difference being that we now consider $\mathrm{tr}(\boldsymbol{\epsilon}) \leq 0$. Similar developments as in Sect. 4 lead to the following set of equations:

$$\sigma_1 = \frac{4G_0}{3}(\epsilon_1-\epsilon_2)+K_0\mathrm{tr}(\boldsymbol{\epsilon})+2\left[\alpha^-\mathrm{tr}(\boldsymbol{\epsilon})+\beta\epsilon_1\right]\mathrm{tr}(\mathbf{D}) + \gamma\left[\mathrm{tr}(\boldsymbol{\epsilon}\cdot\mathbf{D})+\mathrm{tr}(\boldsymbol{\epsilon})D_1\right], \quad (31)$$

$$\epsilon_2 = \frac{\frac{2G_0}{3}-K_0-2\alpha^-(D_1+2D_2)-\gamma(D_1+D_2)}{\frac{2G_0}{3}+2K_0+2\left(2\alpha^-+\beta\right)(D_1+2D_2)+2(2\gamma+\delta)D_2}\epsilon_1, \quad (32)$$

$$Y_1 = -\alpha^-\mathrm{tr}\left(-(\boldsymbol{\epsilon})_+\right)^2-\beta\mathrm{tr}\left(\boldsymbol{\epsilon}^2\right)-\gamma\mathrm{tr}(\boldsymbol{\epsilon})\epsilon_1-\delta(\epsilon_1)^2, \quad (33)$$

$$Y_2 = Y_3 = -\alpha^-\mathrm{tr}\left(-(\boldsymbol{\epsilon})_+\right)^2-\beta\mathrm{tr}\left(\boldsymbol{\epsilon}^2\right)-\gamma\mathrm{tr}(\boldsymbol{\epsilon})\epsilon_2-\delta(\epsilon_2)^2. \quad (34)$$

As in uniaxial tension, the compression parameter a_{0c} is identified by writing that $F(\mathbf{Y},\mathbf{D}=\mathbf{0}) = \|\mathbf{Y}\| - a_{0c} = 0$ at the

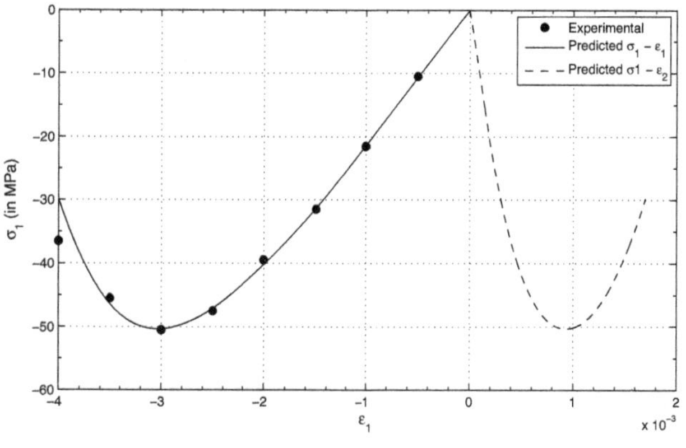

Figure 4. Uniaxial compression. Uniaxial stress-strain curve.

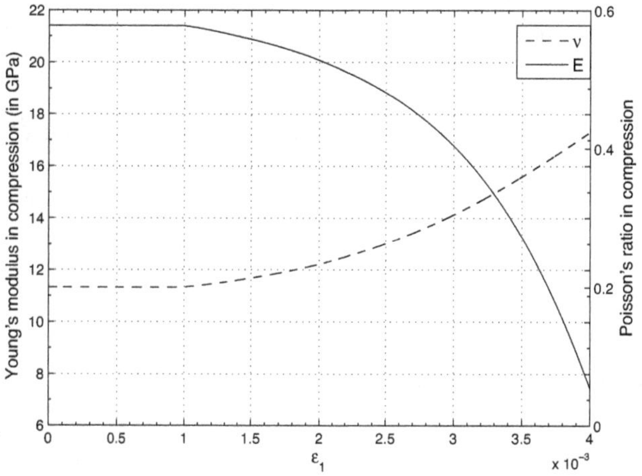

Figure 5. Uniaxial compression. Evolutions of Young's modulus and Poisson's ratio.

onset of damage in compression, resulting in:

$$a_{0c} = \left(\chi_{1c}^2+2\chi_{2c}^2\right)(\epsilon_1^2)_c \quad (35)$$

where

$$\chi_{1c} = \alpha^-(1-2\nu)^2+\beta\left(1+2\nu^2\right)+\gamma(1-2\nu)+\delta,$$

$$\chi_{2c} = \alpha^-(1-2\nu)^2+\beta\left(1+2\nu^2\right)+\gamma(1-2\nu)\nu+\delta\nu^2,$$

and $(\epsilon_1)_c$ corresponds to the threshold compressive strain.

Parameters β, γ and δ have the same values as for tensile loading. The parameters associated to compression loading have been obtained by applying the optimization procedure to the experimental compression curve: $\alpha^- = -0.647$ GPa and $a_{1c} = 0.126$ MPa.

Figures 4 and 5 display respectively the stress-strain, Young's modulus and Poisson's ratio evolutions under uniaxial compression, as predicted by Eqs. (31–32) and a comparison with experimentaldata publishedby Wang and cited by Murakami and Kamiya (1997). Figure 5 shows a decrease in Young's modulus and an increase of Poisson's ratio from

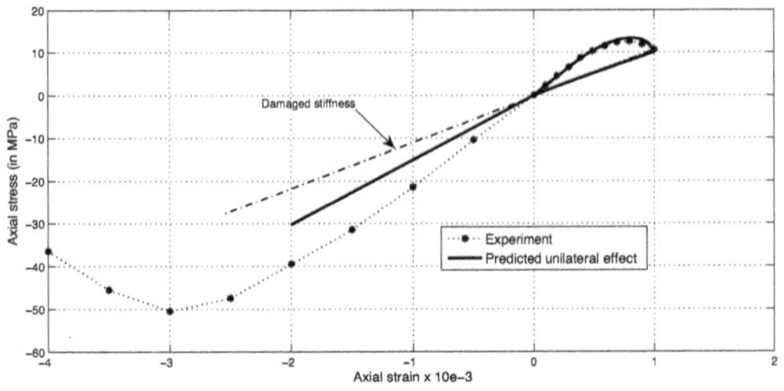

Figure 6. Unilateral effect in tension-compression loading.

the damage threshold strain in compression ($(\epsilon_1)_c = 10^{-3}$). The evolution of these parameters indicates an increase of damage in the transverse direction leading to a diminution of Poisson's ratio. The existence of these microcracks contributes to the weakening of the material stiffness.

Figures 2 and 4 illustrate the dissymmetry existing between the behaviors in tension and compression for this class of material. We observe a ratio of about 4 between the ultimate tensile strength and ultimate compressive stress. This ratio is directly dependent on the values adopted for the paramcters of the model.

6 Unilateral effect

As mentioned previously, the transition from a tensile loading to a compressive loading leads to partial or complete closure of cracks. This change in microstructure causes a recovery of the stiffness of the damaged material. There are relatively few experimental results published on this subject. Ramtani et al. (1992) showed an unilateral effect with total recovery of the stiffness for the compression phase. The simulation we produce in Fig. 6 clearly indicates that the initial stiffness is not completely recovered, and therefore does not reproduce the behavior described by Ramtani. Nevertheless, we may note that the loading path is quite complex to achieve experimentally. It is likely that some cracks remain open during the compression phase and prevent the total recovery of stiffness.

7 Conclusion

A unified model based on continuum damage theory has been developed for quasi-brittle materials such as concrete and rocks. The proposed model relies on a second order tensorial variable to describe anisotropic damage, and has been specifically formulated to be continuously differentiable at the transition between the open and closed cracks configuration. The model has proved to be effective in describing the mechanical behavior of concrete and geomaterials under static loading, more specifically we are able to express:

- the concurrent decrease of Young's modulus and increase of Poisson's ratio with damage development,
- damage-induced anisotropy,
- dissymmetric behavior between tension and compression,
- strain softening behaviour under uniaxial tension-compression loads.

This model requires the identification of a limited set of five parameters that have been extracted from tension-compression experiments through an original and systematic identification procedure based on genetic algorithms. A satisfying agreement is obtained between the experimental tension-compression tests and the theoretical simulations.

Future work concerns the investigation of the predictive capacities of the anisotropic damage model with unilateral effects, by confronting Finite Element simulations with available experimental results, including conventional triaxial tests and real case studies.

References

Alliche, A. and Dumontet, H.: Anisotropic model of damage for geomaterials and concrete, Int. J. Numer. Anal. Met., 35, 969–979, doi:10.1002/nag.933, 2011.

Badel, P., Godard, V., and Leblond, J.-B.: Application of some damage model to the prediction of the failure of some complex industrial concrete structure, Int. J. Solids Struct., 44, 5848–5874, doi:10.1016/j.ijsolstr.2007.02.001, 2007.

Chaboche, J.-L.: A new unilateral condition for the description of material behavior with anisotropic damage, C.R. Acad. Sci. II, 314, 1395–1401, 1992.

Chaboche, J.-L.: Development of continuum damage mechanics for elastic solids sustaining anisotropic and unilateral damage, Int. J. Damage Mech., 2, 311–329, doi:10.1177/105678959300200401, 1993.

Challamel, N., Lanos, C., and Casandjian, C.: Strain-based anisotropic modelling and unilateral effects, Int. J. Mech. Sci., 47, 459–473, doi:10.1016/j.ijmecsci.2005.01.002, 2005.

Cormery, F. and Welemane, H.: A critical review of some damage models with unilateral effect, Mech. Res. Commun., 29, 391–395, doi:10.1016/S0093-6413(02)00262-8, 2002.

Curnier, A., He, Q. C., and Zysset, P.: Conewise linear elastic materials, J. Elasticity, 37, 1–38, doi:10.1007/BF00043417, 1995.

Desmorat, R., Gatuingt, F., and Ragueneau, F.: Nonlocal anisotropic damage model and related computational aspects for quasi-brittle materials, Eng. Fract. Mech., 74, 1539–1560, doi:10.1016/j.engfracmech.2006.09.012, 2007.

Halm, D. and Dragon, A.: A model of anisotropic damage by mesocrack growth; Unilateral effect, Int. J. Damage Mech., 5, 384–402, doi:10.1177/105678959600500403, 1996.

Halm, D. and Dragon, A.: An anisotropic model of damage and frictional sliding for brittle materials, Eur. J. Mech. A-Solid., 17, 439–460, doi:10.1016/S0997-7538(98)80054-5, 1998.

Krajcinovic, D.: Constitutive equations for damaging materials, J. Appl. Mech., 50, 355–360, doi:10.1115/1.3167044, 1989.

Marigo, J.-J.: Modeling of brittle and fatigue damage for elastic material by growth of microvoids, Eng. Fract. Mech., 21, 861–874, doi:10.1016/0013-7944(85)90093-1, 1985.

Murakami, S. and Kamiya, K.: Constitutive and damage evolution equation of elastic-brittle materials based on irreversible thermodynamics, Int. J. Mech. Sci., 39, 473–486, doi:10.1016/S0020-7403(97)87627-8, 1997.

Ortiz, M.: A constitutive theory for inelastic behaviour of concrete, Mech. Mater., 4, 67–93, doi:10.1016/0167-6636(85)90007-9, 1985.

Pigeon, M.: The process of crack initiation and propagation in concrete, PhD thesis, Imperial College, London, 1969.

Rabier, P. J.: Some remarks on damage theory, Int. J. Eng. Sci., 27, 29–54, doi:10.1016/0020-7225(89)90166-3, 1989.

Ramtani, S., Berthaud, Y., and Mazars, J.: Orthotropic behavior of concrete with directional aspects: modeling and experiments, Nucl. Eng. Des., 133, 97–111, doi:10.1016/0029-5493(92)90094-C, 1992.

Torrenti, J. M. and Djebri, B.: Constitutive laws for concrete: an attempt of comparison, Proceedings of the Second International Conference on Computer Aided Analysis and Design of Concrete Structures, 4–6 April 1990, Zell am See, Austria, 871–882, 1990.

Vincenti, A., Ahamadian, M. R., and Vannucci, P.: BIANCA: a genetic algorithm to solve hard combinatorial optimisation problems in engineering, J. Global Optim., 48, 399–421, doi:10.1007/s10898-009-9503-2, 2010.

Welemane, H.: Une modélisation des matériaux microfissures. Application aux roches et aux bétons, PhD thesis, Université Lille 1, France, 2002.

Yazdani, S. and Schreyer, H. L.: An anisotropic damage model with dilatation of concrete, Mech. Mater., 7, 231–244 doi:10.1016/0167-6636(88)90022-1, 1988.

22

Guaranteed detection of the singularities of 3R robotic manipulators

R. Benoit[1], **N. Delanoue**[1], **S. Lagrange**[1], **and P. Wenger**[2]

[1]Laboratoire Angevin de Recherche en Ingénierie des Systèmes (LARIS), 62 avenue Notre Dame du Lac, 49000 Angers, France

[2]Institut de Recherche en Communications et Cybernétique de Nantes (IRCyN), 1 rue la Noë, 44321 Nantes CEDEX 03, France

Correspondence to: R. Benoit (romain.benoit@etud.univ-angers.fr)

Abstract. The design of new manipulators requires the knowledge of their kinematic behaviour. Important kinematic properties can be characterized by the determination of certain points of interest. Important points of interest are cusps and nodes, which are special singular points responsible for the non-singular posture changing ability and for the existence of voids in the workspace, respectively. In practice, numerical errors should be properly tackled when calculating these points. This paper proposes an interval analysis based approach for the design of a numerical algorithm that finds enclosures of points of interest in the workspace and joint space of the studied robot. The algorithm is applied on 3R manipulators with mutually orthogonal joint axes. A pre-processing collision detection algorithm is also proposed, allowing, for instance, to check for the accessibility of a manipulator to its points of interest. Through the two proposed complementary algorithms, based on interval analysis, this paper aims to provide a guaranteed way to obtain a broad characterisation of robotic manipulators.

1 Introduction

Algorithms and methods described in this article are applied to the study of a family of robotic manipulators: *3 revolute-jointed manipulators with mutually orthogonal joint axes.* Those manipulators are first studied because they can be regarded as the positioning structure of a 6R manipulator with a spherical wrist. A main point is that they can be *cuspidal*, which means that they can change their posture without having to meet a singularity, as detailed in Baili et al. (2004) and Wenger (2007). It may or may not be the desired behaviour.

To help the reader understand the notion of non singular posture changing that motivate the study of cuspidal manipulators, two videos showing, respectively, a non-singular and a singular posture changing trajectory, are proposed alongside the online version of this paper (see Supplement). For a robot with only revolute joint axis, checking that a configuration is singular can be done through a geometrical method. Indeed, a configuration is singular if *the end effector is in a revolute joint axis* or if *the end effector is on a line that cross all of the actuated revolute joint axis* (see Baili, 2004).

A cuspidal robot has at least one cusp in a planar cross section of its workspace. On the other hand, the existence of nodes in this section is intimately related to the existence of voids in the robot workspace. Thus, cusps and nodes are important points of interest (Husty et al., 2008). A classification based on the number of such points can be established (Corvez and Rouillier, 2004; Baili et al., 2004).

Cusp points ans nodes points are named after the local form admitted by the image of the singular set at such point. Indeed, a cusp point, in the workspace, is a mathematical cusp point for at least one cross section of the image of the singular set. Similarly, in a cross section of the workspace, a node is located at the crossing of two branches of the image of the singular set.

Formally, a node is defined as a workspace point with two singular inverse kinematic solutions (IKS). In a similar fashion, a cusp can be defined as a workspace point with three equal singular IKS. These definitions are the one used in Baili et al. (2004), so as to define a formal condition for the presence of cusps and nodes through a characteristic polynomial.

Studying the nature of singular points instead of only isolating them to avoid unstable behaviour is relatively recent (Wenger, 2007). However, this approach is quite complementary to the common objective to detect the singular set of a robot, providing useful information on the properties of the robot, particularly for novel design. Methods for detecting the singular set include the brute force method of evaluating numerically the norm of the determinant of the Jacobian (det(**Df**)) and extracting the set of points minimizing this quantity. At the opposite of the spectrum, the equation $\det(\mathbf{Df}) = 0$ is formally or implicitly solved and numerical solution may be extracted from this resolution. For formally complicated kinematic function, a middle ground is needed in the form of methods returning precise constraints on the singular points. This middle ground usually implies a general scheme synergistic with interval analysis methods and will be detailed in Sect. 3.

The main point of the algorithm and methods we are detailing here is to use Interval Analysis to enclose, in a guaranteed way, the cusps and nodes in the generator plane section of the manipulator workspace. To find these points, we use two systems of equations, whose roots are joint space points yielding the cusps and nodes. To enclose the roots of those systems of equations, the *Interval Newton* method is used.

We will verify that, for manipulators with no internal motion, and with some imprecision in their geometric parameters, it is possible to find their cusp and node points, with the formerly introduced algorithms.

Complete studies of manipulator families, as done in Baili et al. (2004), allow one to choose a manipulator within a large range of geometric parameters, when a precise behaviour is needed. Alternatively, algorithms presented in this article make it possible to study manipulators with geometric parameters between chosen bounds. It makes them a first step in guaranteeing the behaviour of a manipulator, given its geometric parameters, and the precision affordable for building the actual manipulator.

2 Studied manipulators

The studied manipulators have three unlimited revolute joints. Thus, it is sufficient to restrict the analysis to their last two joints. Since the workspace is symmetric about the first joint axis, it is enough to restrict its analysis to a planar half cross-section in the plane defined by $\left(\sqrt{x^2+y^2}, z\right)$, that we will identify to $\left(x^2+y^2, z\right)$ for computational purposes.

Figure 1 shows the studied manipulator and its geometric parameters. Note that, for a matter of convenience in our algorithms, angles β_i have been used instead of the standard α_i, where $\beta_i = \pi/2 - \alpha_i$.

We will first consider the same manipulators as in Baili et al. (2004) that is, *manipulators with orthogonal rotations and no offset along their last joint*. With conventions chosen in Fig. 1, these manipulators are defined by $\beta_2 = \beta_3 = r_3 = 0$.

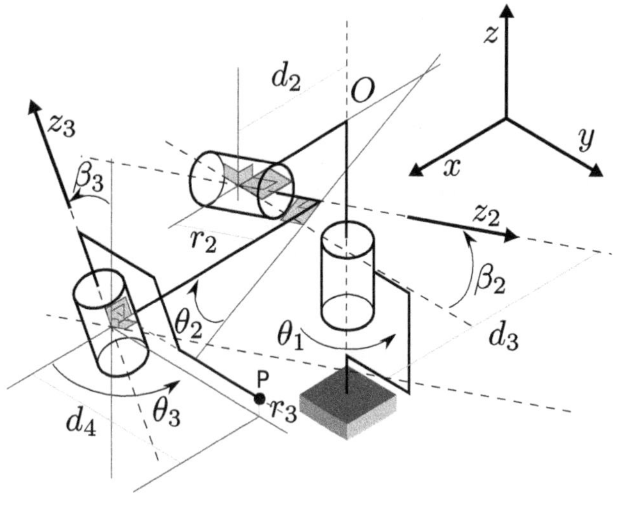

Figure 1. Kinematic diagram of a general 3R manipulator with $\theta_1 = 0$.

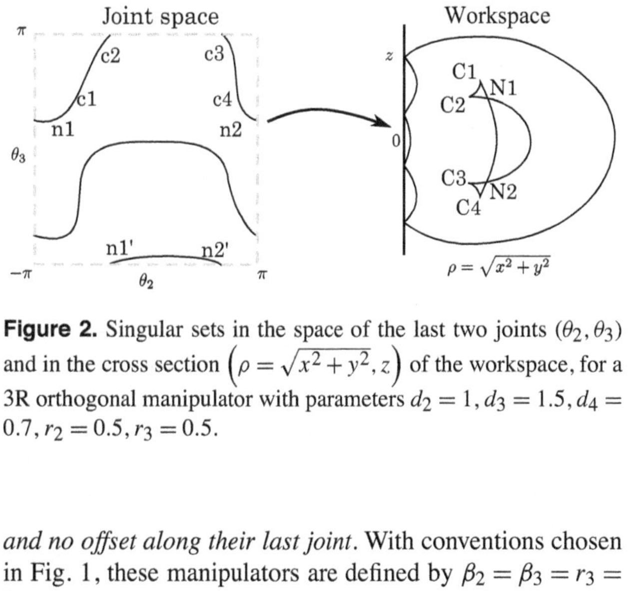

Figure 2. Singular sets in the space of the last two joints (θ_2, θ_3) and in the cross section $\left(\rho = \sqrt{x^2+y^2}, z\right)$ of the workspace, for a 3R orthogonal manipulator with parameters $d_2 = 1, d_3 = 1.5, d_4 = 0.7, r_2 = 0.5, r_3 = 0.5$.

Figure 2 shows, for an instance of orthogonal 3R manipulator, the singular sets of its kinematic function, in the joint space (θ_2, θ_3) and the workspace cross section $\left(\rho = \sqrt{x^2+y^2}, z\right)$. Figure 2 also illustrate the nature of the cusps and nodes, as the cusps (C1, C2, C3, C4) and the nodes (N1, N2), are pictured in the workspace cross section. Their inverse kinematic solutions, on the singular set in the joint space, which are respectively (c1, c2, c3, c4) and ({n1, n1′}, {n2, n2′}), are also pictured, as they are the points we are effectively searching in this paper.

We will show that our methodology is able to provide the same results as in Baili et al. (2004). Furthermore, our approach can also be used for *manipulators with an offset along their last joint* and always returns an exact enclosure of the searched singular joint space points.

3 Application of Interval analysis

3.1 Interval analysis

Interval analysis is a computing method, that operates on intervals instead of operating on values. The point of this is mainly for numerical computation because it allows one to guarantee values to be in intervals (see Jaulin et al., 2001; Moore, 1996) whose bounds can be *exactly* stored by a computer. *Interval analysis is a simultaneous computation of numbers and errors.*

In this article, boxes will be vectors of intervals. The notion of interval can be extended by Cartesian product, so Interval analysis can be extended to boxes by the use of inclusion maps.

Let f be a map. An inclusion map of f is a function $[f]$ that associates to a box D, a box $[f](D)$ such that $f(D) \subset [f](D)$. Note that $(x \in D \Rightarrow f(x) \in [f](D))$.

In practice, the *inclusion map* $[f]$ of f is chosen to minimize the boxes $[f](D)$ with respect to inclusion.

This computing method is useful for its usability when a limited set of values can be exactly represented, as for numerical computations. In this case, a point P is represented by the smallest box D containing P and $f(P)$ is represented by $[f](D)$, the smallest box in the image space containing $f(D)$.

3.2 Interval analysis in Robotics

Interval analysis is a tool that, due to its properties seen in Sect. 3.1, can be used for many applications in Robotics (see Merlet, 2011) such as computing the kinematics of manipulators, including parallel ones.

One of the robotic applications of Interval Analysis is *singularity analysis*, that is, finding singular points of the kinematic map of a manipulator. To find those singular points, a general scheme is used, which consists of a subdivision and shrinking process on the box of study. The main idea is that the searched points are defined as roots of an equation. Then, any box whose image by the map associated with the equation does not contain 0, does not contain any searched point. If a box may contain a root, then an operator is used to shrink the box to smaller ones containing the roots in the initial box. Ultimately, when the box cannot be reduced this way, it is cut into several sub-boxes that are studied again. An instance of this scheme, to enclose the singular points of manipulators, can be found in Bohigas et al. (2012) and Bohigas et al. (2013). What makes the general scheme synergistic with Interval Analysis, is that they both operate on boxes and have the purpose to enclose computed values.

As stated previously, several methods, using Interval Analysis or not, exist to enclose the singular points of a manipulator. But, *it is also necessary to verify the nature of those singular points.* For instance, suppose you succeeded in finding the enclosure of the singular set in the workspace as in

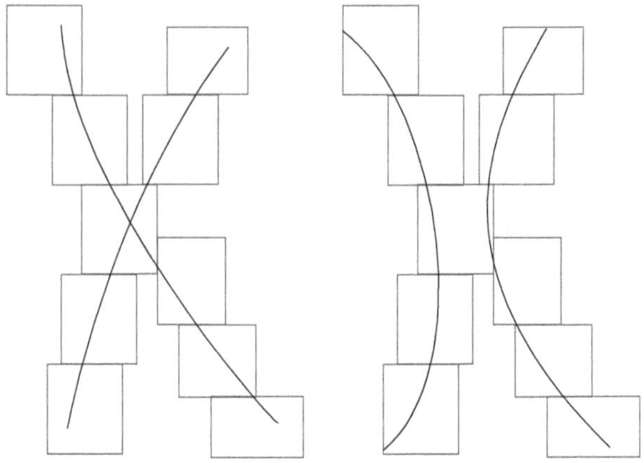

Figure 3. Two identical box coverings with two possible couples of covered curves.

Fig. 3. The real singular set can be either one of the two instance depicted in this Fig. 3. To conclude on the behaviour of the manipulator, it is necessary to verify if the two curves intersect or not.

In this paper, we propose an algorithm to enclose *specific singular points* that define the behaviour of a manipulator, using Interval analysis. Accordingly, next subsection proposes a method to enclose numerically roots from a system of equations, through Interval Analysis: *the Interval Newton method.*

3.3 The Interval Newton algorithm

Given a square system of equations described by $f = 0$, we can define an operator over *boxes*. This *Interval Newton operator N_f associated to the map f* is defined by:

$$N_f : D \longmapsto x - \left((\mathbf{df}(D))^{-1} \times f(x)\right), \qquad (1)$$

where D is a box and $x \in D$.

$\mathbf{df}(D)$ is the matrix of intervals enclosing all the matrices associated to the linear map of the differential of f at a point in D and $(.)^{-1}$ is the operator of matrix inversion. In practice, in our algorithm, $(.)^{-1}$ is computed applying the formulae of the inverse of a matrix. It should be noted that, in Eq. (1), instead of $(\mathbf{df}(D))^{-1} \times f(x)$, any set $\Sigma(D, f(x))$ could be used, as long as it encloses the solutions w of $Aw = f(x)$ where $A \in \mathbf{df}(D)$.

The main point is that the topological relation between D and $N_f(D)$ depends on the presence of a root in D:

1. if $N_f(D) \subset D$ then $\exists! x \in D$ such as $f(x) = 0$ and $x \in N_f(D)$,

2. if $N_f(D) \cap D = \varnothing$ then $\nexists x \in D$ such as $f(x) = 0$,

3. if $N_f(D) \cap D \neq \varnothing$ then (if $\exists x \in D$ such as $f(x) = 0$ then $x \in N_f(D) \cap D$).

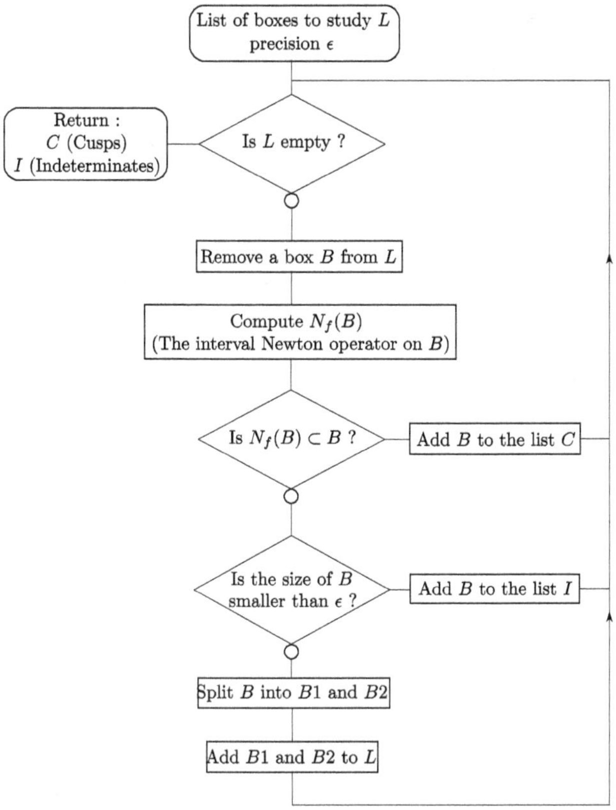

Figure 4. Interval Newton Algorithm.

The *Interval Newton method applied with f* is defined (see Neumaier, 1990) as being the Algorithm following the flow diagram of Fig. 4.

The Interval Newton algorithm is able to find the roots of a square system of equations if the Jacobian matrix associated with it is invertible for the roots of the studied system, implying that the Interval Newton method can only find isolated roots.

The Interval Newton method can also fail if the chosen precision is not small enough. For instance it can allow a studied box with a size smaller than the precision to contains several roots. One then has to choose a smaller precision, such as no box can contain several roots.

4 Finding cusps and nodes

4.1 Kinematic map and singularity concepts

We should first recall that *Cusps points and nodes points in the workspace* are singular positions of the end effector satisfying some additional properties: a cusp admits three equal inverse kinematic solutions and a node admits two distinct pairs of equal inverse kinematic solutions. Instead of searching for those points in the workspace, we are searching for their inverse kinematic solutions. Our points of interest are then what we are defining as *Cusps and nodes in the joint space* which are *the sets of the singular inverse kinematic solutions of the cusp points and node points in the workspace, respectively*.

In the case of 3R orthogonal manipulators, using the conventions of Fig. 1, due to their invariance along the rotation of parameter θ_1 we consider the joint space JS defined by $\mathrm{JS} = \{(\theta_2, \theta_3) \mid -\pi \leq \theta_2 < \pi; -\pi \leq \theta_2 < \pi\}$. Similarly, instead of considering the entire workspace, we consider a generator cross section of the workspace, SWS, that can be easily defined as the cylinder coordinates around the z axis, minus the angular coordinate. A position (x, y, z) in the workspace is then converted as $(\rho = x^2 + y^2, z)$. From these consideration, the manipulator kinematic map can be expressed as $f = (f_1, f_2) : \mathbb{R}^2 \supset \mathrm{JS} \longmapsto \mathrm{SWS} = \mathbb{R}^2$ with $f_1(\theta_2, \theta_3) = (\cos(\theta_2)(d_4 \cos(\theta_3) + d_3) + d_2 - r_3 \sin(\theta_2))^2 + (d_4 \sin(\theta_3) + r_2)^2$ and $f_2(\theta_2, \theta_3) = \sin(\theta_2)(d_4 \cos(\theta_3) + d_3) + r_3 \cos(\theta_2)$.

An *internal motion* occurs when the end tip point P reaches a joint axis. In this case, the inverse kinematics admits a continuum of solutions, which forms a line in the joint space. *On any box that intersect an internal motion line, the proposed algorithm cannot conclude.*

4.2 Applying the Interval Newton algorithm

Applying the Interval Newton algorithm to find cusps and nodes requires to define those points and pairs of points as roots of square systems of equations. We will then consider the same properties and characterisation of cusp and nodes points, in the joint space, that were developed in Delanoue and Lagrange (2014). Additionally, the situations where the defining systems are degenerated will be handled in a non-trivial manner to allow a quicker execution of the constructed algorithm.

4.2.1 Application to the cusps

In the following, $\mathrm{d}f$ refers to the differential of f and **Df** refers to the Jacobian matrix of f, which is the matrix associated to $\mathrm{d}f$.

Geometric considerations: we consider that a joint cusp point, C, is a point for which the orthogonal of $\mathrm{Ker}(\mathrm{d}f(C))$ is collinear with the gradient of the singular curve, defined by $\det(\mathbf{Df}) = 0$. It is worth noting that in $\mathbb{R}^2$, being collinear with a vector $\boldsymbol{v} = (v_1; v_2) \neq 0$ is the same as being orthogonal to the vector $\boldsymbol{w} = (-v_2; v_1) \neq 0$. Also, if $\mathbf{Df}(P) \neq 0$, the rows of **Df** are a base of the orthogonal of $\mathrm{Ker}(\mathrm{d}f(P))$ and as long as $\mathbf{Df}(P)$ is invertible, the orthogonal of $\mathrm{Ker}(\mathrm{d}f(P))$ is of dimension 2 and thus it cannot be collinear with $\mathrm{grad}(\det(\mathbf{Df}))(P)$. Putting all of this together, we can conclude that if $\mathrm{grad}(\det(\mathbf{Df}))(P)$ is not the null vector and $\mathbf{Df}(P)$

is not the null matrix, then P is a cusp point if:

$$\left\{\begin{array}{l} \frac{\partial f_1}{\partial \theta_2}(P)\cdot\left(-\frac{\partial \det(\mathrm{d}f)}{\partial \theta_3}(P)\right)+\frac{\partial f_1}{\partial \theta_3}(P)\cdot\frac{\partial \det(\mathrm{d}f)}{\partial \theta_2}(P)=0 \\ \frac{\partial f_2}{\partial \theta_2}(P)\cdot\left(-\frac{\partial \det(\mathrm{d}f)}{\partial \theta_3}(P)\right)+\frac{\partial f_2}{\partial \theta_3}(P)\cdot\frac{\partial \det(\mathrm{d}f)}{\partial \theta_2}(P)=0 \end{array}\right. . \tag{2}$$

Specificities for the algorithm: system (2) is square, which allows one to use the Interval Newton Method to find its isolated roots. The roots of system (2) that we are searching are singular points. Then, we will apply the Interval Newton Method only if a studied box contains a singular point, that is, if det(**Df**) may be null on the box. The final point is that grad(det(**Df**))(P) and **Df**(P) must not be null for the searched roots P, in order to detect those. Then, we will always verify that the components of grad(det(**Df**)) and **Df**(P) are not null on the boxes that should contain a cusp-root. If it is not the case on one of the isolated box, it will be cut into pieces that will be studied again.

4.2.2 Application to the nodes

Geometric considerations: node points are much simpler than cusp points for transcription in roots of a map. Indeed, let ΔE be the diagonal of E, that is $\Delta E = \{(a,a) | a \in E\}$. Then, we are searching for couples $(x_1, x_2) \in \mathbb{R}^2 \times \mathbb{R}^2 - \Delta\mathbb{R}^2$, satisfying:

$$\left\{\begin{array}{ll} f(x_1) & = f(x_2) \\ \det(Df(x_1)) & = 0 \\ \det(Df(x_2)) & = 0 \end{array}\right. \tag{3}$$

Specificities for the algorithm: to apply the Interval Newton method to the system (3), this system needs to be a square one, which is the case here, with 4 joint variables and 4 equations. We search the roots in JS × JS $\subset \mathbb{R}^2 \times \mathbb{R}^2$ while avoiding the roots in ΔJS $\subset \Delta\mathbb{R}^2$, because on this last subset, the Jacobian matrix associated with the system (3) is not invertible while having roots and the Interval Newton method fails.

Let S_j be the singular set of f (in the joint space JS). Instead of applying the time consuming process of verifying that a studied box does not intersect ΔJS and verifying the injectivity of f, restricted to a subset of S_j each time the intersection occurs, one can build a covering of S_j verifying a well chosen property. Indeed, if the covering is done so that any intersecting boxes admit a hull on which f, restricted to S_j, is injective, then, it suffices to apply Interval Newton algorithm with system (3) to couples of disjoint boxes, in this last covering.

Note that *the covering, built along with the process, is a guaranteed covering of the singular set.*

5 Performances of the Algorithms

5.1 Implementing and running the cusp and node algorithms

All results in this section are valid for any value, or interval of values, of r_3.

To implement, in C++, the algorithms defined in Sect. 4.2, for 3 revolute-jointed manipulators with mutually orthogonal joint axes, formal expressions of the derivatives and matrices derived from f, needed in the algorithms, were calculated. The algorithms evaluate the needed expression on the required boxes, replacing the standard functions and operators by corresponding inclusion maps. To handle intervals and operations on them, the library "Filib++" is used.

The application to more general 3 revolute-jointed manipulators, with $\beta_2 \neq 0$ or $\beta_3 \neq 0$, can be done by calculating their kinematic map. But, as the formal expressions increase in length, the running time of the algorithm may increase and the precision needed to enclose the interest points may need to be higher.

In the implemented algorithms, the initial box of study for (θ_2, θ_3) can be defined using any box or list of boxes, in $\mathbb{R}^2$. The box of geometric parameters can also be chosen. Our algorithms can also be coupled with a procedure enclosing the usable joint space, given a simple volumetric model of the manipulator. The returned enclosure may also be chosen as the boxes of study.

Table 2 shows results returned by the algorithms, applied to examples of classes of studied parameters for 3 revolute-jointed manipulators with orthogonal axes, reported in Table 1, and with an initial box of study for (θ_2, θ_3) of $[-3.1415, 3.1415] \times [-3.1415, 3.1415]$ close to the $[-\pi, \pi] \times [-\pi, \pi]$ full range for the joint angles.

5.2 The cusp enclosing Algorithm

Manipulator inducing no indeterminate (cases a, b, d and e of Table 1): the algorithm has been applied to every example of geometric parameters sets in Baili et al. (2004). When the manipulator does not have an internal motion, for a moderate precision, the algorithm needs little time to find the rigorous enclosures of the cusps, and does not return any indeterminate box.

Manipulator inducing indeterminate (case f of Table 1): when the algorithm is applied to a robot that has internal motions, it finds the cusps outside the internal motions, with the same running time as before. The algorithm then has to run for some time until it encloses the lines associated with the internal motions with boxes whose size is the chosen precision. The running time is then dependant of the chosen precision.

Table 1. Some studied cases of robotic manipulators.

Characteristics	Geometric parameters					Properties of manipulator		
Designation	d_2	d_3	d_4	r_2	r_3	Internal motion	Cusps	Nodes
a	1	2	1.5	1	0	no	4	0
b	1	2	1.5	1	0.5	no	4	0
c	$[1, 1.001]$	$[2, 2.001]$	$[1.5, 1.501]$	$[1, 1.001]$	0	NA	4	NA
d	1	$[0.7]$	$[0.3]$	$[0.2]$	0	no	0	0
e	1	1.5	$[0.7]$	0.5	0	no	4	2
f	1	0.5	$[1.3]$	$[0.2]$	0	yes	0	2

Table 2. Algorithms performances on the robotic manipulators of Table 1.

	Cusp algorithm				Node algorithm				
Case	Precision	Cusps	Indeterminate	Time	Precision	Nodes	Indeterminate	Time	Improved time
a	10^{-4}	4	no	32 s	2.5×10^{-10}	0	no	10 h	23 min
b	10^{-4}	4	no	46 s	2.5×10^{-10}	0	no	18 h	45 min
c	10^{-4}	4	no	35 s	2.5×10^{-10}	NA	yes	NA	out of memory
d	10^{-4}	0	no	12 s	2.5×10^{-10}	0	no	52 s	16 s
e	10^{-4}	4	no	52 s	2.5×10^{-10}	2	no	35 h	5 h and 42 min
f	10^{-2}	0	yes	12 min	10^{-2}	2	yes	42 s	42 s
f	10^{-3}	0	yes	90 min	10^{-3}	2	yes	41 s	41 s

The running times are given for a computer with a 64 bits operating system and an Intel® Core™ i7 CPU.
When the parameter p is not computer storable, then it is replaced by the smallest interval containing it, noted $[p]$.

5.3 The nodes enclosing Algorithm

On boxes where there is no cusps and no internal motion lines (case d of Table 1) the nodes enclosing algorithm concludes after a running time close to the one needed for the cusp enclosing algorithm with no internal motion. However, when the box includes a cusp (cases a, b and e of Table 1) the running time of the algorithm increases quite significantly, because, near cusps, f restricted to S_j, is injective only on small boxes. In the same way, the Interval Newton method can conclude, only on small boxes when the hull box of its two components is close to a cusp point.

5.3.1 Performance improvement using contraction methods

As it has been formerly noted, the main drawback of the algorithm is its relatively slow check of the absence of nodes near cusps. To improve on this, we decided to rely on the contraction method library *Ibex*, available freely at http://www.ibex-lib.org/, with documentation.

A *Contractor* is an operator on Boxes, associated to a set, that reduce the box to a smaller box without removing any element of the associated set. Contraction methods are used in Interval Analysis to enclose a set. It relies on contractors, associated to the chosen set, and may use subdivisions, so as to get a enclosure of the chosen set. The main interest of those methods is that reducing a box using contractors is a lot less time consuming than bisecting it until a chosen precision.

An Ibex contraction procedure is included in the algorithm as an additional check before applying an iteration of the node Interval Newton method on a couple of disjoint boxes. The procedure is based upon a contractor using the Interval Newton method with the system dedicated to the node as parameter. As the Ibex procedure's contractor reduce quite efficiently the studied boxes, we use it as a quick way to check the absence of node in a couple of boxes (see as a box of double dimension). If the procedure return an empty box as a result, then, there is no node in the initial couple of boxes and it is not needed to apply any subdivision process or interval Newton iterations further.

As a result of including the Ibex calling test in the node searching step, the performances of the algorithm toward the length of checking the absence of nodes have been greatly improved. For instance, the time needed to execute the node searching step, for a 3R manipulators with nodes and cusps is *decreased to less than a fifth of its value* (case e of Table 1).

5.4 Application with boxes of geometric parameters

Our algorithms have been implemented to handle intervals of geometric parameters, so to use intervals of parameters (as for case b of Table 1) it is only needed to define a box of geometric parameters which is not restricted to a point.

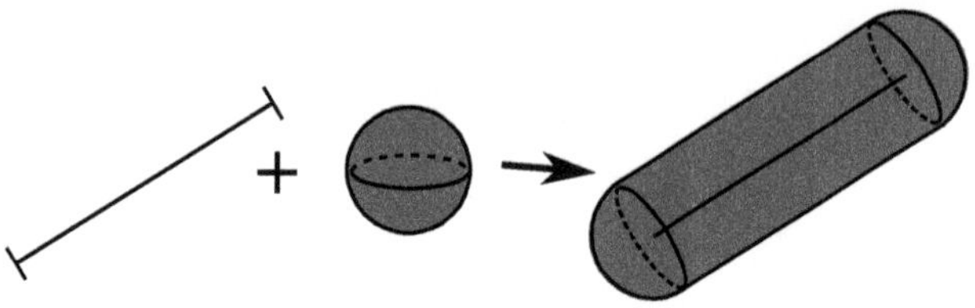

Figure 5. Oblong solid model as a Minkowski sum of a segment and a ball.

If the algorithms find a solution box, then, *for any set of geometric parameter in the defined box of parameters*, there is a single interest point in the solution box. There will be no interest point in any box that is neither a solution box nor an indeterminate box *for any set of geometric parameter, in the defined box of parameters*. Ultimately, it can exist interest points, for any set of geometric parameter in the defined box of parameters, only in solution boxes and in indeterminate boxes. For a manipulator with an internal motion, the algorithms return, at least, enclosures for a subset of the interest point and a covering of the research space that can contain interest points.

6 Collisions detection through Interval Analysis

An complementary procedure have been added to our algorithm, allowing the user to get an enclosure of the set of parameters inducing collisions and of the set of parameters inducing no collisions at all.

6.1 Used model

Solids that may collide (either elements of the manipulator's kinematic chain or environment obstacle) are considered oblong object defined by a segment and a radius, where the oblong object is the set of all points distant to the segment from at most the defining radius, see Fig. 5. With this model, two objects collide if and only the distance between the respective defining segments is equal or less than the sum of the two defining radius.

6.2 Implemented procedure

The implemented procedure is based upon the SIVIA inversion algorithm, and consists in applying it for the distance between every pair of defined segments. As it implies computing the minimum of the distance between a point in one segment and a point in a second one, the two segments are split until a limit size and the distance between each couple of sub-segment is checked if greater than the sum of the radius.

As the distance varies with the articular parameters, the former process is applied for sub-boxes of the initial list of boxes of articular parameters. To sum up, the procedure is

Algorithm 1 Set Inversion Via Interval Analysis (SIVIA) algorithm

```
Require: A set S, a function, and a real number ε (a limit of size)
  and a list of boxes of research L
  return 3 lists of boxes I, O and U
  while L is not empty do
    extract B from L
    evaluate D = f(B) through Interval Analysis
    if D = f(B) ⊂ S then
      add B to I
    else if D = f(B) ∩ S = ∅ then
      add B to O
    else if size(B) > ε then
      split B in B1 and B2 and add them to L
    else
      add Bi to U
    end if
  end while
  In the end (∪_{B∈I} B) ⊂ S ⊂ (∪_{B∈(I∪U)} B) and S ∩ (∪_{B∈O} B) = ∅
```

applying a list of consecutive double-SIVIA for each couple of solids that may collide, the user defined to be studied.

6.3 Joint use with cusp and node detection

The interest of the collision detection procedure in itself is to control the feasibility of given paths in the joint space, by the studied manipulators. Combined with the the knowledge of an enclosure of the singular set, returned by the preprocessing step of the node enclosing procedure, one can also check for the possibility to join two posture by a non singular feasible path.

The joint use of the procedure with the main detection algorithm also allows, quite naturally, to check for the manipulator access to the chosen interest points. In our case, the inaccessibility to the cusps and nodes may not change the behaviour of the manipulator. However, for instance, a wide collision zone around a cusp, may imply, that the manipulator is, in practice, *not cuspidal* if no articular path can link two IKS without crossing a singularity or inducing collision.

7 Conclusions

The main interest of the proposed method is that it can be used to find any isolated point of interest for the evaluation of the behaviour of any manipulator, provided it can be defined by a root of a square system of equations. Then, this methodology constitutes a possible way of describing a robotic manipulator singular set, allowing for the guaranteed detection of isolated specific singular points of interest.

It is to be noted that most of the running time of the algorithm is used to treat boxes where the Interval Newton algorithm fails to conclude. To increase the performance of the algorithm, alternate methods for splitting and localized tests need to be used and are still searched.

As for a lot of Interval Analysis algorithms, our algorithm can be time consuming when dealing with complicated kinematic functions or high dimension boxes of study, especially for the nodes enclosing algorithm, due to the doubled dimension of the box of study, although attenuated by a pre-subdividing in the joint space. However, provided that the algorithm runs for the time needed with a sufficient precision, it is able to find enclosures for the searched points without errors, or at least a subset of those enclosures and a covering of the searched points.

With the joint use of the collision procedure, the algorithm aims to provide efficient and guaranteed information about the manipulator's kinematic properties. The algorithm could provide additional information that may be relevant to the user's interests with additional procedures to, for instance, enclose the singular positions in the workspace or enclose the non-singular IKS of the singular positions. As such, the reader may found the source code to the algorithms at http://perso-laris.univ-angers.fr/~benoitr/contenu/thom_2d_online.zip.

References

Baili, M.: Analyse et classification de manipulateurs 3R à axes orthogonaux, PhD thesis, Ecole Centrale de Nantes (ECN), Université de Nantes, France, 2004.

Baili, M., Wenger, P., and Chablat, D.: A Classification of 3R Orthogonal Manipulators by the Topology of their Workspace, in: Robotics and Automation, Proceedings, ICRA '04, IEEE International Conference on, 26 April–1 May 2004, New Orleans, LA, USA, 2, 1933–1938, 2004.

Bohigas, O., Zlatanov, D., Ros, L., Manubens, M., and Porta, J. M.: Numerical computation of manipulator singularities, in: Robotics and Automation (ICRA), IEEE International Conference on, 14–18 May 2012, Saint Paul, MN, USA, 1351–1358, 2012.

Bohigas, O., Manubens, M., and Ros, L.: Singularities of non-redundant manipulators: A short account and a method for their computation in the planar case, Mech. Mach. Theory, 68, 1–17, 2013.

Corvez, S. and Rouillier, F.: Using Computer Algebra Tools to Classify Serial Manipulators, chapter: Automated Deduction in Geometry, in: Lecture Notes in Computer Science, edited by: Winkler, F., Springer, Berlin, Germany, 2930, 31–43, 2004.

Delanoue, N. and Lagrange, S.: A numerical approach to compute the topology of the Apparent Contour of a smooth mapping from $\mathbb{R}^2$ to $\mathbb{R}^2$, J. Comput. Appl. Math., 271, 267–284, 2014.

Husty, M., Ottaviano, E., and Ceccarelli, M.: A Geometrical Characterization of Workspace Singularities in 3R Manipulators, in: Advances in Robot Kinematics, Analysis and Design, edited by: Lenarcic, J. and Wenger, P., Springer, 411–418, 2008.

Jaulin, L., Kieffer, M., Didrit, O., and Walter, E.: Applied Interval Analysis with Examples in Parameter and State Estimation, Robust Control and Robotics, Springer-Verlag, 2001.

Merlet, J.-P.: Interval analysis and robotics, in: Robotics Research, edited by: Kaneko, M. and Nakamura, Y., Springer Berlin Heidelberg, 66, 147–156, 2011.

Moore, R. E.: Interval Analysis, Prentice-Hall, Englewood Cliffs, NJ, USA, 1996.

Neumaier, A.: Interval methods for systems of equations, vol. 37 of Encyclopedia of mathematics and its applications, Cambridge University Press, Cambridge, UK, 1990.

Wenger, P.: Cuspidal and noncuspidal robot manipulators, Robotica, 25, 677–689, doi:10.1017/S0263574707003761, 2007.

Permissions

All chapters in this book were first published in MS, by Copernicus Publications; hereby published with permission under the Creative Commons Attribution License or equivalent. Every chapter published in this book has been scrutinized by our experts. Their significance has been extensively debated. The topics covered herein carry significant findings which will fuel the growth of the discipline. They may even be implemented as practical applications or may be referred to as a beginning point for another development.

The contributors of this book come from diverse backgrounds, making this book a truly international effort. This book will bring forth new frontiers with its revolutionizing research information and detailed analysis of the nascent developments around the world.

We would like to thank all the contributing authors for lending their expertise to make the book truly unique. They have played a crucial role in the development of this book. Without their invaluable contributions this book wouldn't have been possible. They have made vital efforts to compile up to date information on the varied aspects of this subject to make this book a valuable addition to the collection of many professionals and students.

This book was conceptualized with the vision of imparting up-to-date information and advanced data in this field. To ensure the same, a matchless editorial board was set up. Every individual on the board went through rigorous rounds of assessment to prove their worth. After which they invested a large part of their time researching and compiling the most relevant data for our readers.

The editorial board has been involved in producing this book since its inception. They have spent rigorous hours researching and exploring the diverse topics which have resulted in the successful publishing of this book. They have passed on their knowledge of decades through this book. To expedite this challenging task, the publisher supported the team at every step. A small team of assistant editors was also appointed to further simplify the editing procedure and attain best results for the readers.

Apart from the editorial board, the designing team has also invested a significant amount of their time in understanding the subject and creating the most relevant covers. They scrutinized every image to scout for the most suitable representation of the subject and create an appropriate cover for the book.

The publishing team has been an ardent support to the editorial, designing and production team. Their endless efforts to recruit the best for this project, has resulted in the accomplishment of this book. They are a veteran in the field of academics and their pool of knowledge is as vast as their experience in printing. Their expertise and guidance has proved useful at every step. Their uncompromising quality standards have made this book an exceptional effort. Their encouragement from time to time has been an inspiration for everyone.

The publisher and the editorial board hope that this book will prove to be a valuable piece of knowledge for researchers, students, practitioners and scholars across the globe.

List of Contributors

G. Palmieri
Università degli Studi eCampus, Via Isimbardi 10, 22060 Novedrate (CO), Italy

H. Giberti
Politecnico Di Milano, Dipartimento di Meccanica, Campus Bovisa Sud, via La Masa 1, 20156, Milano, Italy

A. Pagani
Fpz S.p.a., Via Fratelli Cervi, 18, 20049 Concorezzo (MB), Italy

D. Trimble, H. Mitrogiannopoulos, G. E. O'Donnell and S. McFadden
Department of Mechanical and Manufacturing Engineering, Trinity College Dublin, Dublin 2, Ireland

G. Hao and J. Mullins
School of Engineering-Electrical and Electronic Engineering, University College Cork, Cork, Ireland

F. Klocke, M. Brockmann, S. Gierlings and D. Veselovac
Laboratory for Machine Tools and Production Engineering (WZL), Aachen, Germany

N. Khan
University of Engineering and Technology Peshawar, Pakistan

I. Ullah and M. Al-Grafi
College of Engineering, Taibah University, Yanbu, Saudi Arabia

W. Kraus, P. Miermeister, V. Schmidt and A. Pott
Fraunhofer Institute for Manufacturing Engineering and Automation IPA in Stuttgart, Germany

C. M. C. G. Fernandes, P. M. T. Marques and R. C. Martins
INEGI, Universidade do Porto, Campus FEUP, Rua Dr. Roberto Frias 400, 4200-465 Porto, Portugal

J. H. O. Seabra
FEUP, Universidade do Porto, Rua Dr. Roberto Frias s/n, 4200-465 Porto, Portugal

S. G. Khan
Department of Mechanical Engineering, College of Engineering Yanbu, Taibah University, Al Madinah, Saudi Arabia

J. Jalani
Department of Electrical Engineering Technology, University Tun Hussein Onn Malaysia, Batu Pahat, Malaysia

B. Li and Y. M. Li
Faculty of Science and Technology, University of Macau, Taipa, Macau, China
Tianjin Key Laboratory for Advanced Mechatronic System Design and Intelligent Control, Tianjin University of Technology, Tianjin, China

X. H. Zhao and W. M. Ge
Tianjin Key Laboratory for Advanced Mechatronic System Design and Intelligent Control, Tianjin University of Technology, Tianjin, China

Y. F. Liu1, J. Li
Complex and Intelligent System Laboratory, School of Mechanical and Power Engineering, East China University of Science and Technology, Shanghai, China

W. J. Zhang
Complex and Intelligent System Laboratory, School of Mechanical and Power Engineering, East China University of Science and Technology, Shanghai, China
Department of Mechanical Engineering, University of Saskatchewan, Saskatoon, Canada

Z. M. Zhang and X. H. Hu
Department of Mechanical Engineering, University of Saskatchewan, Saskatoon, Canada

M. Díaz-Rodríguez
Departamento de Tecnología y Diseño, Facultad de Ingeniería, Universidad de los Andes, Mérida, 5101, Venezuela

J. A. Carretero
Department of Mechanical Engineering, University of New Brunswick, Fredericton, NB, E3A 5A3, Canada

R. Bautista-Quintero
Department of Mechanical Engineering, University of New Brunswick, Fredericton, NB, E3A 5A3, Canada
Departamento de Ingeniería Mecánica, Instituto Tecnológico de Culiacán, Sinaloa, 80220, Mexico

R. Pakrouh and A. A. Ranjbar
Department of Mechanical Engineering, Babol University of Technology, P.O. Box 484, Babol, Iran

M. J. Hosseini
Department of Mechanical Engineering, Golestan University, P.O. Box 155, Gorgan, Iran

A. A. Jomartov and S. U. Joldasbekov
Institute Mechanics and Mechanical Engineering, Almaty, Kazakhstan

Yu. M. Drakunov
al-Farabi Kazakh National University, Almaty, Kazakhstan

Y. F. Liu, J. Li and X. H. Hu
Complex and Intelligent System Laboratory, School of Mechanical and Power Engineering, East China University of Science and Technology, Shanghai, China

W. J. Zhang
Complex and Intelligent System Laboratory, School of Mechanical and Power Engineering, East China University of Science and Technology, Shanghai, China
Department of Mechanical Engineering, University of Saskatchewan, Saskatoon, Canada

Z. M. Zhang
Department of Mechanical Engineering, University of Saskatchewan, Saskatoon, Canada

L. Cheng
Institute of Automation, Chinese Academy of Sciences, Beijing, China

Y. Lin
Department of Mechanical and Industrial Engineering, Northeastern University, Boston, USA

A. V. Perig
Manufacturing Processes and Automation Engineering Department, Donbass State Engineering Academy, Shkadinova Str. 72, 84313 Kramatorsk, Ukraine

N. N. Golodenko
Department of Water Supply, Water Disposal and Water Resources Protection, Donbass National Academy of Civil Engineering and Architecture, Derzhavin Str. 2, 86123 Makeyevka, Ukraine

F. Romero and F. J. Alonso
Department of Mechanical, Energy and Materials Engineering, University of Extremadura, Avda. de Elvas s/n, 06006 Badajoz, Spain

Jin Li, Chang Jun Liu and Xin Wen Xiong
The Complex and Intelligent System Research Center, School of Mechanical and Power Engineering, East China University of Science and Technology, Meilong Road 130, Shanghai, 200237, China

Yi Fan Liu
Robotic Systems Laboratory, Ecole Polytechnique Fédérale de Lausanne (EPFL), C/o Nicolas Cantale Avenue de prefaully 56, 1020 Lausanne, Switzerland

Wen Jun Zhang
Department of Mechanical Engineering, University of Saskatchewan, Saskatoon, S7N5A9, Canada

X. Ding
School of Mechanical Engineering and Automation, Beihang University, Beijing, China

W. Zhang
School of Mechanical Engineering and Automation, Beihang University, Beijing, China
State Key Laboratory of Robotics, Shenyang Institute of Automation, Chinese Academy of Sciences, Shenyang, China

J. Liu
State Key Laboratory of Robotics, Shenyang Institute of Automation, Chinese Academy of Sciences, Shenyang, China

Y. S. Du, T. M. Li, Y. Jiang and J. L. Zhang
Manufacturing Engineering Institute, Department of Mechanical Engineering, Tsinghua University, Beijing, 100084, China
Beijing Key Lab of Precision/Ultra-precision Manufacturing Equipment and Control, Tsinghua University, Beijing, 100084, China

A. Alliche
Sorbonne Universités, UPMC Univ Paris 06, CNRS, UMR 7190, Institut Jean Le Rond d'Alembert, 75005 Paris, France

R. Benoit, N. Delanoue and S. Lagrange
Laboratoire Angevin de Recherche en Ingénierie des Systèmes (LARIS), 62 avenue Notre Dame du Lac, 49000 Angers, France

P. Wenger
Institut de Recherche en Communications et Cybernétique de Nantes (IRCyN), 1 rue la Noë, 44321 Nantes CEDEX 03, France

Index

O

P

R

S

T

U

Printed in the USA
CPSIA information can be obtained
at www.ICGtesting.com
JSHW051444221024
72173JS00006B/1577

9 781632 385338